Arthur Hill Hassall

The Inhalation Treatment of Diseases of the Organs of Respiration

Including Consumption

Arthur Hill Hassall

The Inhalation Treatment of Diseases of the Organs of Respiration
Including Consumption

ISBN/EAN: 9783337182199

Printed in Europe, USA, Canada, Australia, Japan

Cover: Foto ©berggeist007 / pixelio.de

More available books at **www.hansebooks.com**

THE INHALATION TREATMENT

OF

DISEASES

OF THE

ORGANS OF RESPIRATION

INCLUDING

CONSUMPTION

BY

ARTHUR HILL HASSALL, M.D. LOND.

MEMBER OF THE ROYAL COLLEGE OF PHYSICIANS OF ENGLAND
LATE SENIOR PHYSICIAN TO THE ROYAL FREE HOSPITAL, LONDON
FOUNDER OF, AND CONSULTING PHYSICIAN TO, THE ROYAL NATIONAL HOSPITAL
FOR CONSUMPTION AND DISEASES OF THE CHEST
PHYSICIAN TO THE NORTH BRITISH AND MERCANTILE INSURANCE COMPANY
AUTHOR OF 'THE MICROSCOPIC ANATOMY OF THE HUMAN BODY IN HEALTH AND DISEASE'
'THE URINE IN HEALTH AND DISEASE'
'SAN REMO CLIMATICALLY AND MEDICALLY CONSIDERED'
AND VARIOUS OTHER WORKS

With numerous Illustrations

LONDON
LONGMANS, GREEN, AND CO
1885

All rights reserved

DR. HASSALL'S INHALATION CHAMBER.

THIS Chamber is for the Inhalation Treatment of Diseases of the Organs of Respiration, including Maladies of the Throat, Larynx, Bronchi, and Lungs.

The air of the Chamber is charged to any required extent by the Method and with the appliances described in Dr. Hassall's work on Inhalation, the medicaments employed being equally diffused in a gaseous or aëriform state, and this object is effected at a temperature of about 64° Fahrenheit.

The medicaments used consist chiefly of Carbolic Acid or Phenol, and all preparations containing this ; Creasote, and products of which this is a constituent ; Thymol ; Oil of Eucalyptus ; and Oil of Pinus Sylvestris.

VILLA BRACCO,
    SAN REMO.

PREFACE.

ALTHOUGH several books have been published abroad, chiefly in America and Germany, on Inhalation in Disease, especially of the Organs of Respiration, no original English publication has yet appeared, notwithstanding that there is great need of such a work.

There are several reasons why a treatise on Inhalation is required; one is, the importance of the subject, and another, that the laws and principles which govern Inhalation have been but little studied and very imperfectly understood, much error even prevailing in consequence; the knowledge of the subject being so incomplete, the practice has been faulty and inefficient in the extreme. That this statement is amply justified will be proved in the course of this work.

Another result of this want of knowledge of the principles on which Inhalation is founded is, that many of the ingenious contrivances which have been devised for facilitating Inhalation have fulfilled the objects intended in a very imperfect manner. A striking example of this is furnished by the Oral and Oro-nasal Inhalers, which are chiefly employed in this country in the treatment of Lung Affections. The great majority of the substances used in these are so little volatile, and the quantities volatilized are so small, that it is impossible the treatment carried out by their means can be effective. In the case of such substances as carbolic acid, creasote, and some others, four-fifths of the very small quantities with which the Inhalers are usually charged are recoverable from the sponge or cotton-wool at the end of the Inhalation.

The above circumstance alone, independent of many other facts and considerations described and noticed in the course of the work, goes far to explain the disappointment which has been expressed by many at the results hitherto obtained of treatment by Inhalation. The fact is, that this method has not, especially in this country, had a fair trial, and its real capabilities have not yet been put to the test. With extended knowledge of the subject, and with the new and improved appliances now at our disposal, the method admits of being tested in a much more satisfactory manner, and doubtless far more encouraging and important results will be obtained.

So far as I have yet been able to carry out the Inhalation treatment of Lung Affections in accordance with the principles and by the means described in this work, I have much reason to be satisfied. For the treatment to be successful, however, it is necessary that it should be conducted systematically and thoroughly, and that it should also in the majority of cases be constitutional as well as local. In fact, when antiseptic remedies such as carbolic acid and creasote are employed, the system should be brought under their influence to such an extent that the renal secretion gives evidence of their presence.

The treatment of Diseases of the Organs of Respiration by Inhalation may now be said to rest on a scientific and secure basis. The method has always been a favourite one, both with the profession and the public, and there can be no question but that, when efficiently carried out, it is capable of yielding results of the highest importance.

The subject of Inhalation is treated of in this work in seven chapters, under the following heads :—

The Entrance of Medicaments into the Organs of Respiration.

The Principles concerned in the Volatiliztion and Inhalation of the Medicaments.

The Apparatus employed.

Inhalation Chambers.

The Quantities of the Medicaments, the Manner, Frequency, and Duration of the Inhalations.

The Medicaments employed in Inhalation.

The Inhalation Treatment of Diseases of the Organs of Respiration.

Under this last heading a brief description is given of each disease. This has been done to indicate more clearly the several stages and conditions of each malady requiring Inhalation treatment. Further, I have not confined my remarks to Inhalation, since this is not to be regarded as an exclusive method of treatment, but is rather to be viewed as a valuable, and indeed indispensable, addition to the other and more usual methods.

I would now acknowledge the obligation I am under to the work of Dr. J. Solis Cohen, of Philadelphia, on 'Inhalation in the Treatment of Disease.' The first edition of this work was published in 1867, and the second in 1876, and it is from the latter that I have obtained many hints and much useful information.

I am equally indebted to the voluminous work of Dr. M. J. Oertel, published in 1882, entitled, 'Handbuch der Respiratorischen Therapie,' and especially to that portion of it which treats of the Inhalation of Compressed and Rarefied Air—a subject but little studied and still less practised in this country, and which is very fully described and considered in the work in question.

For the translation of Dr. Oertel's book, and for other help, I am greatly indebted to my Wife. My thanks are also due to my friend Dr. J. de Tymowski, who kindly explained certain passages.

I have further to express my obligations to Mr. Edwy G. Clayton, F.C.S., who at much trouble has most kindly procured for me certain information from books which, here in San Remo, were inaccessible to me; also for carrying out at my suggestion some experiments in evaporation, and especially those embodied in Tables I. and II.

Lastly, I would thank Mr. Evans for the pains he has taken in the execution of the Wood Engravings.

VILLA BRACCO, SAN REMO, ITALY (*winter*);
54 HOLBORN VIADUCT, LONDON, E.C.

February 1885.

CONTENTS.

CHAPTER		PAGE
I.	On the Entrance of Medicaments into the Organs of Respiration .	1
II.	The Principles concerned in the Volatilization and Inhalation of Medicaments .	21
III.	The Apparatus employed in Diseases of the Organs of Respiration .	30
IV.	On Inhalation Chambers .	76
V.	On the Quantities of the Medicaments and on the Manner, Frequency, and Duration of the Inhalations . . .	91
VI.	The Medicaments employed in Inhalation in Diseases of the Organs of Respiration	107
VII.	The Inhalation Treatment of diseases of the Organs of Respiration	227
INDEX . . .		359

INHALATION
IN
DISEASES OF THE ORGANS OF RESPIRATION.

CHAPTER I.
ON THE ENTRANCE OF MEDICAMENTS INTO THE ORGANS OF RESPIRATION.

THE various medicaments, using the word in its widest sense, employed for inhalation in diseases of the organs of respiration include gases, vapours, liquids, and powders.

No difficulty whatever is encountered in the application, by means of inhalation, of any of the above substances to the nares, throat, fauces, and pharynx, indeed to all the parts situated above the epiglottis, and which also may be reached with great facility by means of gargles and brushes.

Neither is any considerable difficulty experienced in the entrance of gases, providing these are not of an irritating character and do not produce spasm, into the proper respiratory track; that is to say, into the channels and parts which lie below the epiglottis, namely, the larynx, trachea, bronchi, and air cells of the lungs. It is obvious that where atmospheric air can enter freely such gases must also find a ready entrance.

The difficulty of the entrance of powders into the deeper parts of the organs of respiration is somewhat greater; but that these can and do enter, when persistently respired, is sufficiently demonstrated by what occurs in the case of millers, glass and stone cutters, and others engaged in occupations attended with the suspension of fine dust in the air breathed.

B

It is chiefly in the case of liquids and vapours, that any difficulties arise; and here the question is not whether these substances are physically capable of passing into the air passages and of entering the lungs, for this of course is undoubted, but as to the circumstances under which and the extent to which many of them actually do enter, more particularly by means of the different kinds of apparatus which have hitherto for the most part been employed.

This is a very important part of our subject, and one by no means thoroughly worked out and determined as yet, notwithstanding the confident assertions of some writers on inhalation in diseases of the lungs. I propose therefore to consider this question somewhat fully, as until we know what becomes of the medicaments we employ, and how much of them reaches the seat of the disease, we are working in the dark, and we shall be apt to deceive ourselves as well as others. Dr. Solis Cohen, of Philadelphia, in his work 'On Inhalation, its Therapeutics and Practice,' the first edition of which was published in 1866, has dealt with this subject in much detail and in a most interesting manner.

Fournié, as the result of a great many apparently well-devised and carefully conducted experiments, by means of sprays, made both by himself and others, chiefly on animals, and in many of which the medicaments failed to reach the air passages, arrived, on a consideration of the whole question, at the following conclusions:—

'First, it is nevertheless possible by this method to induce a toxic effect upon the larynx, but one of short duration only. Second, the fluid can certainly penetrate into the air passages, but not with facility nor in sufficient quantity, and this only when the inhalation is performed with peculiar care, such as opening the mouth well, separating the base of the tongue from the soft palate, with the head inclined backwards to destroy as much as possible the rectangular curve that the windpipe makes with the oral cavity.'

The comparative failures of the earlier experimenters led subsequently to the employment of more exact and rigorous methods of investigation.

The first more successful experiments were those of Demarquay, surgeon to the Maison Municipale de Santé, Paris. Some of these were upon animals and one on a hospital nurse. In the case of the experiments with rabbits and dogs the mouths were kept forcibly dilated and the tongues drawn forward.

The nurse was the same person who had been previously unsuccessfully experimented on by Fournié, and upon whom the operation of tracheotomy had been performed in consequence of stenosis of the larynx. Demarquay repeated on her the experiment of Fournié. A piece of paper moistened with chloride of iron was introduced through the opening in the trachea, which was then completely closed with the finger. A 1 per cent. solution of tannic acid was then sprayed into the throat by Matthieu's spray apparatus. On removal of the finger the paper was found to have become quite black from the formation of tannate cf iron, thus showing that the tannin had penetrated at least as far down as the opening in the trachea.

A somewhat similar experiment on the human subject was made by Tavernier. A spray containing chloride of iron was first inhaled, and afterwards one of ferrocyanide of potassium. On examination with the laryngoscope, a layer of ferrocyanide of iron or Prussian blue was detected, and which extended even below the vocal cords. The sputa were at first irregularly and afterwards more uniformly coloured.

Bataille inhaled a spray containing extract of rhatany; this produced a redness of the larynx and trachea, which after some hours entirely disappeared, but the sputa continued red for some time longer.

Schnitzler and Störk made some experiments upon the man servant of the latter, who was accustomed to have his throat examined with the laryngoscope; they used coloured solutions of rhatany, Campeachy wood, and saffron, and the colours could be detected below the larynx and in one case in the trachea. They also succeeded, as did Dr. Frederick Fieber, of Vienna, after experiencing some failures, in the case of two individuals upon whom the operation of laryngotomy had been performed. They employed a solution of tannin, and on applying to the opening a strip of linen moistened with a solution of perchloride of iron and closing it with the finger obtained the inky coloration.

A successful result was likewise obtained with iodine, although this failed in one of the three experiments made, as did also some of the trials with the other substances.

Again, Zdekauer has made known the particulars of a case of hæmoptysis which terminated fatally, and which was treated by sprays of chloride of iron. After death, increased quantities of iron were found in the hæmorrhagic spots and in the surrounding lung tissue.

A second fatal case of hæmoptysis has been communicated by Lewin, in which at the post-mortem examination a sac containing a blackish fluid was found in the right lung, and in this were detected granules of free iron in small quantity.

But though these and other experiments sufficiently establish the fact of the entrance of medicaments into the air passages, under favourable circumstances, by the use of sprays, particularly steam sprays, it must not be supposed that these results are easily arrived at or are obtainable in all cases. Unless the patient holds himself erect, and inspires deeply with the mouth open and the tongue depressed, the spray will not pass beyond the glottis. Nor do they, as will be seen presently, by any means settle the question in all its bearings. The results above recorded were all obtained by the spray apparatus, and not by the use of oro-nasal inhalers. If, therefore, it can be shown that these inhalers are comparatively ineffective for the introduction of medicaments into the lungs, as at present used, then their employment must be abandoned and other less faulty methods substituted. The majority of cases in which inhalation is resorted to are cases of consumption and bronchitis, and these are almost invariably treated in Great Britain by some form or other of oral or oro-nasal inhaler, and not by the spray apparatus. Now, if by these inhalers the medicaments employed enter the air passages and lungs only in small quantities—in amounts, in fact, wholly insufficient—then it follows, that hitherto the treatment of the above affections by inhalation, has not been practised in such a way as to afford a fair test of the efficacy or otherwise of this method. The question of the efficiency or otherwise of oro-nasal inhalers will be considered further on.

With respect to the evidence furnished by the sputa of the entrance of inhaled substances into the track of the organs of respiration, Oertel, in his ' Handbuch der respiratorischen Therapie,' simply states that the sputa are sometimes noticed to have the odour of carbolic acid, and to lose the offensive and putrid smell which they occasionally acquire; but he nowhere affirms that these results have been noticed to follow the use of an ordinary oro-nasal inhaler. When these results are observed they are the consequence of the free use of the acid in the form of sprays. At all events I have never met with a case in which the presence of the carbolic acid in the sputa from the use of the ordinary oro-nasal inhalers was to be detected either by the smell of the acid or by the characteristic chemical tests. This statement of course applies to sputa coming really from the bronchial tubes and not from some part of the mucous surfaces above the epiglottis.

Again, in the case of the disappearance of the offensive smell of the sputa, it would be necessary to determine whether the odour might not proceed from putrid discharges from the tonsils, throat, or teeth, and not from parts more deeply seated. At the same time the possibility is not denied of the sputa being found to contain in some cases traces of carbolic acid after the use of oro-nasal inhalers, but it is affirmed that this result is an uncommon one when these respirators are of the usual construction and are charged with the ordinary small amount of carbolic acid, viz. from two to five grains.

Even when the offensive smell of the sputa disappears, this result is sometimes brought about by wholly different means, and admits of an explanation entirely independent of the entrance of the acid into the lungs either by sprays or oro-nasal inhalers. Thus we learn from Ringer, that Da Costa administers by the stomach salicylic acid in five-grain doses to correct the foul breath and offensive expectoration sometimes occurring in phthisis. Berthold of Dresden relates a case which yielded promptly to internal administration of salicylic acid after the failure of turpentine inhalations and large doses of quinia. Lastly, Dr. Starkey has detected carbolic acid in the serum of a blister, and even in the expectoration, although it had not been inhaled.

Now, what has been stated with respect to carbolic acid is equally applicable to creasote, thymol, and other important antiseptic and medicinal substances. That the more volatile medicaments, when oro-nasal inhalers are used—as turpentine, chloroform, ether, &c.—do make their way into the lungs to some extent must be admitted; but even of this little satisfactory evidence has been produced. If with oro-nasal inhalers it were a fact, that such substances as carbolic acid, creasote, and thymol enter the lungs in quantities sufficient to give even a chance of their proving beneficial, then it ought to be very easy to demonstrate their presence in the sputa by appropriate tests.

With respect to the proofs of the entrance of carbolic acid into the lungs, furnished by the characteristic discoloration of the urine, the facts stand thus : This discoloration does undoubtedly sometimes occur, either from the administration of the acid by the stomach or by the use of sprays, which is nearly the same thing, since much of the acid in the atomized liquid is swallowed and does not enter the lungs at all; so that even in these cases the dark green tint of the urine affords no sufficient proof that the acid has entered the system through the lungs. The author believes further, and certainly this is the result of his own somewhat extended experience, that the alteration in the colour of the urine is never brought about by the use of the oro-nasal inhalers as hitherto employed.

The physical composition and characters of vapours vary according to their nature. The vapours from oro-nasal inhalers are for the most part insensible; those of hot water and steam approximate in their physical condition to atomized sprays; the vapour of iodine is molecular; that of arsenious acid and chloride of ammonium crystalline; they all travel but a short distance, and speedily become condensed, mainly by the abstraction of heat and by dispersion in the colder surrounding atmosphere.

The vapour of hot water is often employed as the vehicle for the diffusion and conveyance of medicaments to the organs of respiration, and when these are of a sufficiently volatile character they are no doubt more or less effective, but any little or non-volatile substances contained in the liquid from which the vapour is generated remain in the liquid, and hence are useless. This is a fact which is constantly overlooked, and many prescriptions for inhalation are thus rendered worthless.

The constitution of the several smokes or fumes derived from the ignition, chiefly of different vegetable substances, also varies much; but they mostly consist of watery vapour, certain gases, and volatile empyreumatic and medicinal substances; and when they do not possess irritating properties and produce spasm they find a more ready entrance into the lungs than the vapour of hot water or steam, but this only when certain precautions are adopted.

The results of some observations and experiments may now be quoted, which throw much further light on the question of the entrance of medicaments into the organs of respiration.

The first series of these have reference to inhalation by means of the oro-nasal inhalers now so much in use; they were first made known in a communication by the author, 'On the Comparative Inutility of Antiseptic Inhalations as at present Practised in Phthisis and other Diseases of the Lungs,' and which appeared in 'The Lancet' of May 5, 1883.

The substances experimented with were phenol or carbolic acid, the most important and frequently employed in lung inhalation; creasote, thymol, and iodine, all much used.

One-half gramme of phenol, or 500 milligrammes, equal to about 7·70 grains, was dissolved in water, and placed on the sponge of one of Mayer and Meltzer's cellulose oro-nasal inhalers, and the inhalation was continued for one hour. Of the 500 milligrammes taken, 412 were recovered from the sponge, showing a deficiency of only 88 milligrammes. In the second and third similar experiments the amounts recovered were 413 and 458 milligrammes respectively, showing a loss of only 87 and 42 milligrammes. Thinking that the volatility of the carbolic acid would be aided by dissolving it in some more volatile substance than water, experiments were next made with spirits of wine and spirits of chloroform as the solvents. In three trials with half a gramme of carbolic acid, dissolved in alcohol, the amounts still retained in the sponge at the end of the inhalation were 442, 485, and 480 milligrammes respectively, showing deficiencies of 58, 15, and 20 milligrammes only. The results of two experiments with spirits of chloroform as the solvent were similar, the amounts recovered being 485 and 495; in the one case 15 milligrammes had disappeared, and in the other 5 only.

It may now be stated, that the phenol was quantitatively determined by the very delicate and accurate process of M. Chandelon. It is difficult to suppose that the small quantity of carbolic acid which disappeared during the several inhalations could exercise any marked effect in the alleviation or cure of such a malady as phthisis, even if the whole quantity made its way to the seat of the disease, but this it certainly does not do; some is lost in the respirator itself, more in the manipulations necessary to remove the remaining phenol from the sponge, but a still larger amount is absorbed by the skin of the lips and the mucous membrane of the nose, mouth, and fauces; so that it becomes extremely doubtful whether any portion of this antiseptic ever reaches the air cells of the lungs. This doubt is greatly strengthened by the fact, that the sputa in cases of phthisis, expectorated shortly after inhalation by means of oro-nasal inhalers, never, so far as my experience goes, smell of carbolic acid; neither have I ever found that acid present in the sputa in any notable quantity. Another fact corroborative of this view may here be cited. The air expired after each inhalation of the carbolic acid was passed through distilled water, which was afterwards tested for the acid, the faintest trace only being discovered.

The next antiseptic experimented with was creasote. No volumetric chemical process being known for the quantitative estimation of this compound, the gravimetric method had to be pursued, which, however, furnishes results sufficiently precise for the purpose.

From half a gramme, or 500 milligrammes, of creasote, inhaled with water for one hour, 430 milligrammes were recovered, the creasote being dissolved out of the sponge by means of alcohol; from the same quantity of creasote dissolved in spirits of wine there were recovered in the first experiment 406 and in the second 410 milligrammes; while, lastly, from the same quantity of creasote dissolved in spirits of ether there were obtained 416 and 406 milligrammes respectively. These experiments show, as was the case with the phenol, that upwards of four-fifths of the creasote used remain in the sponge of the inhaler, while of the remaining small quantity some is lost in the inhaler itself, some is absorbed by the mucous membrane of the nose and mouth, while part is lost during the evaporation of the spirits of wine and ether employed as the solvents of the creasote.

The results of the experiments with thymol may now be detailed.

There were recovered after an hour's inhalation, in the first experiment, 412, and in the second 466, out of the 500 milligrammes employed, and which were dissolved in ether. From the same amount of thymol dissolved in spirits of wine, there were obtained 450 and 466 milligrammes. It is therefore scarcely possible to conceive, that this antiseptic can, as at present employed, exert any beneficial effect.

The last antiseptic of which trial was made was iodine. The experiments with that substance were the following, and, as will be seen, they were very conclusive:—200 milligrammes of iodine were dissolved in 10 cubic centimetres of spirits of wine, and of this mixture 1 cubic centimetre, containing ·020 of iodine, was placed on the sponge of the inhaler, with a little alcohol added, and inhaled for an hour. A similar experiment was made with spirit of ether as the solvent. The vapour given off was at first pungent and stinging to the mucous membrane of the nose and mouth, and on examining the sponge at the end of the hour all but a trace of the iodine was found to have disappeared. In iodine, therefore, we really have a sufficiently volatile agent to deal with, and hence it might be presumed that it did, in fact, make its way into the lungs. This conclusion, though apparently warranted by the disappearance of the antiseptic during inhalation, is not confirmed by further observation.

When testing the saliva and mucus of the mouth and throat on the completion of the inhalation, with a solution of starch, I noticed that the colour of the starch was unchanged, proving the absence of free iodine. On applying, however, an acid to the mucus, the blue colour was abundantly developed, showing that very much of the iodine inhaled, and possibly the whole of it, had become converted into an iodide, in which transformation it loses a great part of its antiseptic properties. This is an interesting and important fact, not only in relation to the subject now under consideration, but in other ways. Thus, for one thing, it seems to show, that it is almost useless, when oro-nasal inhalers are used, to administer free iodine as a medicine. Again, it should be known, that when, as is frequently the case, carbolic acid and iodine are inhaled together, a strong chemical action is set up between them, whereby possibly the properties of the iodine are impaired. In the case of iodine, then, evidence is also wanting to show that this antiseptic does really make its way into the lungs as such, when oro-nasal inhalers are employed.

Now it may be urged, that if the inhalation of the several antiseptics had been continued for a longer period than an hour, the result would have been different—that is to say, that more of them would have been inhaled. In order to test this point, the inhalation of carbolic acid was continued for two hours in three experiments, with the following results: Of the 500 milligrammes taken there were recovered by Chandelon's process 410, 400, and 390 milligrammes respectively, thus showing only a very moderate increase, quite insufficient to affect in any material manner the general results previously arrived at. Even had the amount inhaled been much greater, it would by no means have followed that a proportionate increase of the antiseptic was to be found in the lungs. Supposing a small quantity of any of the antiseptics really reaches those organs, it is not to be supposed that it remains there for an indefinite time, and goes on accumulating as long as the inhalation is continued. The action of the absorbents would doubtless come into play, and the antiseptic which was inhaled the first hour would become removed by absorption during the second hour. Again, it might be urged that if smaller quantities of the antiseptics were placed in the inhaler, the proportionate volatilization would be greater. 250 milligrammes of phenol in water were sprinkled on the sponge of the inhaler, and inhalation continued for an hour, at the end of which time 234 milligrammes were recovered from the sponge, showing rather a smaller and not a larger proportionate loss.

It thus appears that in the case of phenol or carbolic acid, creasote, and thymol, more than four-fifths of the quantities originally taken, were still present in and were recovered from the sponge at the end of the inhalation, and further that the small amounts of those substances which had actually disappeared were in great part to be accounted for in other ways than by supposing that they had entered the lungs—namely, by loss in extraction from the sponge, by loss in the inhaler, on the skin of the chin and cheeks, and on the mucous membrane of the mouth and fauces of the person inhaling. This result is in no respect surprising, and is only what naturally might have been expected by anyone conversant with the chemical properties of the antiseptics experimented with. The volatility of carbolic acid, creasote, and thymol at ordinary temperatures is but slight, and in winter often nil, and it is nearly always necessary in using oro-nasal inhalers that some means should be adopted whereby an increase of temperature is obtained sufficient to ensure the necessary degree of volatilization.

A second series of observations and experiments differed considerably from those just detailed. They were communicated to the British Medical Association at its meeting in Liverpool in August 1883, and were afterwards published in the journal of the

Association of November 3, 1883, in a paper which bore the following title : ' On Inhalation, more particularly Antiseptic Inhalation, in Diseases of the Lungs.' The author contrived the following arrangement in imitation, as near as was practicable, of natural respiration :—A syringe was prepared, having a capacity of 200 cubic inches, and provided with a hollow piston having a valve at each end. The lower valve prevented the escape of the air in that direction when the piston was pushed down, while the upper one at the same time of course allowed of its escape upwards. A Woulfe's bottle, filled with water, or a bulb tube filled with spirit, as the case might be, was next attached to the distal end of the syringe ; this bottle or bulb in its turn was joined on to the termination of the trachea of an unskinned sheep's head and neck. The mouth and nares were covered with a well-fitting papier-mâché respirator, furnished with the usual cribriform receiver and sponge. The syringe of course was intended to take the place of the lungs, and the Woulfe's bottle to intercept any of the substances used, which might pass from the respirator down the trachea, with a view to their subsequent determination. The syringe was capable of being worked at the rate of about 250 times per hour.

The first experiment was with half a gramme, equal to a little more than 7½ grains of carbolic acid. After the rapid action of the syringe for an hour and a half, the water in the Woulfe's bottle was tested by Chandelon's process, with the result of finding in it only ·003 gramme of carbolic acid—that is to say, a mere trace. A second experiment furnished a similar result.

Such being the results of the experiments with carbolic acid, it is not probable that any with creasote and thymol would have been different, having regard to the relative volatility of these substances. Thymol, indeed, at ordinary temperatures, is far less volatile than carbolic acid.

In the next experiments much more volatile substances were selected, namely, eucalyptol and oil of turpentine, both also possessing strongly penetrating odours. In the case of the eucalyptol, the faintest possible odour only, recognizable with difficulty, was perceptible in the alcohol through which the air was passed. With the turpentine, however, the spirit was found to possess a decided smell of this oil, although it was obvious that the quantity actually present was very minute.

Lastly, an experiment was made with iodine; of this two cubic centimetres of an alcoholic solution were taken, containing ·06 gramme of iodine (equal to about nine-tenths of a grain), a much larger quantity than that ordinarily employed, in consequence of the irritating nature of the fumes. The alcohol through which the air issuing from the trachea was passed, did not become coloured in the slightest degree, nor did it furnish any evidence whatever of the presence of iodine. This experiment was performed a second time with another sheep's head and neck. The saliva of the mouth, and the mucus, were also tested for free iodine; not a trace was present; but, after the addition to the mucus of a little dilute nitric acid, it was freely liberated. This shows that the iodine had really entered the mouth, but had become converted into an iodide, of course losing in the process much of its antiseptic properties. In this respect the experiment accorded exactly with what has elsewhere been shown to take place in the living human subject.

Of course it may be urged in objection that the results of the experiments just recorded would have been different in the case of the human subject; and this is to some extent possible, although scarcely probable, seeing that the construction of the larynx of the sheep is less complicated than that of the human subject, and that the passage is much more open and free. However this may be, the experiments are confirmatory of those previously quoted, and they show that substances of little volatility at ordinary temperatures, such as carbolic acid, creasote, and thymol, make their way into the lungs, when oral or oro-nasal respirators are used, with difficulty, and in very minute quantities only, while even substances with a considerable volatility, as eucalyptol and oil of turpentine, do not always find a ready access.

But the results of a variety of other observations and experiments designed to test the efficacy of inhalation in other forms were described in the communication already referred to. One of these forms embraced inhalation by the vapour of hot water.

The British Pharmacopœia contains five formulæ for the inhalation of medicaments by the aid of the vapour of hot water; two of these are Vapor Coniæ and Vapor Creasoti. Considerable discrimination has been exercised in the choice of these two substances, and in the methods described for their use, seeing that the alkaloid of conium is volatile, as of course is also the creasote to some extent.

For the Vapor Coniæ, The Pharmacopœia directs that twenty minims of a mixture of one part of the extract of hemlock with one part of a solution of potash and ten parts of water should be put on the sponge of a suitable apparatus, so that the vapour of hot water passing over it may be inhaled. It will be observed that the twenty minims of the mixture will contain less than two grains of the extract of conium. The quantity of the active principle, conia, contained in the two grains is infinitesimal, and may be said to be homœopathic; it may be taken to be rather over one-tenth of a grain per 100 grains of the extract. Two cubic centimetres of the mixture referred to above, equal to about thirty-one grains, and containing ·16 gramme of the extract, or rather more than two grains and a half, were placed on a piece of sponge, suspended in the neck of a retort containing 250 cubic centimetres of distilled water. Twenty-five cubic centimetres were distilled over and tested with Mayer's general reagent for alkaloids (potassio-iodide of mercury), but not a trace of conia could be discovered. After slow evaporation to dryness at a very low temperature, a faint reaction only was obtained with Mayer's reagent. Subsequently a similar experiment was made with ·01 gramme of the alkaloid itself, or rather more than one-tenth of a grain; but in this case the alkaloid was put into the water, and not placed on the sponge, as in the previous experiment. The first twenty-five cubic centimetres distilled over gave a distinct reaction with Mayer's reagent, but the second only after concentration almost to dryness. It thus appears that, in Mayer's reagent, we have a most delicate test for conia; and, that point being determined, an experiment was made with the extract of conium, in the quantity and under the conditions set forth in The Pharmacopœia. The result was that not a trace of the alkaloid was to be detected in the alcohol through which the vapour was passed, even after careful evaporation, almost to dryness.

With respect to the employment of the Vapor Creasoti, the following instructions are given in The British Pharmacopœia. Twelve minims of creasote are directed to be mixed with eight ounces of boiling water, in an apparatus so arranged that air may be made to pass through the solution for inhalation. With a view of testing how far this method of employing the creasote is effective, the two following experiments were made:—Half a gramme, equal to 7·7 grains, was added to 300 cubic centimetres of water, and the vapour, mixed with air, was drawn through alcohol for fifteen minutes; this of course retained whatever creasote passed over. The alcohol was afterwards found to contain only ·007 gramme of creasote, or about the $\frac{1}{72}$nd part of the amount originally taken. In the second experiment the same quantity of creasote was taken, but in a somewhat differently constructed apparatus, with tubes of a greater diameter, so as to allow of a freer passage for the vapour. In this case ·026 gramme of the creasote was recovered, equal to about the $\frac{1}{19}$th part of the original quantity.

It may be safely affirmed, therefore, that the method of inhaling conium and creasote by the vapour of hot water, as prescribed by The British Pharmacopœia, is most ineffective, and almost, if not quite, valueless.

The pharmacopœia of The Hospital for Diseases of the Throat contains a variety of formulæ for medicated inhalation by the vapour of hot water. I will select two of the most important of these and put them to the test of experiment. One of them is the Vapor Acidi Carbolici. Twenty-one drachms, or 1260 grains, of the acid are directed to be dissolved in three drachms of water, and of this mixture a teaspoonful, say equal to a drachm, and containing the very large quantity of 52·5 grains of phenol, is to be put into an eclectic inhaler, containing a pint of water at a temperature of 150° Fahr., and maintained at that temperature by the aid of a spirit lamp. A quantity of the mixture containing exactly three grammes and a half of carbolic acid, equal to 53·9 grains, was put into the inhaler, the inhalation being continued for the space of twenty minutes; after which 3·44 grammes of the acid, equal to 53·08 grains, were recovered from the inhaler, the loss thus amounting to ·06 gramme, equal to only ·82, or four-fifths of a grain. It must not be supposed, however, that the whole of even this small amount was actually inhaled; part, no doubt, was dissipated in the uninhaled vapour, while of that which really entered the mouth, some was absorbed by the mucous membrane of the mouth and fauces.

The air expelled from the lungs at each expiration was also tested for carbolic acid, the quantity found being excessively minute, namely, ·0076 gramme, equal to a little over one-tenth of a grain.

The second formula selected from The Throat Hospital Pharmacopœia for experiment was that for Vapor Thymolis, or thymol. The directions in this case are, to dissolve twenty grains of thymol by means of three drachms of spirit, and to make up with water to twenty-four drachms, or three ounces; a teaspoonful of the mixture to be added to a pint of water at 150° Fahr. Contrast for a moment the very large quantity of carbolic acid employed in the first inhalation—namely, 52·5 grains—with the minute amount of the thymol, less than one grain, employed in the second inhalation. Thymol, though little volatile at ordinary temperatures, melts readily in hot water, and then becomes very diffusible. Of course, the greater part of the portion of a grain contained in the 8750 grains of the pint of water was volatilized in the vapour; but what possible curative effect could be expected to result from so minute a quantity of thymol, even if the whole were really inhaled?

When the substances added to hot water possess a high degree of volatility, and are employed in considerable quantity, and when, at the same time, the temperature of the water is maintained by means of a lamp, vapour inhalation may be practised in some cases with benefit, especially in affections of the throat. In some instances the warm vapour of the steam itself, unmedicated, proves serviceable, although it is surprising how little water really passes over in most cases, except the inhalation be continued for a long time and the temperature be maintained by the aid of a lamp. In the two experiments made with the vapor creasoti of The British Pharmacopœia, the loss of water amounted to only 10 and 12 cubic centimetres respectively, equal to about $2\frac{1}{2}$ and 3 drachms.

We may now pass on to treat of inhalation by steam. Steam, of course, does not differ essentially from the vapour of hot water, only that the vapour is generated faster and the temperature is higher. This temperature, however, is rapidly reduced by contact and intermixture with the air in which it becomes diffused, the vapour or steam with equal rapidity becoming condensed and reduced to particles or atoms of sensible dimensions. In fact, by the time the steam reaches the air passages it is for the most part reduced to the condition of an atomized liquid or spray.

It will be well to refer here to the fact, which, however, is constantly ignored or forgotten in many prescriptions for inhalation, that no substance is volatilized, and passes over in the vapour of steam, which is not itself more or less volatile at the temperature of the water or steam. Substances which are volatile at ordinary temperatures have, of course, their volatility greatly increased at the boiling point of water. Whatever substances, therefore, are contained in or added to water and which are not volatile at 100° C. or 212° F., will not pass over by distillation, but will be found in the residue of the retort or still. It is thus useless to prescribe for inhalation by the vapour of hot water or by means of oro-nasal inhalers, as is frequently done, such remedies as preparations of opium, cannabis Indica, stramonium, hyoscyamus, and many other medicinal substances.

To show the effect of the temperature of boiling water in increasing the volatility of substances which, at ordinary temperatures, are but little volatile, the results of the three following experiments may be given. Carbolic acid or phenol, notwithstanding the strong odour which it emits, is, under ordinary circumstances, but little volatile. Half a gramme of this, or rather more than $7\frac{1}{2}$ grains, was placed in a retort, and distilled with 250 cubic centimetres of water; and of this one-fifth part, or 50 cubic centimetres, was distilled off, ·12 gramme or 1·85 grains being found in the distillate by Chandelon's process. In a similar experiment with creasote, which is more volatile than phenol, ·17 gramme, or 2·6 grains, passed over; while, lastly, in the case of thymol, which is scarcely at all volatile at ordinary temperatures, no less than ·267 gramme, or 4·1 grains, was obtained. Had the distillation been carried further, the quantities recovered would have been proportionately increased.

Steam, therefore, especially when given off in a concentrated form, does carry over a very considerable amount of the antiseptic substances referred to above.

In some steam inhalers, as in that of Dr. Lee, the substances used are added to the water itself, prior to its being boiled; but usually, as in Siegle's steam inhaler, the medicament is not put into the receiver itself, but into a separate receptacle or bottle. In the latter case, the steam and medicated liquid come into contact at the points of the capillary glass tubes, the hot and rapidly moving steam producing a vacuum in the ascending tube, which causes the liquid to flow up it. The force of the jet of steam

ENTRANCE OF MEDICAMENTS INTO THE LUNGS. 17

atomizes the medicated liquid, and the contact of the cold air condenses it as well as the steam itself. Whether the medicament used be added to the water before boiling, or be contained in a separate vessel, the actual result is very similar, although the quantity of liquid used is, of course, greater in those cases in which two receiving vessels are employed. In Dr. Lee's spray producer, the medicaments contained in the water in the boiler pass over bodily with the steam till the receiver is emptied, and this to some extent whether they are volatile or not.

I will now proceed to state the results of experiments with Siegle's steam inhaler.

The vapour of steam, as already stated, coming into contact with the colder air, very rapidly cools and condenses; so that, if the mouth be applied within a few inches of the spray, the temperature becomes bearable, while the condensation is shown by the rapid deposition of moisture in the track of the jet of steam. Attentively watching the action of a steam inhaler, it is seen, first, that much of the steam is deposited in the vicinity of the inhaler; secondly, that a considerable portion of the steam spray does not enter the mouth at all; thirdly, that part of that which enters the mouth is returned during the act of expiration: indeed, it has appeared to me that not one-third of the steam generated is actually retained, and most of that which is so, there is good reason to believe, settles upon and is absorbed by the mucous membrane of the mouth, cheeks, and fauces. But another and fourth great cause of waste is, that the steam spray is always in operation, whereas the act of inspiration probably does not occupy much more than one-third of the whole time, some ten or fifteen minutes, consumed in the inhalation; so that, from this cause alone, it may safely be affirmed that fully two-thirds of the whole quantity of the medicament employed is lost, and can be of no utility whatever for the purpose in view. Lastly, it may be pointed out, that in using steam inhalers, respiration is carried on almost entirely by the mouth.

The following experiments with steam inhalers will show, in a measure, how great the loss is from the causes above mentioned.

It may be remarked, at the outset, that it is most difficult to determine the actual loss which takes place in using a steam inhaler, owing to the rapid diffusion of the vapour and the extreme difficulty of confining and condensing it all. With a view of estimating the loss, an apparatus was constructed, consisting of a large hood, fastened round the neck and furnished with a long chimney, which was made to pass through a Liebig's condenser and to terminate in a flask surrounded with ice, the upper part of the chimney also being packed round with ice. It will thus be seen that the surface over which the vapour of the steam spray was necessarily spread was very large, and the difficulty of recovering the whole of the material taken proportionately great.

Taking 0·750 gramme of carbolic acid, = 11·55 grains, there was recovered, as the mean of two experiments, 0·368 gramme, = 5·67 grains, of the acid. In an experiment made, using the same quantity of carbolic acid, by allowing the steam jet simply to expend itself in the hood, without any inhalation, there was recovered 0·470 gramme, = 7·24 grains. This experiment is important, the difference in the amount recovered when the steam spray was inhaled and without inhalation being but 0·102 gramme, or 1½ grain.

There is still another form of spray often employed, namely, the cold spray, the motor power being air. In this case the liquid condenses still more readily, and the moisture is deposited in a more limited area. This spray may be used either intermittently—that is to say, the jet may be thrown out only during inspiration—or continuously, when of course there will be the same loss of material as in the case of the continuous steam spray. In two experiments with the air spray, the same quantity of carbolic acid being taken—namely, 0·750 gramme—the mean amount recovered was 0·438 gramme, = 6·74 grains; while in an experiment with the hood empty, there was recovered 0·451 gramme, = 6·94 grains, being only a very little more than was obtained after inhalation.

With a hood and other apparatus in every way perfect, I entertain no doubt, that I could succeed in recovering larger quantities of the carbolic acid than those obtained in the above experiments with the steam and air sprays.

That a quantity, small in comparison with the amount originally taken, of such medicinal substances as carbolic acid, thymol, &c., really makes its way into the lungs, when steam or other sprays are used, would appear to be further shown by the result of experiments with those sprays and the apparatus

ENTRANCE OF MEDICAMENTS INTO THE LUNGS. 19

referred to already, namely, the large syringe and the head and neck of a freshly-killed sheep. Thus half a gramme of carbolic acid, dissolved in water, was placed in the bottle of a Siegle's inhaler, and the steam generated in the usual manner; after the syringe had been in operation for fifteen minutes the water in the Woulfe's bottle through which the air was passed was tested, but not a trace of the acid was found to be present.

Such, then, are the results of the author's experiments on inhalation, especially antiseptic inhalation, as at present ordinarily practised : 1. with the vapour of hot water, medicated or otherwise ; 2. with oro-nasal inhalers ; and 3. with steam and air spray producers.

The following conclusions may now be deduced, founded partly on the foregoing observations and experiments :—

1. That substances of a gaseous nature and unirritating character pass, as might indeed be assumed would be the case, readily into the air passages and lungs.

2. That the fumes derived from the burning of certain mineral and organic substances, when unirritating and inhaled with certain precautions, also enter with comparative facility.

3. That the same may be affirmed of the vapours of many volatile substances, and particularly that of hot water, although much of this is apt to be condensed and deposited prior to its entrance into the lungs. Except in the case of the more volatile substances, and when the temperature is maintained by the aid of a spirit lamp, the inhalation of medicaments by the vapour of hot water is, as we have seen, but little effective; these, even when vaporized in any considerable quantity, are expended rather on the mouth and fauces than upon the lungs, and by far the greater part of the substances used remains, in the majority of cases, in the inhalers themselves.

4. That the medicinal substances, whether volatile or nonvolatile, contained in sprays, either warm or cold, also reach the lungs, when proper precautions are observed; though in greatly diminished quantities, part being lost before the entrance of the spray into the mouth, and part being swallowed, or absorbed by the pharyngeal mucous membrane. No doubt, by the employment of sprays, either hot or cold, any quantity of a given substance may be thrown into the mouth, applied to the fauces, and even introduced into the stomach; and, so far, they are

c 2

effective; but, from the evidence adduced, there is reason to believe that but a proportionately small amount of the substances employed, as the sprays are at present used, really reaches the lungs, or even the larynx and trachea.

5. That the employment of oral and oro-nasal inhalers for the volatilization and inhalation of such substances as carbolic acid, creasote, thymol, and many others which are, in fact, but little volatile at the ordinary temperatures of the air, or which may even be entirely non-volatile, is of but little utility in general, and is in many cases, entirely delusive. When such inhalers are used for the diffusion and inhalation of the more volatile medicaments, as alcohol, chloroform, ether, or turpentine, they are no doubt more effective. When it is remembered, that inhalation in affections of the organs of respiration, is carried on in Great Britain, chiefly by means of oral and oro-nasal inhalers, the practical importance of the foregoing conclusion will be at once apparent. I would again recall to mind the fact, that four-fifths of the more important antiseptics, such as carbolic acid, creasote, and thymol, are, with the ordinary oro-nasal inhalers in use, recoverable after the completion of the inhalation.

But the question may be asked : cannot the patient himself tell whether the medicaments inhaled really enter the air passages? The answer is, that the mucous membrane of those parts is but little sensitive, and that sometimes he can and at others he is unable to say whether they have effected an entrance or not. If the inhalations are either decidedly hot or cold, or if the substances employed are powerful, they may give rise to sensations and feelings of a positive character, as of cold or warmth, of constriction or soreness, and these would be first felt high up behind the sternum, and as the inhalations penetrated farther and deeper, the area of the sensations would become proportionately increased. If, however, the substances inhaled were mild and unirritating, and of the ordinary temperature, the patient would be unconscious of their entrance.

CHAPTER II.

THE PRINCIPLES CONCERNED IN THE VOLATILIZATION AND INHALATION OF MEDICAMENTS.

THERE are several principles and circumstances which govern and affect the volatilization of the medicaments employed in inhalation. The nature of some of these is either not understood, or in some cases altogether overlooked. In fact, the chemistry and physics of inhalation have been but little studied by medical men.

The chief circumstances to be taken into consideration are: the relative volatility of the substances employed, temperature, relative humidity of the air, the motion of the air, the extent of the surface of exposure, the physical condition of the medicaments themselves, whether in solution or not, whether wholly or only partially dissolved, the nature of the media by which they are held in solution, and, lastly, the manner in which they are inhaled.

With respect to volatility, medicaments vary greatly; some pass into a state of vapour rapidly at the ordinary temperature of the air, as alcohol, chloroform, and ether, while others do so only slowly and with difficulty, as carbolic acid, creasote, and thymol.

The expression, ordinary temperature of the air, is a very wide one, and may include all degrees of temperature above 32° F., the freezing point, up to 76° F., or summer heat, a range of no less than 44°. Now the rate of evaporation of any given substance will vary greatly between this range; it will be much diminished as the lower limit is approached, and still more augmented in proportion as the higher limit is reached. Thus taking carbolic acid as an example, the antiseptic on which so much reliance is placed in this country, the evaporation at or near the freezing point will be almost nothing, while at summer heat it will be greater, though still small.

This is well shown in the following table, kindly prepared at my suggestion by Mr. Edwy G. Clayton, F.C.S.

TABLE I.

Half a gramme, = 0·500, of pure crystallized phenol was taken in each experiment, with 4 c.c., or as nearly as possible 1 drachm, of rectified spirit. The specific gravity of the latter was 0·842.

The dishes used were all similar in size and shape, and the diameter of the *surface of the liquid exposed* was in each case $1\frac{7}{8}$ inch. The diameter across the *top* of each dish was $3\frac{1}{8}$ inches.

A free current of air was allowed to pass through the room during all the experiments, two windows and a ventilator being open the whole time.

In the first experiment the temperature was obtained by the aid of ice, the dish floating in water at 32° F.

No. of Experiment	Temperature at which Volatilization took place	Amount of Phenol recovered, estimated by Chaudelon's method
	° F.	
1	32	0·497
2	40	0·492
3	45	0·489
4	50	0·486
5	55	0·477
6	60	0·473
7	65	0·468
8	70	0·457
9	75	0·446
10	80	0·431

N.B. The time of exposure was in each case *two hours*.

These experiments show, in the first place, a steadily increasing volatilization with increase of temperature, and in the second, that the extent of the loss even at 80° F., and after two hours' exposure, was but small, it amounting to only 0·069 gramme, = 1·38 grain. This last result confirms in a very conclusive manner the accuracy of the statement, founded on experiments, already more than once referred to—namely, that fully four-fifths of the carbolic acid are still retained in the oro-nasal respirator at the completion of the inhalation.

The second table, also prepared by Mr. Clayton at a later date, is equally instructive, and shows the rate of evaporation at temperatures between 50° and 100° F., in the case of some medicaments having a higher degree of volatility than carbolic acid.

THE VOLATILIZATION OF MEDICAMENTS.

TABLE II.

Tabulated Results of some Experiments made with the View of Determining APPROXIMATELY *the Quantities Lost of Certain Solids and Liquids during their Evaporation for One Hour at each of the following Temperatures:* 50°, 60°, 70°, 80°, 90°, *and* 100° *Fahrenheit.*

Name of Substance	Boiling Point °F	Specific Gravity	Quantity of Substance taken for each Single Experiment	Percentage Proportion by Weight Lost during Evaporation at					
				50° F.	60° F.	70° F.	80° F.	90° F.	100° F.
			Grains						
Ol. eucalypti	347	0·905	40	1·8	3·7	6·0	8·8	9·9	13·2
Ol. juniperi	320	0·855	40	2·2	3·9	6·3	9·1	10·3	12·9
Ol. pini sylv.	312	0·877	40	2·3	3·7	5·9	9·2	10·7	13·3
Ether	95	0·713	40	95·9	—	—	—	—	—
Chloroform	141	1·491	40	58·0	61·4	62·9	64·7	67·0	70·1
Spirit. vini rect.	174	0·838	40	5·5	7·8	10·2	14·1	19·3	26·7
Benzene (coal tar)	176	0·85	40	7·1	9·4	11·4	14·9	19·9	28·7
Iodine	347–356	4·948	40	4·8	6·2	7·9	10·0	13·8	16·8
Thymol	428–446	1·028	40	0·3	0·5	0·9	1·4	2·2	2·9
Creasote	422	1·071	40	0·4	0·7	1·4	1·9	2·6	3·5
Phenol	358	1·065	40	0·6	0·9	1·7	2·2	2·7	3·8

Then there are other medicaments which are termed non-volatile, because they do not evaporate at ordinary temperatures of the air. It is obvious, that this last large class can only be employed in inhalation in the form of atomized liquids or sprays, and not in the vapour of hot water, nor by means of oro-nasal inhalers.

While the rate of evaporation varies greatly within the limits of the ordinary temperature of the air, it augmenting in proportion as the upper limit is approached, the rate is, with certain substances, still further very greatly increased when this is passed—that is to say, at temperatures between 76° and 212° F. This fact will be demonstrated by certain experiments, the particulars of which will be hereafter narrated. The effect of temperature in augmenting or retarding the volatilization of many of the medicaments employed in inhalation was, up to the time of the publication of my various communications on inhalation in the 'Lancet' and in the 'British Medical Journal,' but little understood.

The relative humidity of the air makes also a very considerable difference in the rate of evaporation; a dry air, as is well known, promotes it, and a moist condition of the atmosphere exerts a retarding effect.

Again, the condition of the air as to movement exerts a considerable effect. If the air be still, vaporization proceeds slowly, and if it be both still and humid, then the rate is still more retarded; if, on the other hand, the air be in active movement, if a light breeze or strong wind be blowing, then the ratio is surprisingly accelerated. Ordinarily when a strong wind is blowing the air is very dry. If to the actively moving air increased temperature be superadded then the maximum effects are produced.

But there is still a most important principle to notice in connection with the subject of volatility, and that is the effect produced by augmenting the surface of exposure or evaporation. This principle in its application to the subject of inhalation was not at all understood till the author drew attention to it, first, in a paper 'On the Principles of the Construction of Chambers for Inhalation in Diseases of the Lungs,' which was communicated to the British Medical Association at its meeting in Liverpool in August 1883, and which afterwards appeared in the Journal of the Association of January 12, 1884; and secondly, in two articles in the 'Lancet,' the titles of which will be given later on, and which were published in the numbers of that journal for October 6, 1883, and January 19, 1884.

It occurred to the author, that since the surface of exposure of most medicaments employed in inhalation, by means of oro-nasal inhalers, was exceedingly small, if this surface could be very greatly increased there would be a proportionate augmentation of volatilization. Thus, say the surface of exposure in an ordinary oro-nasal inhaler was 2 inches, and that this surface could be extended to 2000 inches; the results obtained would be a thousand times enhanced.

The results of actual experiments were most surprising, and proved that the principle was a sound one, even in the case of substances whose volatility at ordinary temperatures is comparatively so slight, as carbolic acid, creasote, thymol, &c.

This principle was at first thus tested :—

Fifty grammes of carbolic acid, $= 771$ grains, were dissolved in water, two Turkish towels being saturated with the solution and exposed to the air of a closed room at a temperature of $22°$ C., $= 72°$ F.

After the lapse of forty-eight hours the towels had become

quite dry, and the author was not a little surprised on examining them to perceive that they did not possess the slightest smell or taste of the carbolic acid. One of the towels was then tested, and found to contain only 0·07 gramme of the acid, equal to 1·08 grain; that is to say, the whole of the carbolic acid employed, namely, 771 grains, had disappeared, with the exception of the small quantity above referred to.

This result appeared so surprising, although the author was prepared for a considerable reduction in the amount of the acid taken, that he thought the towels must have been inadvertently washed. He therefore repeated the experiment in a more precise manner, taking one of the Turkish towels only. It measured 48 by 18 inches, thus presenting a surface of 1728 square inches, and it was saturated with an aqueous solution containing 25 grammes, or 385 grains, of carbolic acid. The experiment was commenced at 10 A.M., the towel being moistened in the course of the day once or twice with a little water. At 10 A.M. the next day the towel was still somewhat wet, but at 4 P.M. it was only just damp. It was then tested for the acid, 0·09 gramme only being obtained, equal to 1·39 grain. Thus in the space of 30 hours the whole of the acid used, 385 grains, had practically disappeared. Had the same quantity of the acid been exposed to the air in any vessel presenting only a small surface, the loss would have been but trifling.

This result is of the highest importance, as it places in our hands the means of charging inhalation chambers to almost any desired extent with all those medicinal substances, which possess even a feeble volatility, at the ordinary temperature of the air.

The following experiments were also made with carbolic acid. Turkish towels were cut into pieces, measuring 18½ inches in one direction and 9½ inches in the other. Each of these was charged with 5 grammes, = 77 grains, of carbolic acid, dissolved in about 100 cubic centimetres of water, each piece being tested at intervals of 4 hours.

There were recovered out of the 5 grammes, or 77 grains, taken—

After 4 hours 2·77 grammes = 42·66 grains
,, 8 ,, 1·82 ,, = 28·03 ,,
,, 12 ,, — ,, = — ,,
,, 16 ,, 0·49 ,, = 7·55 ,,
,, 20 ,, 0·20 ,, = 3·08 ,,

These experiments were conducted in a closed chamber and at a comparatively low temperature. Had the chamber been freely open to the air, or had the temperature been raised to even 62° F., the results would have been far more striking. In place of 20 hours being required for the almost complete volatilization of the acid, much less than half that time is necessary.

As illustrating the combined effects of extension of surface and free exposure to the air, the following experiment may be cited:— A portion of a Turkish towel, 18 by 9 inches, and charged with a watery solution containing 100 grains of carbolic acid, was exposed for 4 hours in the outer air, with a light breeze blowing and the sun shining at the time, the temperature being 90° F. in the sun. At the end of that time the towel was found to be quite dry, and the whole of the carbolic acid had disappeared, with the exception of 4·74 grains.

The following experiment stands in marked contrast with the above. In this case the portion of Turkish towel, charged with the same amount of carbolic acid, was placed in the still air of a room without a fireplace, having a temperature of 58° F., the door, window, and shutters being closed, and allowed to remain for 10 hours, at the end of which time no less than 58·2 grains of the acid were recovered.

Further experiments showing the increased volatilization of certain medicaments, resulting from the extension of the surface of evaporation, will be found detailed in Chapters III. and IV., those treating of the apparatus employed in inhalation and of the construction of true inhalation chambers.

It will also be proved that this principle holds good in the case of creasote and thymol and some other medicaments.

It is thus seen that the idea or principle on which the foregoing and many other experiments were founded was correctly conceived, but when the principle is combined with increase of temperature some further surprising results were obtained. The reader should be reminded, that the experiments just narrated were made at the ordinary temperatures of the air.

The increase of temperature should vary with the volatility and other properties of the substances employed, and should be carefully regulated.

One of the best and most practical methods of applying this increased temperature, is by means of hot water spread over a comparatively large surface. Owing, however, to the increase of volatility obtained by the augmentation of temperature, the extent of surface required will be very many times less than that which is necessary when Turkish towels or other suitable fabrics are placed in the air at any ordinary temperature, while the time occupied will also be proportionately reduced—two great advantages.

The effect of the combination of a somewhat large surface with increase of temperature is well illustrated by what takes place in the case of thymol. 40 grammes of thymol, = 616 grains, were melted in hot water in a dish having a superficies of 64 inches, the temperature being maintained at 72° C., = 161·6° F. At the end of 12 hours it was found that the whole of the thymol had disappeared, a substance which, as already remarked, possesses but little volatility at ordinary temperatures, and which, even when spread over the large surface of a Turkish towel, is far less volatile than carbolic acid, creasote, and a variety of other substances.

It is difficult to overrate the practical importance of these principles. By means of one or other of them, or of both combined, chambers may be quickly charged with the most powerful antiseptic substances, and indeed with a great variety of other medicaments which possess even a very low volatility at the usual temperatures of the air, and this to any extent required. Guided by these principles, I have been led to devise a number of improved forms of apparatus for inhalation.

I may also point out that the principles above described may be applied with great effect to disinfection as well as inhalation.

But there are still other circumstances to be noticed which exert a considerable effect in promoting volatilization; one of these is that the substance should be either in a very fine state of division, or, better still, entirely liquid or held in complete solution.

Another circumstance which has to be considered, is the nature of the media in which the substance is dissolved, and which are often added to the prescription under the expectation that the volatility of certain not very volatile medicaments is thereby promoted. This has appeared a point of so much importance, that I have had it put to the test by certain experiments, which Mr. Clayton has also kindly carried out at my suggestion.

Five separate quantities of pure crystallized phenol, weighing exactly 0·5 gramme each, = 7·7 grains, were exposed for *four hours* to a temperature of 21° C., = 69·8° F., in five dishes of similar size and shape. To four of the weighed quantities of phenol were added 4 c.c., = 61·7 grains, of certain solvents, namely, 4 c.c. of water, 4 c.c. of rectified spirit, 4 c.c. of chloroform, and 4·4 c.c. of ether. The fifth quantity was not treated with any solvent, so that the evaporation of the substance alone could be compared with its volatilization in the presence of solvents. The results were as follows :—

0·5 gramme alone	0·4637
,, water	0·4610
,, rectified spirit	0·4603
,, chloroform	0·4321
,, ether	0·3746

All the estimations were made by titration, according to Chandelon's process, the solutions used being carefully standardised.

The figures above given show, that under the conditions of the experiments, the volatilization is increased somewhat, in accordance with the volatility of the dissolving menstrua, the loss in each case being, with water 0·0027, chloroform 0·031, and ether 0·089. It will be seen that the loss is the greatest with ether, but even in this case it does not amount to one-fifth of the quantity of phenol originally taken.

Now these experiments, while they are conclusive as to the general principle, are not entirely applicable to the case of oro and oro-nasal inhalers.

In the first place, the ordinary periods during which the inhalation lasts when these inhalers are used is from one to two hours; so the above small losses would have to be divided by 2 or 4, according to the length of time. Again in the experiments, the solvents in the dishes were in the fluid state the whole time, whereas in the sponge of the inhaler the solvents are more or less atomized, whereby the rate of their volatilization is so greatly enhanced that the alcohol, chloroform, or ether are dissipated in a very short time, usually in a few minutes. The volatilisation of the phenol is not, however, accelerated in the same proportion as is sufficiently established by the experiments with oro-nasal inhalers already recorded, in which in every case, no matter what the solvent employed was, four-fifths of the amount of the carbolic acid originally taken were recovered from the inhaler at the termination of the inhalation.

The general conclusion in using oro-nasal inhalers therefore is, that the device of adding volatile solvents to the carbolic acid and other substances, as is now so generally done, with a view to increase their volatility, is of extremely little practical utility.

Lastly, the mode of inhalation exerts some, but by no means a considerable, effect in increasing the volatility of such slightly volatile substances as carbolic acid, creasote, and thymol; that it is not great is shown by the fact, so often adverted to, that four-fifths of these substances are recoverable from the oro-nasal inhalers at the completion of the inhalation, although these inhalers were used by people of average vigour and lung capacity.

The method of inhalation does, however, make some difference in the amount of any substance which, after its volatilization, is carried into the lungs. If the inspiration be feeble and shallow, little or none will penetrate into the air passages; but if deep, prolonged, and forcible, the quantity will be more considerable. This constitutes another difficulty in the use of oro-nasal inhalers on the part of feeble patients with extensive lung disease, and these are just the persons on whom inhalation, to be at all effective, should be carried out in the most efficient manner possible.

In smoking tobacco in the ordinary manner, scarcely any of the fumes enter the lungs; they do not, in fact, usually pass beyond the epiglottis, and smokers are well aware of the fact that if they wish the smoke to pass into their lungs they must inspire deeply and continuously.

CHAPTER III.

THE APPARATUS EMPLOYED IN DISEASES OF THE ORGANS OF RESPIRATION.

The various kinds of apparatus employed in inhalation have been designed and constructed for several different purposes—
1. For the inhalation of atmospheric air and gases.
2. For the inhalation of medicated vapours.
3. For the inhalation of atomized liquids in the form of spray.
4. For the inhalation of hot aqueous vapour, or steam, medicated or non-medicated.
5. For the inhalation of various kinds of fumes or smoke, arising from the combustion of certain mineral and vegetable substances.
6. For the inhalation of powders.

These several purposes are so different, that the apparatus required must necessarily vary greatly according to the object to be fulfilled. It is not my purpose to describe all the various forms of apparatus which have been constructed and used, but chiefly those which are practically useful and generally employed. A great many of the forms devised are quite useless and obsolete, while many more are merely unimportant modifications of each other.

APPARATUS FOR THE INHALATION OF AIR AND GASES.

The inhalation of atmospheric air, either compressed or rarefied, and of gaseous substances, is in this country but little practised, indeed not so much as is desirable.

A great multiplicity of contrivances, many of a very elaborate character, have been devised for the inhalation of gases. To the great majority of these it will not be necessary even to allude in this work; some of them are very costly, while others are of little real utility and are rarely, if ever, employed. Those who desire very full information on this subject are referred to the treatises of Dr. Cohen and Dr. M. J. Oertel.

The principal and most natural gas inhalation is that of the air itself, its physical condition being modified in accordance with certain requirements.

Cooled and Warmed Air.

It is sometimes of advantage that the air be made cool before being inspired, at others that it should be warmed.

The air may be cooled by causing it, before it is inhaled, to pass through a coiled pipe or worm surrounded with ice, such as that used in ordinary refrigerators. The apparatus represented in fig. 1 is constructed on this principle.

Substituting hot for cold water, we have an apparatus for warming the air, which might be so constructed that the degree of warmth could be regulated and maintained at any point desired.

Dr. Joscelyn Seaton's Respirator.

Some years ago Mr. Jeffrey invented an oral respirator, which he designed for the purpose of warming the air on its passage through the mouth to the lungs. In it he utilized the caloric contained in the air expired; this air was made to pass through a series of perforated metal plates. The metal, being a good conductor of heat, absorbed a certain amount of caloric, with which it readily parted again at each inspiration of the outer cold air. This

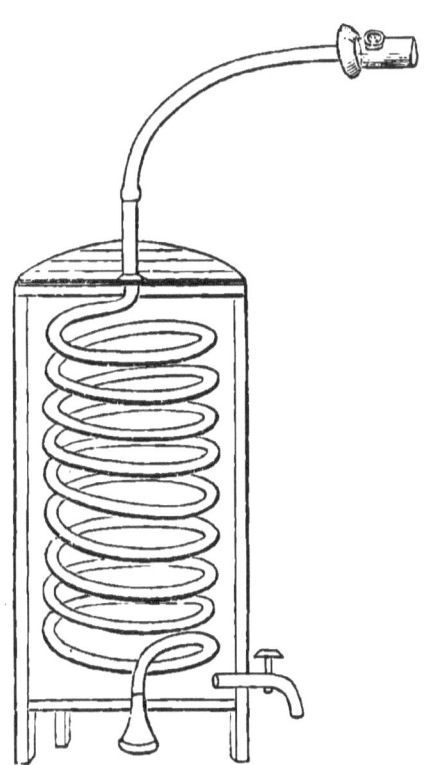

FIG. 1.—APPARATUS FOR COOLING THE AIR.

respirator was much used for many years, and is still so to some extent; it is doubtful, however, whether it fulfilled in a satisfactory manner the purpose for which it was designed. The shawls or woollen 'clouds' now so much worn act partly on the same principle.

A very ingenious and portable contrivance for warming the outer air to such a degree as to render it suitable for inhalation in cold weather by delicate persons has been devised by Dr. Joscelyn Seaton, who, like Jeffrey, makes use of the natural heat of the body to effect his purpose. It consists essentially of a belt formed of a series of elastic pipes, which are coiled round the waist next the skin, one end of the series communicating with the outer air and the other terminating in an oro-nasal inhaler. In the course of the pipe leading to the mouth, a vulcanite box is inserted, containing a small piece of sponge, in which carbolic acid or some other suitable medicament may be placed. Dr. Seaton writes—

'The temperature of the air inhaled will of course depend upon the length and calibre of the tubing and the temperature of the air supply. With an air temperature of 57° F. and twelve feet of a quarter of an inch tubing an elevation to 80° may be procured, the average result of my experiments being that a minimum rise of from 15° to 20° may be depended upon at all seasons for purposes of outdoor respiration, whilst for inhalation indoors, the temperature in the medicament box may be always maintained at or near 80°, sufficient for the vaporization of carbolic acid, most of the terebinthinæ, and other medicaments in common use. By the extension of the free afferent extremity, the closeness of the sick chamber may be readily counteracted by allowing the tubing to communicate with the outer atmosphere through a small hole in the window frame.'

A full description of this apparatus will be found in 'The Lancet' of April 19, 1884. It is made by Messrs. Maw, Son, and Thompson.

Compressed and Rarefied Air.

Many different mechanical arrangements have been designed for the inhalation of compressed or condensed, and rarefied or thinned air. Some of these are very expensive and complicated, while the cost of others is so reasonable as to placeth em within the reach of most persons.

Compressed and rarefied air are employed in the treatment of disease and for several purposes, as will be fully explained hereafter. Sometimes the patient inspires only compressed air, the degree of compression varying according to the necessities of the case, and expires into ordinary air; at others, and more frequently, he expires into rarefied air, or more rarely into compressed air; or the process may be reversed and he may inspire rarefied air. Again, the pressure may be partial or complete, one or both sided. When the air pressure is confined to the lungs only, it is partial or one-sided; but when it is applied to the whole surface of the body as well, it is complete, or general. This last form of increased pressure is only to be obtained in properly constructed pneumatic chambers.

Of the contrivances for one-sided pressure, some are for compressed air only, others for rarefied air, and others again for both combined.

The inhalations ordinarily practised are those of compressed air only, followed by expiration into ordinary air or into rarefied air.

The different kinds of apparatus devised and employed are very fully treated of by Oertel, as already stated; much more so indeed than is necessary in the present work. I shall therefore only refer to the more useful and complete forms.

One of the earliest forms was that by Waldenburg, constructed on the same principles as Hutchinson's spirometer. Waldenburg's single apparatus may be used for either compressed or rarefied air separately, but when both are employed consecutively two of these must be joined. It is not well to use the same gasometer or receptacle for both inspiration and expiration, as the air expired of course contains many impurities which would speedily foul the receiver—the tubercle bacillus, for example. Thus, scrupulous cleanliness is essential in the employment of these appliances; the like remark applies to the mouthpieces, of which there should always be two, the one to be used for inspiration and the other for expiration. The apparatus of Waldenburg may also be employed as an ordinary spirometer.

An advance on the above is Cube's double apparatus; but that of Tobold is smaller, cheaper, and more simple, and Oertel remarks that there is little to be urged to its disadvantage.

D

34   THE APPARATUS EMPLOYED IN INHALATION.

Schnitzler's single apparatus, again, is an improvement on Waldenburg's, and he has also devised a double continuously working apparatus; but this last he seldom uses, as he is of opinion

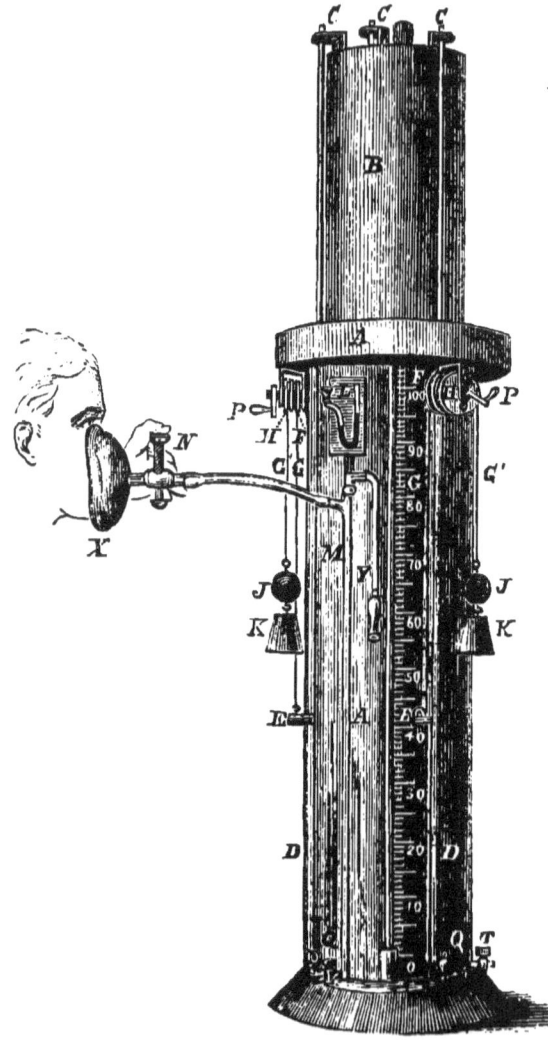

A, outer receptacle; B, bell receiver; C, guiding rods; D, grooves for rods; E, projection on which the chain G is hung, which passes over roller F; H, second roller, on which, by means of a second chain, the equilibrium weight J is fastened; K, overweight. If the outside cylinder be filled to a certain height with water and the overweight K hung on to the equilibrium weight of the bell receiver, B rises and the enclosed air is thinned. The degree of rarefaction can be read off on the mercurial manometer attached to the pipe M. If both cocks, N and T, or only one, are opened, the air enters the bell receiver till it is quite full. If it is desired to compress the air, then you push forward the peg in roller H, by which F is set free, in consequence of which the equilibrium weight is rendered powerless and the bell receiver compresses the air by means of its own weight. If it be desired to compress the air still more it is only necessary to lay the chain of roller F over roller Q at the foot of the apparatus, draw up the weight, and push back the peg in roller H, by means of which the former traction is reversed.

FIG. 2.—SCHNITZLER'S SINGLE APPARATUS FOR COMPRESSED OR RAREFIED AIR. Copied from Oertel.

that the alternate inspiration of compressed air and expiration into rarefied air is in many cases hurtful, owing to the great variation of the pressure.

Geigel and Mayer have also devised a small single apparatus, which, however, weighs 30 kilos.

When cost has to be considered, the apparatus of Waldenburg, or preferably that of Schnitzler, may be chosen. In both, the air or gas used may be measured, and the pressure is constant.

The simple apparatus of Biedert, in which the receiver is on the accordion principle, costs only about two guineas.

But of all the forms of apparatus for the inhalation of compressed and rarefied air hitherto devised, Geigel and Mayer's, with a double ventilator, is by far the best, allowing of a constant and continuous effect. It is, however, costly, and best suited to public institutions.

The Pneumatic Chamber.

The pneumatic chamber is employed chiefly for the inhalation of compressed air, but may be used also for rarefied air.

The action of the chamber differs in several important respects from that of the transportable appliances; while in the first the pressure is exerted on the surface of the whole body as well as on the interior of the lungs, in the latter it is confined to the lungs only.

The pneumatic chamber is constructed on the principle of the diving bell; indeed, it was this which mainly furnished the idea of such a chamber, the form even of the diving bell being still retained in some of the chambers.

One of the first pneumatic chambers was devised and constructed in 1838 under the direction of M. Tabarié. It consisted of an iron chamber of an elliptical form, capable of accommodating from four to twelve persons, and to this was attached a small ante-room. Into the chamber the air was forced by a steam pump until the required pressure was obtained, while the object of the ante-room was to allow of ingress and egress without any disturbance of the air pressure in the chamber. There were also arrangements for the ingress of fresh air and for the egress of the expired air, a mercurial manometer for measuring the pressure of the air, and a thermometer. Each sitting was supposed to last two hours. During the first half-hour the pressure was gradually increased to the required extent, at which it was maintained for the next hour, while during the last half-hour it was carefully lessened.

The chamber devised by Tabarié has since undergone various

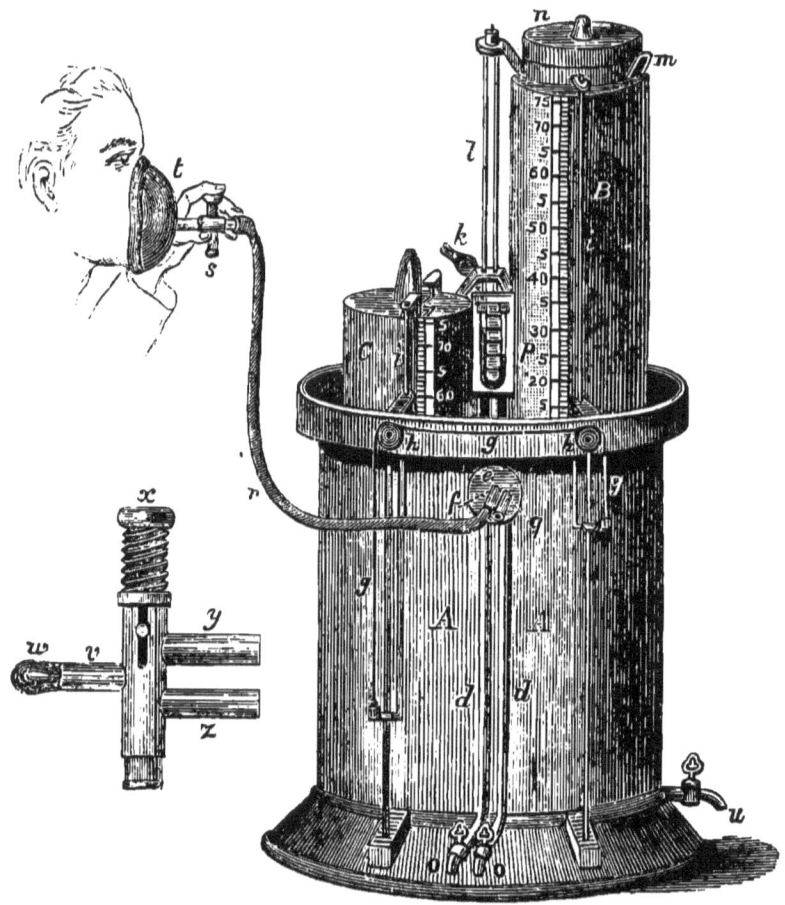

Fig. 3.—Schnitzler's Double Apparatus. Copied from Oertel.

This apparatus may be used for the inhalation of either compressed or rarefied air, or for the inhalation of compressed air followed by expiration into rarefied air. In the latter case two breathing tubes are required, which may be applied by means of the arrangement shown at *x*.

modifications; first by Lange, who aimed at simplifying it and reducing the cost; he also made better arrangements for the ventilation and for warming and cooling the air. Again, Lange's chamber is cylindrical instead of elliptical, like Tabarié's, and is for four persons only ; it has also a regulator for preventing the too sudden entrance of the air, and an arrangement for charging the air with volatile medicaments, as, for instance, pinewood oil. For cooling the air, cold water is sprinkled upon the air pump, the pipe, and even on the exterior of the chamber ; while for warming it the room in which the apparatus is placed is heated by a stove. There is also an arrangement whereby the air can be rarefied.

It is unnecessary in this practical work, either to give engravings or to describe minutely the mechanical arrangements needed in order to fulfil the various requirements of a well-appointed pneumatic chamber. It will be sufficient merely to enumerate the principal of them ; as the means of producing compression and rarefaction of the air, the method of entering and leaving the chamber without disturbing the air pressure, arrangements for the admission of fresh air and for filtering the same, for the escape of the expired air, for warming and cooling the air and for preventing it becoming overladen with moisture ; instruments for determining the degree of pressure, the temperature, and the humidity ; as a manometer, thermometer, and psychrometer. For those who require to be made acquainted with all the necessary structural details, the reader is referred to the work of Dr. Oertel, who devotes nearly 100 pages to the subject of pneumatic chambers.

The material of which the chambers consist varies as much as their form and construction, and this may be of iron, tiles, or stone, it being claimed for the latter that the temperature of the chamber may be more easily regulated in consequence of the less conducting power of the stone for caloric.

We learn from Oertel's work that pneumatic chambers, constructed after Tabarié's method, exist at the following places : Lyons, Montpellier, Nice, Stockholm, London, and St. Petersburg.

There is one of Lange's chambers at Johannisberg and another at Ems.

Upon one or other of the preceding models, pneumatic chambers have been constructed at Neuschöneberg near Berlin, at Vienna, Hanover, Wiesbaden, and other places.

The chambers at Reichenhall, devised many years since by Dr. G. von Liebig, are amongst the most complete hitherto constructed

A brief account of them was given by Dr. J. Burdon Sanderson in the 'Practitioner' for October 1868, and from which the following description is taken :—' It consists of three air-tight chambers, all of which open into the central antechamber. Each chamber is eight feet high and seven feet wide, so that three persons can sit with comfort. The pressure employed is equal to $1\frac{1}{2}$ atmosphere, i.e. about 45 inches of mercury, or about 22 lbs. on every square inch of surface. The patient remains in the chamber about an hour and forty minutes, of which time forty minutes is occupied in gradually increasing and diminishing the pressure, of which processes the latter demands the greatest caution on the part of the engineer. To those who are not conversant with the phenomena which attend rapid changes of density of elastic fluids the problem appears simple enough; in practical reality it is attended with perplexing difficulties. These arise principally from the absolute necessity which exists of maintaining a moderately equable temperature and of preventing the air from becoming saturated with moisture; for a failure in either of these particulars would certainly be detrimental. . . . At the beginning of the sitting, while the pressure is gradually rising, the temperature of the air tends to increase in exact proportion to the mechanical work converted into heat in the act of compression. For a similar reason, as the pressure diminishes at the end of the sitting, the air tends to become cooler, and consequently to become saturated with moisture. To guard against this contingency, it is necessary to watch the psychrometer with the utmost care, for the moment the readings of the two thermometers coincide a cloud of mist appears in the chamber, which is most disagreeable to the patients and would probably materially interfere with the beneficial results.'

But the most perfect pneumatic chamber is Simonoff's, in St. Petersburg; this is made of stone, is most luxuriously furnished, and so complete in every respect that patients may reside in it for days.

In all the chambers above mentioned the pressure of steam is employed, but Dr. S. A. Fontaine has devised a chamber worked by water pressure, the cost of working being thereby greatly reduced.

APPARATUS FOR THE INHALATION OF OXYGEN.

M. Limousin, pharmaceutist of the Rue Blanche, Place de la Trinité, Paris, has devised a very convenient and portable apparatus for the inhalation of oxygen gas, mixed with variable quantities of atmospheric air. It consists essentially of two parts, that for the liberation of the gas and that for its storage and admixture with air. The mechanism will be sufficiently explained

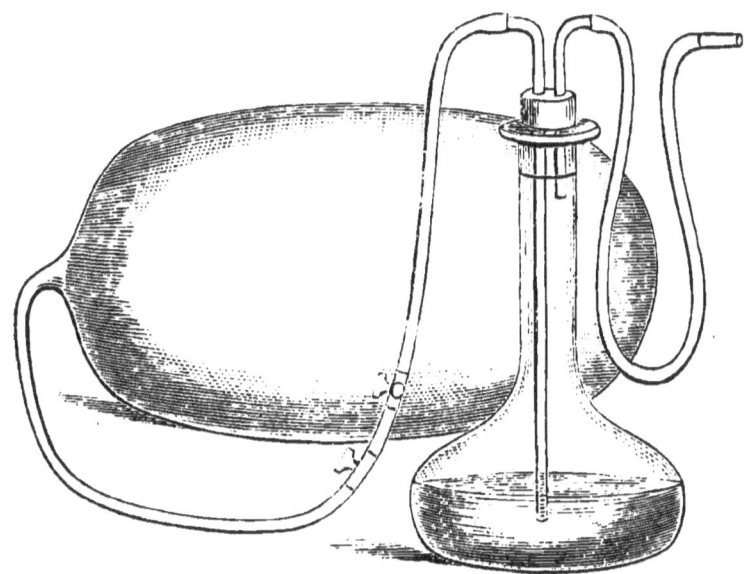

FIG. 4.—LIMOUSIN'S OXYGEN RECEIVER.

by an examination of the figure. The oxygen is liberated either from chlorate of potash or peroxide of manganese ; the gas should be always passed through water before being inhaled.

The best plan is, for the pharmaceutist to prepare the gas from time to time as may be required, and to convey the balloon containing it, either to the house of the medical man or to that of the patient.

The proportion of atmospheric air with which the oxygen is mixed, must vary according to circumstances.

The patient after each inspiration should retain the oxygen in the lungs for a few seconds, so as to give time for absorption.

40 THE APPARATUS EMPLOYED IN INHALATION.

The cost of the gas is about a penny a litre, and as much as twenty to thirty litres may be inhaled in the day. The apparatus may be hired or purchased, and is kept in readiness for use by many foreign pharmaceutical chemists.

Chloride of Ammonium Inhaler.

When liquid ammonia and fuming hydrochloric acid are brought into close proximity the vapours of each intermingle, and

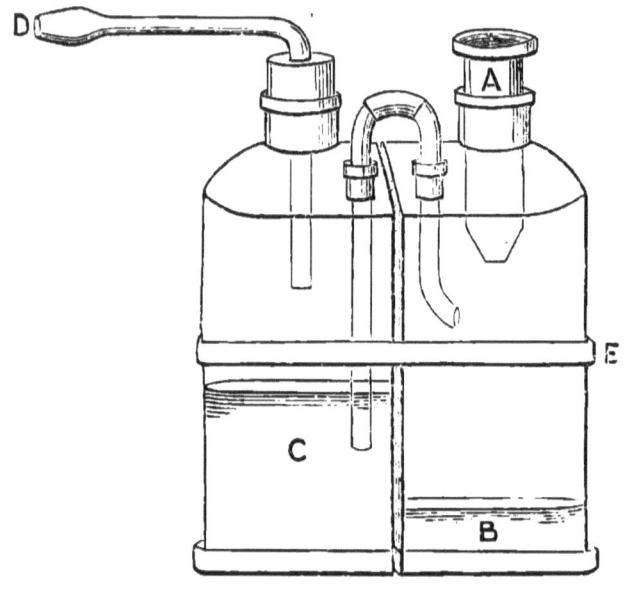

FIG. 5.—CHLORIDE OF AMMONIUM INHALER OF MESSRS. BURROUGHS.
A, tube for the ammonia; B, the hydrochloric acid; C, water; D, mouthpiece; E, connecting band.

dense white fumes of ammonium chloride are abundantly formed, mixed with a little free hydrochloric acid, from which the vapours are freed before inhalation by passing them through water. The apparatus, of which there are several modifications, consists usually of a double bottle united; one of the divisions contains hydrochloric acid and the other water. Into that with the acid, a tube is inserted, filled with cotton-wool or sponge, but which does not touch the acid, and is charged with liquid ammonia of the strength of the British Pharmacopœia. To the other division a tube or mouthpiece is adapted, the lower end of the tube terminating at some distance above the surface of the water. Finally, the two divisions are made to communicate through a small bent tube inserted through apertures in the upper parts of each division (fig. 5).

When the inhaler is in action the fumes of the ammonium chloride are drawn through the connecting tube into the water, from which it escapes in great part into the vacant space above it, to be drawn through the mouthpiece into the mouth. These fumes readily condense in part in the small connecting tube, which is apt to become quickly blocked up, and in part in the mouth, but a further considerable portion is ejected during the act of expiration through the mouth and nose. It has been already stated that tobacco smoke as a rule does not pass beyond the cavity of the mouth, and that the art of passing the smoke into the lungs is one which only some smokers acquire, and this chiefly with cigarettes. It seems highly probable, therefore, that the same may happen with the fumes of ammonium chloride, and that it is only in certain cases that these reach the lungs in any considerable quantity. Other very simple inhalers are those of Dr. Felton and Mr. Kerr; Messrs. Savory and Moore are the agents for the former, and Messrs. Maw and Son for the latter.

APPARATUS FOR THE INHALATION OF MEDICATED VAPOURS.

The inhalation of medicated vapours is effected mainly by the well-known oral and oro-nasal inhalers.

The motor power, when these are used, is derived from the patient, whose inspiratory and expiratory powers vary greatly in certain diseases and in accordance with his strength. This is so reduced in some cases, and the inspiratory effort is so feeble, that little or no beneficial result can possibly ensue from their employment.

Now to inhale effectually, so that the vapours may pass more or less freely into the air passages, it is necessary to inspire deeply and continuously, as I have shown in the preceding chapter. If this be not done, or be not possible, owing to the patient's weakness, the vapour will not pass beyond the epiglottis.

There are two channels for the entrance into the lungs of the medicaments employed in inhalation: by the nose and by the mouth. If an oral inhaler be used, the air will enter the lungs partly by the nose, and then the medicated vapour passing by the mouth, will become diluted and weakened. But if an oro-nasal inhaler be employed, this entrance and dilution will still occur to a less extent, for it is the exception to meet with an oronasal inhaler which is accurately adapted to the contour of the face.

But both the oral and oro-nasal inhalers, as constructed and used to the date of the publication of the author's communication in the 'Lancet' of May 5, 1883, possessed certain other very serious defects, and these have not hitherto been remedied.

One of these is, that the cotton-wool or sponge for the medicaments is so small, that it will only hold a few drops of the solutions employed.

Another is, that some of the most important substances used, as the antiseptics, carbolic acid, creasote, and thymol, are so little volatile at ordinary temperatures that about four-fifths of them are still retained in, and may be recovered from, the inhalers after two hours' inhalation; one-fifth only of the quantity originally taken having disappeared, and even of this it is certain that part only enters the lungs. For the evidence on which this statement is based, the reader is referred to Chapters I. and II.

A third fault is, that the quantity of the medicaments with which the sponge is usually charged, is far too small to prove of service even if a larger proportion were inhaled; see Chapter V. It is to be noticed, however, that since the publication of my observations and experiments, the quantities of the medicaments have in some cases been increased; but still it must not be forgotten that this addition does not make up for the low volatility of such substances as carbolic acid, creasote, and thymol.

To be effective these inhalers, both oral and oro-nasal, require to be entirely remodelled; but on this subject some further remarks will be made later on.

Of the two forms of inhaler, the oral is the least fatiguing to use, as it allows of greater freedom of respiration.

The oro-nasal inhaler has been strongly objected to by many medical men, on the ground that it impedes and limits the entrance of the atmospheric air, which is so much needed, especially in those cases in which the natural capacity of the lungs is greatly lessened by disease. This objection is no doubt a valid one; still, it would not be sufficient to justify the rejection of these inhalers, provided they were really effective and afforded a well-grounded prospect of benefit from their use.

In their present form, and with all their defects, it is certainly extremely questionable whether one is justified in ordering patients to wear them constantly day and night, as do some physicians.

Finally, then, I would state, that I have little confidence in the antiseptic action in diseases of the lungs themselves, of the majority of oral and oro-nasal inhalers, as at present constructed, charged, and used, because amongst other reasons the quantities of phenol, creasote, and thymol employed are usually far too small, and their volatility too low. In affections of the throat, and even of the upper part of the larynx, they may possibly be more efficient.

For the more volatile substances, as alcohol, chloroform, ether, turpentine, eucalyptol, &c., these inhalers are no doubt much more effective, especially in affections of the throat.

Of the oro-nasal inhalers at present in use, those are the best which are provided with a valve to prevent the escape of the expired air through the sponge; with two nasal valves to allow of its escape from the nostrils, and with a flexible rim or border. A piece of indiarubber tubing is perhaps as good as anything to permit of adaptation to the contour of the face. The respirators of Dr. Edward Blake and Dr. G. Hunter Mackenzie answer the above description in most respects.

Dr. G. Hunter Mackenzie's Oro-nasal Respirator.

One of the best of the oro-nasal inhalers which have been devised is that of Dr. G. Hunter Mackenzie, who, writing in the 'British Medical Journal' of August 30, 1884, thus expresses himself:—' In regard to the use of my oro-nasal respirator, I believe that, from the irritant action of the volatile substances with which it is necessarily charged, upon the windpipe, its use is contra-indicated in most of the inflammatory affections of that region, especially in laryngeal phthisis. Further, I have found that, whilst of service in cases of bronchiectasis and putrid bronchitis, it has no influence upon the bacillus of tubercle, and consequently upon tuberculosis.'

Now this, like many other more or less similar oro-nasal

inhalers, is very frequently employed in cases of phthisis, and hence these limitations and admissions are highly important, although I consider that they require some explanation and modification. Dr. Mackenzie does not give the names of the volatile substances which produced the irritant action to which he refers. If either carbolic acid or creasote, without admixture with alcohol, were used, there could have been no danger of any irritation being set up, and this for the very simple reason that only a very small quantity of these substances becomes vaporized at ordinary temperatures, and this with extreme slowness. Again, it by no means follows, because antiseptics which, when inhaled, have no influence on the bacillus, that they are of no avail in phthisis, or, as Dr. Mackenzie expresses it, in tuberculosis; they may prove beneficial in several other ways, by lessening expectoration, arresting putrefaction, mitigating the night sweats, and lowering the fever.

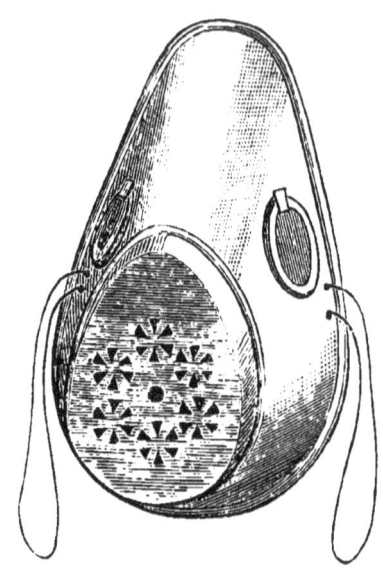

FIG. 6.—DR. G. HUNTER MACKENZIE'S ORO-NASAL INHALER.

I fully concur in the conclusion arrived at, that, charged and used in the ordinary way, his and most other oro-nasal inhalers have no influence on the bacillus of tubercle, and consequently upon tuberculosis.

Dr. Burney Yeo's Respirator.

An inhaler which has been much recommended, partly on account of its cheapness, is Dr. Burney Yeo's oro-nasal respirator. This consists of a piece of ordinary perforated zinc, folded so as to form a covering for the nose and mouth, and with an aperture in front for a small piece of sponge, which is held in position by two folds of the zinc. There are no valves of any kind, and the air enters through the numerous holes which exist on all sides; the expired air, therefore, being unchecked by a valve, escapes in part through the surrounding apertures, but principally through the sponge.

The small piece of sponge in the respirator with which I experimented weighed 4 grains, and it was not capable of holding with safety more than 1 drachm of liquid. I charged it with half a gramme of carbolic acid, = 7·7 grains, dissolved in 1 drachm of water, and inhaled for 1 hour. At the end of that time I recovered from the sponge 0·458 gramme of the acid, = 7·05 grains, so that only about seven-tenths of a grain had disappeared. In a second trial I took 1 drachm of the acid, and after an hour's inhalation recovered 13·99 grains, showing a loss of 1·44 grain.

Dr. Burney Yeo, in a letter which appeared in the 'British Medical Journal' of January 19, 1884, states that with his inhaler, '2·5 grains of carbolic acid, mixed with twice its weight of spirit of chloroform, can be vaporized per hour, so that if worn for 12 hours out of the 24—and many patients wear them all night —30 grains of carbolic acid will be respired.' Dr. Yeo omits, however, to mention the quantity of acid with which the inhaler was charged, and this is a most material point, since the loss of course must be in a great measure proportionate to the amount originally taken.

What the quantity of carbolic acid usually employed in this inhaler is, may be gathered to some extent from the letter of Mr. J. Brindley James which appeared in the 'British Medical Journal' of December 8, 1883. In this Mr. James states, that he prescribed in a very successful case an inhalation having the following composition: ℞ tinct. iodi ætherialis, acidi carbolici a a ℨij, olei thymi ℨj, spirit. vini rect. ad ℨj. 20 drops of this mixture are directed to be added to the sponge night and morning. Now these 20 drops, or say 20 minims, would contain only 5 grains of carbolic acid.

The statement contained in Dr. Burney Yeo's letter, quoted above, requires, then, further explanation, as it is calculated greatly to mislead.

Of the 30 grains said to disappear in 12 hours, a large proportion, probably not far short of one half, is lost by the passage of the warm air through the sponge during the act of expiration, and so is of no avail, since it does not enter the system at all.

Then, the sponge being exposed on all sides to the outer air, a further loss ensues; while, of the little that really enters the nose and mouth, a portion is absorbed by the mucous membrane of the parts with which it comes into contact, so that the most considerable deductions have to be made from the quantity given by Dr. Yeo. But even this result is not all. The loss on inspiration is not a constant quantity, but varies with the weather; when this is cold, the loss is but slight, and when very cold, there is practically little or no loss at all. Then with private patients how difficult it is, if not impossible in some cases, to get them to wear an oro-nasal respirator for 12, and still more for 24, hours. With hospital patients the case is different, as they must follow the instructions given.

In conclusion, then, it may be affirmed that this respirator has not only all the faults and shortcomings of most other oro-nasal respirators, but some special to itself.

Dr. W. Williams's Oro-nasal Respirator.

The object for which this respirator was designed, it is stated, is not so much for the medication of the inspired air, as its purification.

Dr. Williams, in his communication inserted in the 'British Medical Journal' of July 23, 1881, states, 'With regard to the mechanism adopted, the following is a description of the kind of respirator I find answer best:—Over a wire framework shaped like a respirator made to cover both the mouth and nose, two or more layers of ordinary antiseptic gauze are stretched; along the concavity inside, a narrow strip of sponge is placed; and, finally, the whole is fitted accurately to the face by a circumferential pad made of guttapercha tissue, stuffed with cotton-wool, or folded lint, which is more manageable. Antiseptic gauze in the dry state gives off a vapour of carbolic acid. Bearing in mind, however, the large amount of air that would in ordinary breathing pass to and fro, it will be very evident, that this comparatively small piece of gauze must soon become exhausted and require recharging; this is secured by the whole being dipped every half-hour or so at first, afterwards less frequently, into a watery solution of carbolic acid of the strength of 1 in 40; the gauze is, besides, renewed every two or three days. These respirators or dressings are worn as constantly as possible; in fact, the only occasions on which their temporary removal is permitted are during a meal, for the purpose of expectorating, and for that of dipping.'

Dr. Williams states, that the offensive odour in cases of vomicæ in the lungs is quickly removed by the use of this inhaler. 'The quantity inhaled of this vapour may for any single inspiration be quite insignificant, but when multiplied by the number of inspirations made in only a few hours, it does not seem difficult to believe, that the amount would soon be sufficient to accomplish the disinfection of all the purulent cavities already in communication with the bronchial tubes. That this end is actually gained, and even rapidly gained, my experience certainly tends to prove; and I find moreover that, on the disappearance of odour, it requires subsequently the presence of very little carbolic acid on the gauze to keep the expectoration permanently free from fœtor, as though the ulcerated surfaces having been rendered aseptic, all that remained to be done was to ensure against their re-infection by the inspiration of only pure air, and also to obviate the risk of creating irritation by breathing that which contains but a minimum quantity of suspended carbolic vapour.'

The idea on which this respirator is founded is excellent; the close material or gauze with which it is covered must, however, especially in the case of feeble patients, render inspiration more difficult than ordinary; then again the solution of carbolic acid, 1 in 40, in which the respirator is dipped from time to time, is very weak, and the amount of that acid which actually enters the lungs must be very small. It appears to me, therefore, that any beneficial action exerted by this respirator, is due rather to the purification of the air than to the minute quantity of carbolic acid introduced into the lungs.

Oral Inhalers.

As may be gathered from what has been already stated, oral inhalers possess most of the defects of oro-nasal inhalers, as well as deficiencies of their own, with also one or two advantages.

The receptacle for the sponge or cotton-wool is very small, and therefore capable of holding but little liquid, and if these inhalers are charged with the antiseptics usually employed, as carbolic acid, creasote, or thymol, about four-fifths of those substances would be recoverable from the inhaler after the completion of from one to two hours' inhalation.

One of the advantages is, that the channel leading to the air passages is more direct, and another, that the hindrance to respiration is not so great, since this is still carried on in part by the nostrils, unless these be artificially closed.

But the greater freedom of inspiration operates disadvantageously in another way. There is less vaporization of the antiseptic substances named above, in consequence of the whole of the air not passing through the respirator. On the other hand the loss during expiration is greater, because of the absence of any valve to restrain the expired air from passing, at least in part, through the sponge containing the antiseptics.

Then there is still another consideration to be taken into account, applicable alike to oral and oro-nasal inhalers, and which has not yet been specially dwelt on; it is this:—

The inspired air, which passes through the medicated sponge, is at all times a comparatively cold air; its temperature, according to the season of the year, may vary from 32° to 70° or 80° F., and with this variation so will vary the rate of vaporization of the medicaments; but the temperature of the expired air, on the other hand, no matter what the season, is nearly always the same, approximating to that of the body; hence when the air is allowed to pass through the sponge or other material used, the vaporization is always greater during expiration than during inspiration, and the loss of material of course proportionately larger. This is a practical point of which we should not lose sight. In those cases in which oro-nasal inhalers are provided with an effective valve, this loss during expiration is not experienced.

As is the case with oro-nasal, so with oral inhalers, many modifications have been devised, particularly in Germany. The sponge or cotton-wool is usually kept in position by an inner perforated plate, but this plate is apt to become loose and the sponge to be displaced, though this occurrence is easily avoided by fixing the plate to the inhaler by a hinge, as is now often done.

Nasal Inhalers.

Several different forms of nasal inhalers have been devised. One of the simplest of these is that of Dr. Feldbausch, of Strasburg; this consists of two little cylinders, united below by a small band, which are passed more or less completely up the nares, and are filled with blotting-paper or flannel, charged with the medicament to be inhaled, usually two to four drops of carbolic acid. The apparatus is made in three different sizes, but they are all very small and the surface of evaporation wholly insufficient. It is not probable that its employment would be of the slightest use in any maladies of the air passages or lungs, though it might be serviceable in very limited affections of the nasal mucous membrane. Dr. Feldbausch recommends his apparatus for prophylactic purposes, as a protection against infectious maladies; but no proof has yet been given of its efficacy in this respect, and although it closes the channel by the nose, it leaves open that by the mouth.

Dr. Cousins' Nasal Inhaler.

The 'Lancet' of July 19, 1884, contains a description and figure of an inhaler devised by Dr. J. Ward Cousins, and which may be used either as a nasal or oral inhaler. It is made of vulcanite, and resembles somewhat a bulky cigar in form, in the hollow of which is placed 'the little pledget of cotton-wool upon which the inhalant is dropped.' When used as a nasal inhaler, the narrow end is inserted into one nostril, the other nostril being closed by means of a small spring.

After all that has been stated respecting the construction and action of oral and oro-nasal inhalers, it is unnecessary to make any extended remarks on this instrument. The surface of evaporation is but small, and the vaporization of such antiseptics as carbolic acid, &c., would be very limited, especially in cold weather.

The Nose Inhaler of Dr. George Moore.

The nose piece is flexible and adaptable. A box, placed below the nose, is perforated on its upper and lower surfaces, and contains a layer of absorbent wool or lint, on which the medicine to be used is dropped. The under surface of the box moves on a hinge at one end and is fixed by a clasp at the other, so that the wool or lint can be easily changed. When adjusted, the patient can either inspire by the nose and expire by the mouth, or can 'carry on both acts of respiration by the nose alone, as may be found expedient or necessary.'

50 THE APPARATUS EMPLOYED IN INHALATION.

Dr. Moore writes, 'The important practical point appears to me to be, that respiratory treatment, when required, should be conducted through the nasal passages as parts of the natural respiratory thoroughfare, and not wholly or partly through the mouth, which belongs to the alimentary canal. Moreover, it is desirable to bear in mind, that many, if not all, cases of bronchitis, asthma, and emphysema are preceded by, or accompanied with, nasal catarrh, which ought to be included in respiratory treatment. When a nose-mouth respirator is employed, the patient generally uses the mouth as a breathing organ, so that medication begins in

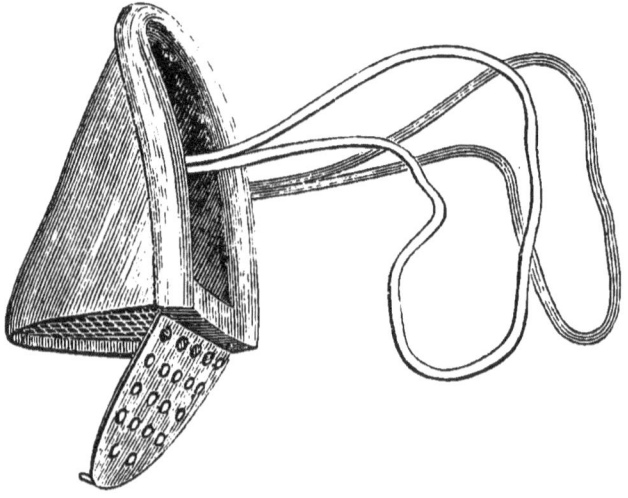

FIG. 7.—DR. GEORGE MOORE'S NASAL INHALER.

the throat, and concurrent disorders of the nasal passages and of the naso-pharyngeal cavity are practically untreated.'

This respirator is certainly superior to the two nasal respirators previously described; it is many times larger than that of Dr. Feldbausch, and, unlike that of Dr. Cousins, it embraces both nostrils. No doubt the most natural respiratory channel is that by the nose; it is much to be doubted, however, whether this is the best for the introduction of medicated vapours into the throat and lungs. Many of these vapours very quickly become condensed, so that the shorter the distance they have to travel, the greater the chance of their reaching their destination. Now the route by the nose is certainly the longer and more tortuous, purposely so indeed, in order that it may act the better as a strainer or filter, depriving the air of the solid particles it contains.

Again, since the warm expired air is allowed to escape through the respirator, a very considerable loss of the carbolic acid or other medicament originally taken, must necessarily ensue. This inhaler may be obtained of Messrs. Wright and Co., of New Bond Street.

Spray Producers or Nebulizers.

A great many different forms of spray producers have been constructed; many of these are obsolete and are no longer employed. The object of them all is to finely divide, or atomize, as it is termed, the medicated fluids employed, and to project through very fine apertures the little streams or sprays of atoms with such force as to aid their entrance into the air passages. The motor power which may be employed varies; sometimes the compression is produced by air, at others by water or steam. When air is employed, the pressure may be exerted either by means of the air pump or more easily and simply by the hand acting on an elastic ball. The spray producers worked by a force pump for either air or water are expensive, and require for the most part the aid of an assistant, so that they are but little used. The forms of apparatus now generally employed are few, simple in their construction, and by no means costly.

The Air Ball Apparatus.

This apparatus usually consists of a bottle or receiver, containing in solution the medicament to be employed; in this is placed the vertical tube of Bergson's arrangement, which is formed of two tubes joined together, placed at right angles with one another and terminating above in fine capillary openings, which are nearly in apposition and almost touch each other. To the horizontal tube is attached a piece of indiarubber piping, in the length of which two elastic balls are introduced. The ball farthest from the receiver acts as a reservoir for the air, which is forced out of it when the apparatus is put into operation by the action of the hand, and is conducted thence by the indiarubber piping, first into the second elastic ball or regulator, and thence into the horizontal glass tube, from the point of which it issues in a fine stream. The force of this creates a vacuum in the vertical tube below, causing the fluid in the bottle to flow up it; when, meeting at its point the current of air from the horizontal tube, it becomes broken and scattered, the two together constituting the atomized spray. This at first is a narrow jet, which soon spreads out in a form resembling somewhat the tail of a comet.

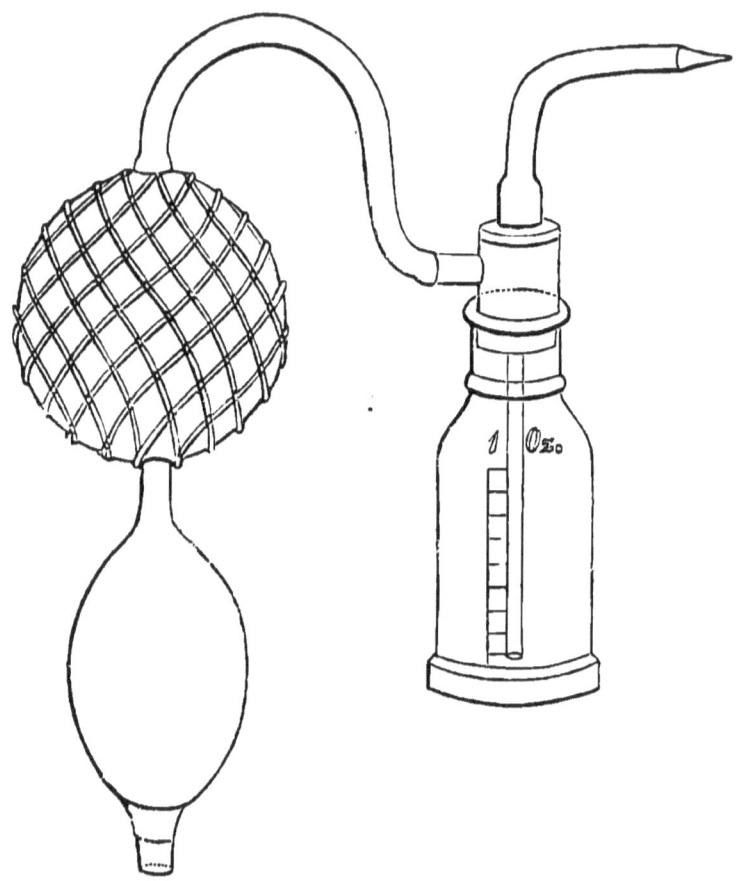

FIG. 8.—IMPROVED AIR SPRAY PRODUCER.

The hand-ball spray apparatus has within the last few years undergone some important modifications, one of which renders it especially serviceable for inhalation purposes, and particularly for spraying the throat. Bergson's tubes, in which the capillary openings are opposed to each other at right angles, are done away with. The air is made to enter through the side of the cork or through the tube as it issues from the receiver; the cork is made of vulcanite or indiarubber, and through its centre passes a glass or vulcanite tube, one end of which terminates in the fluid in the receiver and the other is prolonged to any required length, and may be variously curved and thus adapted to any requirement. The tube may be so prolonged as to reach the back of the throat, and if curved upwards the spray may be delivered into the posterior nares, and if downwards, into the larynx. Again, a curve of a different form may be given to it for spraying through the nostrils. In fact, this and other spray producers should always be furnished with different sets of tubes, suited for each of the purposes referred to above. In some cases the second air ball is done away with; in others the ball is attached to the top of the cork, the discharge tube issuing from near the neck of the receiver.

It will be readily perceived, that the principle of the hand-ball spray with the single tube is quite different from that of Bergson's tubes, in which the medicated fluid is made to rise in the vertical tube by the vacuum produced in its upper part by the current of air which flows through the horizontal tube. In the other apparatus the fluid rises up the tube, in consequence of the pressure on the surface of the liquid in the bottle.

This apparatus will, therefore, doubtless supersede in a great measure the older and more complicated form.

Dr. Sass's Horizontal Nebulizer.

A very convenient cold spray apparatus is the horizontal nebulizer of Dr. Sass, of New York, which is represented in fig. 9. The advantages are, that the receiver can be held in the hand, and that the tubes may be prolonged horizontally, so that they may enter the mouth, when it is desired thoroughly to irrigate the back of the throat. The hand being in this apparatus, and in others on the same principle, the propelling force, the spray is of course intermittent, and corresponds with each action of the hand; in using it, therefore, care should be taken to time each inspiration with the escape of the spray.

Dr. Wright's Atmonemeter.

A very ingenious modification of the ordinary intermittent air spray producer, has been devised by Dr. Wright, of Old Burlington

Street; it is called the atmonemeter, and consists of a bottle to which is attached a syringe; this is worked by pressure of the thumb on a spiral spring fixed round the piston rod. The air thus driven displaces a portion of the contents of the bottle and forces it along a horizontal tube, at the end of which it is dis-

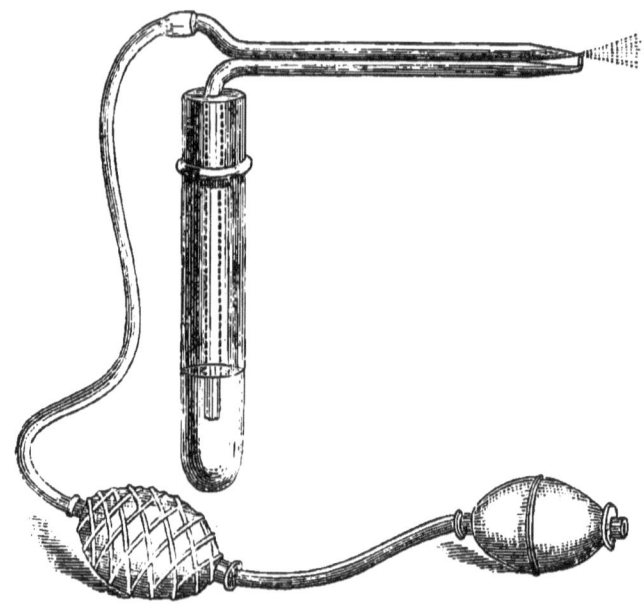

FIG. 9.—DR. SASS'S HORIZONTAL AIR SPRAY PRODUCER WITH TUBULAR RESERVOIR AND DUCK-BILL TUBES. Copied from Cohen's work.

charged through very minute orifices. Though suitable for other purposes, the instrument is primarily intended for use in throat diseases, and to facilitate its employment in these cases the tube is furnished with a plate beneath, which serves as a tongue depresser. It is made by Messrs. Krohne and Sesemann.

STEAM SPRAYS.

In the steam spray apparatus there are usually two receivers, one corresponding to that of the air spray producer, namely, the receiver for the medicaments, and the other being the boiler for the generation of the steam, the two being brought into contact, as in the air spray apparatus, by means of Bergson's tubes.

Siegle's Steam Spray Apparatus.

Siegle's apparatus has been subjected to various modifications and improvements, especially in America.

One of the best and most complete steam spray inhalers is the modification of Siegle's, made by Messrs. Codman and Shurtleff, of Boston, although it has one serious fault; the spray is ejected obliquely, striking the side of the shield, in place of being thrown out at right angles. It has a safety valve, a glass face shield, and a

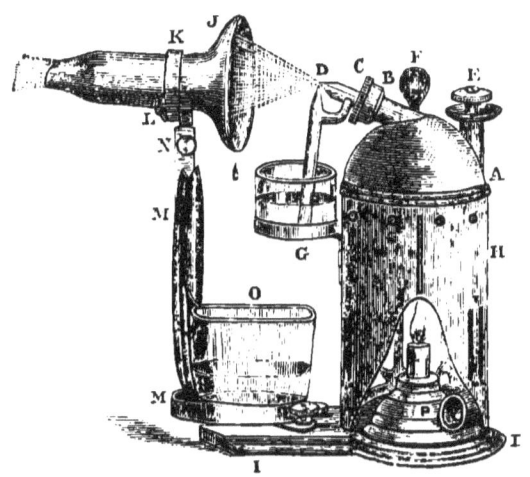

FIG. 10.—CODMAN AND SHURTLEFF'S MODIFICATION OF SIEGLE'S STEAM SPRAY PRODUCER. Copied from Dr. Cohen's work.

A, boiler; B, steam pipe; C, packing box; D, nebulizing tubes; E, safety valve, closing funnel-shaped orifice for pouring in water; F, wooden handle for removing boiler from the stand, &c.; G, socket for medicine cup; H, support for boiler; I I, base; J, glass face shield; M M, support for face shield; O, waste cup.

waste cup. It would be more perfect if provided with a water gauge and if the tubes were arranged on the duck-bill principle.

The shield is a very necessary portion of all forms of spray apparatus, especially when nitrate of silver is used; it prevents too great expansion of the spray, directs this into the mouth, and protects the face.

Dr. Cohen's Modification of Siegle's Steam Spray Producer.

Another very useful modification of Siegle's steam spray producer is that with the duck-bill arrangement of the tubes, which allows of their being so prolonged that they can be made even to enter the mouth.

In the duck-bill arrangement, the tubes being reversed, and that conveying the medicament being the upper of the two, the separate receptacle may be done away with by expanding the termination of the upper tube in the form of a funnel.

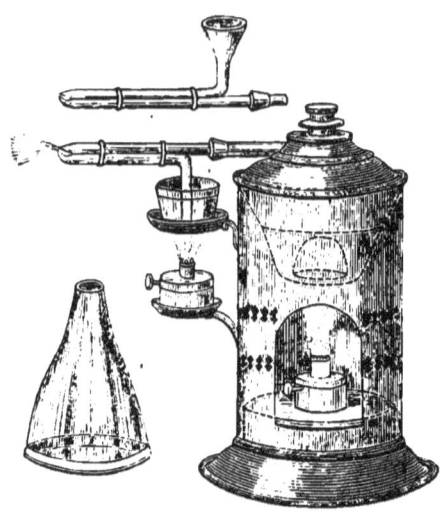

FIG. 11.—DR. COHEN'S MODIFICATION OF SIEGLE'S STEAM SPRAY PRODUCER WITH DUCK-BILL TUBES. Made by Messrs. Wilhelm and Newman, Philadelphia. Copied from Cohen's work.

Dr. Oertel's Modification of Siegle's Steam Spray Producer.

In Oertel's modification, three different kinds of capillary tubes are used; the first correspond in the calibre of the openings with those recommended by Siegle; these are the finest of all, and the spray is cooler, possesses more penetrating power, and is particularly adapted for reaching the smaller bronchial tubes and air cells of the lungs; the quantity of fluid delivered, however, is comparatively small. The second kind of tubes, which Oertel states are more commonly used, have rather larger openings; these when employed, profusely cover the throat and larynx with fluid, which might even pass into the lungs. With the third kind of tubes, having still larger openings, by raising the heat of the steam, a coarser spray is obtained, which is very suitable as a pharynx douche to sprinkle strongly the affected parts of the mucous membrane with large quantities of fluid. These last tubes, Oertel considers, replace to some extent the apparatus of Matthieu and Bergson. Lastly, the sprays are projected into a glass funnel, whereby too great expansion is restrained. This apparatus may be obtained of Messrs. Mayer and Meltzer. Oertel's modified spray producer possesses certain serious defects, and is by no means complete in other respects. The horizontal tube should issue from the side of the boiler, and not from the top, in order to obviate the liability to spurting, which is apt to occur when the tube is bent at right angles. The glass tube itself is too long and frail, and therefore liable to be broken. Again, the receiver for the medicaments is too wide and shallow, and presents too large a surface to the air, although divided into two parts, the one division being intended to catch the waste from the protecting funnel. Furthermore, while the outer case is provided with a handle, there does not appear to be any means of quickly removing the boiler, nor is this furnished with a water gauge.

Mr. Benham's Modification.

Further useful modifications of Siegle's steam spray producer have been devised, as that which bears the name of Mr. Benham. It is furnished with a handle, whereby its manipulation is facilitated, and with a water gauge, so that the amount of water in the boiler can always be ascertained, a matter of some consequence. It may be procured from Messrs. Wright and Co., 108 New Bond Street, London.

Dr. Lee's Steam Draught Inhaler.

But a steam spray producer has been constructed on quite another principle by Dr. Robert Lee. In this the bottle for the medicated fluid and the glass capillary tubes, or syphons, are entirely dispensed with; the medicaments are added to the water in the boiler, and issue from the pin hole with the steam, and this for the most part whether they are volatile or non-volatile.

The steam as it issues is very hot, and with a view to regulate the temperature Dr. Lee has provided the boiler with a funnel-shaped tube, having apertures on its side for the admission of air; these apertures, according to the temperature required, being left open to a greater or less extent. It is stated that by means of the funnel the temperature may be made to range between 80° and 100° F.

If the water be hard and contain much lime, the pin hole might become blocked up, so that the boiler should always be provided with a high-pressure or safety valve. This had not been applied to the apparatus with which I experimented.

In most steam spray producers, there being two vessels and both containing liquids, the quantity of fluid used is usually greater than when there is but one vessel, the medicaments being also proportionately diluted.

FIG. 12.—DR. ROBERT LEE'S STEAM DRAUGHT INHALER.

What takes place while spray producers are in action has already been described in Chapter I., and the reader is now referred to the remarks on that subject therein contained. He should also be reminded, that it was shown in certain experiments, the results of which have been previously recorded, that fully two thirds of the carbolic acid and certain other medicaments taken, do not enter the mouth at all.

Further, it is beyond all question, that of the quantity actually thrown into the mouth, a considerable part is condensed on the extended mucous surfaces of the mouth, fauces, and pharynx—on parts, in fact, above the epiglottis; while much is also swallowed, and thus makes its way into the stomach.

No doubt by the employment of sprays, either hot or cold, any quantity of a given substance may be sprayed into the mouth and thus applied to the fauces, and that so far they are effective; also, that by their means, medicaments may be made to reach the stomach in any quantity, and thus the system at large may be brought under the influence of the remedies employed.

A portion of the medicaments used does certainly make its way into the air passages; of this satisfactory proof has already been advanced; but from all that has been stated the conclusion is fully warranted, that the quantity which thus enters is, in proportion to the whole amount taken, comparatively small; still the quantities which are absorbed by the mucous membrane of the fauces, and which enter the stomach, doubtless contribute in an important degree to the medicinal effects obtained.

All sprays, however, whether cold or hot, possess this great advantage, that they are capable of diffusing non-volatile as well as volatile substances.

The characters of the sprays produced by air and steam, differ greatly in several particulars. The action of the hand-ball spray apparatus is intermittent, and the projecting force much less than with the steam spray, and the spray is coarser. The temperature also of the spray itself is always some degrees less than that of the surrounding air, owing to the expansion of the air causing a portion of the caloric to become insensible. This in some cases is an advantage, in others the reverse. When it is desirable to avoid the cold spray, the medicated solution may be warmed. The hand-ball spray apparatus, therefore, is generally considered more suited for the irrigation of the back of the throat than for penetration into the air passages, but this conclusion has not been established experimentally.

The action of the steam spray apparatus is continuous, and the spray is expansive, is projected with greater force and to a greater distance, is finer and more suitable for deeper inhalation. The medicated solution is, however, admixed and diluted by the steam, and the temperature varies with the distance from the apparatus and the extent of its admixture with air.

In using steam sprays, it is not desirable that they should be approached too closely to the mouth, and this for two reasons: one is that the force of the spray is so great that but little of it becomes deflected during inspiration and really enters the air passages, but it strikes directly against the back of the pharynx; the other is that the temperature is in some cases too high, although it is not nearly so great as might be supposed, the spray having a temperature at a distance of 4 inches from its commencement, of only about 88° Fahr.

A final difference between the improved hand-ball and the ordinary steam spray, is that in the former case one receiving vessel only is needed, whereas in the latter two are required, one for the water and the other for the medicated fluid.

These several differences render air and steam sprays suitable for different purposes. Steam sprays, from their greater penetrating power, are believed to make their way more readily into the air passages.

Perhaps the finest spray is that on the Sales Girons principle, produced by the forcible impinging of a column of water upon a metal or other hard surface. This principle is best adapted for filling the air of inhalatoriums with a fine spray, as is commonly done at many of the foreign water-cure establishments.

APPARATUS FOR THE INHALATION OF VAPOURS.

The vapour of hot water differs from steam chiefly in that it is given off at a lower temperature.

A great many contrivances have been resorted to for the inhalation of the vapour of hot water, either simple or medicated, from the open mouth or spout of a jug to much more elaborate and complete arrangements.

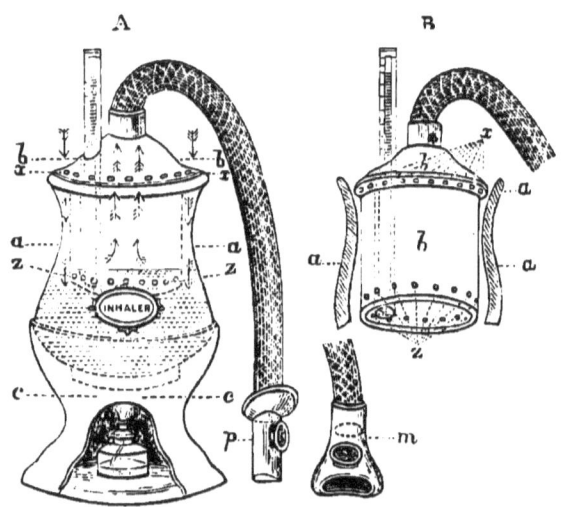

FIG. 13.—DR. MORELL MACKENZIE'S ECLECTIC INHALER.

A, outer receiver, into which the pint of hot water and the medicaments are put; B, the inner receptacle, forming a kind of inverted tumbler, the bottom of which dips down just below the water; it is perforated above and below with a series of apertures, as seen at x and z. During inhalation the air passes through the apertures at x, thence through the air chamber, next through the apertures at z, subsequently through the medicated fluid, until finally it passes through the mouthpiece, as indicated by the arrows.

The essential requirements of a well-constructed vapour inhaler are, a lamp to maintain the temperature of the water, a thermometer to regulate it, an admixture of air to cool the vapour, and a suitable mouthpiece, provided with valves to prevent the expired air from passing into the water or medicated liquid.

Dr. Morell Mackenzie's Eclectic Inhaler.

Such an apparatus is Dr. Mackenzie's eclectic inhaler, made by Maw, Son, and Thompson. Its mechanism will be sufficiently understood on an examination of fig. 13.

Boiling water is placed in the receiver to start with, and the temperature is maintained usually at from 140° or 150° F.

The patient may inhale from five to thirty minutes, according to the nature of the case, but the mouth should be removed from the mouthpiece from time to time. 'About six inspirations should be taken in a minute.' To avoid taking cold, the patient should remain indoors for half an hour after using the inhaler.

Another very useful form is *Martindale's Portable Inhaler*. This is furnished with an earthenware mouthpiece, a woollen covering to maintain the warmth of the water, and it admits of being easily and thoroughly cleansed.

Dr. Spencer Thompson's Vapour Inhaler.

Dr. Spencer Thompson has devised a vapour inhaler, the mouth of which is so shaped as to fit more or less accurately the contour of the face, undoubtedly a considerable advantage. It is not, however, provided with a lamp or thermometer, so that there are no means of maintaining the water at any given temperature. This inhaler may be obtained of Mr. Toogood, Mount Street, Grosvenor Street.

APPARATUS FOR THE INHALATION OF MEDICAMENTS IN THE FORM OF FUMES.

The smoke that results from the igneous combustion of organic substances, chiefly vegetable, is made up of the vapour of water, of certain gases, especially carbonic acid and ammonia, and usually of more or less of the volatile medicinal constituents of the substance burned; it has therefore a very complex composition.

The substances to be burned are often made into pastilles, cigarettes, or are cut into strips like tobacco, and sometimes, as in the case of nitrate of potash, papers are saturated with them.

APPARATUS FOR THE INHALATION OF POWDERS.

Several methods are employed for conveying powders to the throat and air passages. One of the oldest of these is by putting the powder into a hollow quill, inserting one end of this into the mouth, and drawing out the powder by a strong inspiratory movement. One disadvantage of this proceeding is, that the whole of the powder is apt to be brought at once into communication with the back of the throat, and there is but little time allowed for its reaching the deeper air passages; another fault is that the powder is very apt to be blown out of the tube on the slightest expiratory action.

A second method is by means of the well-known insufflator of Rauchfuss. This consists of a tube curved at one end and having an elastic ball at the other, with an opening provided with a cover in the centre of the tube, for the reception of the powder. Ingenious as is this instrument, it often fails. One fault is, that the powder is liable to be more or less drawn back into the elastic ball, in consequence of its expansion, the moment the thumb is taken off.

Other forms of apparatus are the following: *Mr. Bryant's Insufflator* consists of a tube, so curved as to allow one end to be placed in the nostril and the other in the mouth. At the top of this tube is an opening for introducing the medicament, and this is afterwards closed by a cork. When charged, the patient blows gently through the tube, the powder being thus driven up the nostril.

Dr. Andrew Smith's Insufflator resembles a hand ball spray producer, with one receiver and two openings; one of these is in connection, by means of the indiarubber tubing, with the air ball, and the other is furnished with a tube for the escape of the powder; this tube may be made to vary in its curvature and length, according to whether it is desired to apply the powder to the anterior or posterior nares or to the throat. The objection to this apparatus is that it would be most difficult to apply by its means with any exactitude the very small quantity of powder usually required. This objection, as I learn from the second volume of Dr. Mackenzie's recent work 'On the Diseases of the Throat and Nose,' is to a great extent obviated by the substitution, at the suggestion of Clinton Wagner, of a test tube in place of the bottle.

NEW AND IMPROVED FORMS OF APPARATUS.

The four forms of apparatus next to be noticed were first described by me in the 'Lancet' of October 6, 1883, and they have since been manufactured by Messrs. Maw, Son, and Thompson, from whom they may be obtained. In devising them, the author has relied on the two important principles which have already been adverted to in Chapter II., which treats of the principles concerned in the volatilization and inhalation of medicaments. One of these principles is the greatly increased vaporization of certain chemical substances, which at the ordinary temperature of the air are but little volatile, obtained by spreading them over a very considerable surface and exposing them either to the air or on water. The other is the augmented evaporation resulting from increase of temperature in combination with extension of surface. The results of certain experiments showing the extraordinary effects of the action of these principles have already been quoted in Chapters II. and III., but further examples will be found detailed in Chapter IV., which treats of the principles of the construction of inhalation chambers.

Armed with the above principles the author has been led to devise the four following forms of apparatus, the first two being intended for charging the air of chambers for inhalation and disinfection:—

Chamber Inhaler and Disinfector No. 1.

This apparatus consists of a long cotton fabric, woven so as to afford a very large extent of surface for evaporation. The length and breadth of this, vary with the extent of the effect required to be produced. It is spread out in several layers, one above the other, by means of an arrangement consisting of a double series of rails, which are attached to a box, and lift out and in. The greater the distance between the layers, the better. These rails, when not in use, are packed away in the box, which also contains the long cloth. The apparatus may be made of any size, according to the purpose for which it is required. At present it is manufactured in two sizes, namely, 11 by 16 and 8 by 11 inches.

The quantity of carbolic acid for the larger size is 800 grains or 100 for each fold. Of this quantity, upwards of one-half, at a temperature of about 64° F., becomes evaporated during the first six hours, and by the time the fabric is dry the whole of the acid will have disappeared.

A convenient quantity of creasote, if it be desired to use it with this apparatus, is 400 grains, or 50 grains for each fold. Of this amount, about three-sevenths will have become volatilized at the end of the first six hours. If thymol be used, the same

FIG. 14.—CHAMBER INHALER No. 1.

amount may be employed, and more than one-third will have disappeared at the end of the six hours.

A mixture of the carbolic acid in water should be prepared beforehand of the required strength, in readiness for use. Of this the requisite quantity should be placed in a basin of convenient size, into which the dry cloth should be put. This should absorb the whole of the mixture, so that when suspended there is no dripping from the cloths. The exact amount of liquid absorbed will, of course, depend on the composition and texture of the fabric employed.

If either creasote or thymol be used, mixtures of these should also be prepared beforehand, but the solvent in these cases should consist of water, and as little spirit as is required to keep them in solution. In consequence of the presence of the spirit, the cloths become dry in much less time than when water only is used, and when once dry, scarcely any further volatilization of the substances takes place.

A great variety of inhalants and disinfectants may be employed with this apparatus, besides those specially referred to. Some of these are the several preparations and solutions which contain carbolic acid or creasote as the basis, firwood oil, eucalyptol, camphor, &c., but preparations containing permanganates or chlorides of soda and lime are not suitable. Indeed, the volatilization of thymol, camphor, and some other substances is best effected by the Chamber Inhaler and Disinfector No. 2, as the use of spirit, which is costly, is thereby rendered unnecessary.

The No. 1 Chamber Inhaler, 11 by 16 inches, is best suited to a room of a capacity of about 800 cubic feet. For rooms of greater dimensions, or for hospital wards, a larger apparatus should be employed.

The smaller apparatus, 8 by 11 inches, is adapted for a room of a capacity of 500 to 600 cubic feet, and only half the quantity of the several substances and mixtures employed in the large apparatus should be taken.

When these inhalers, which are really vaporizers, are being used, the doors and windows of the chamber should be closed, and in some cases even the register of the stove. It is not necessary to re-charge the cloths for at least six hours, at the end of which time, however, they will still contain half the quantity of the substances originally taken. If the inhaler be employed for only two or three hours at a time, the cloths should be folded and put away in readiness for use on successive days until the six hours have expired. Care should be taken that the fabric is perfectly dry before it is re-charged, and it will be well to rinse it out occasionally with clean water. Each patient should not remain in the room or chamber at first, for more than an hour or an hour and a half at a time.

Chamber Inhaler and Disinfector No. 2.

This consists of an outer water bath, to which a thermometer is attached, and an inner porcelain dish, divided into four equal parts. The divisions of this are also filled with water, the temperature of which, by means of the outer bath, can be regulated and maintained to a nicety. With this apparatus, if desired, no less than four antiseptic or other substances may be used at the same time, either for inhalation or disinfection. Although it is usually advisable to add the medicaments to the water in the dishes, yet in some cases they may be placed in the divisions without water. This apparatus combines the advantages of moderate extent of surface, with a temperature regulated according to the nature and composition of the substances used—a very valuable combination.

The outer bath should be filled to the extent of about two-thirds with water, and the lamp of course applied; the inner bath should be floated on the water, and the divisions of this also filled with water to a like extent. A thermometer should be placed in the centre of one of the divisions and watched until the necessary temperature is reached; this in the case of carbolic acid, thymol, and creasote should be about 74° C., or 165° F. When more volatile substances are used, the temperature required will be much less.

As soon as the requisite temperature has been attained, the substance or substances intended to be employed should be added

FIG. 15.—CHAMBER INHALER No. 2.

to the water in the dishes. If carbolic acid be used, 75 to 100 grains should be added to each of the divisions, and it will be found that at the end of six hours nearly four-fifths have become dissipated.

If creasote or thymol be employed, half the quantity may be taken. At the end of the six hours the whole of these substances will practically have disappeared, particularly if the creasote, which falls in part to the bottom of the water, be disturbed occasionally by stirring.

A great variety of other inhalants and disinfectants may be used with this apparatus : as eucalyptol, firwood oil, camphor, &c.; but in these cases the temperature of the water must be raised only very slightly.

All mixtures, or preparations containing carbolic acid or creasote as the basis, may likewise be advantageously employed with this apparatus.

The apparatus, 8 by 7 inches and $4\frac{1}{2}$ inches deep, is suited to rooms or chambers having a capacity of from 800 to 1000 cubic feet. It may be made of any size required, so as to admit of the use of large quantities of a variety of medicinal substances.

No hospital should be without the Chamber Inhalers and Disinfectors Nos. 1 and 2, especially the former, which is a very powerful vaporizer. They will also be found of great service in private houses in cases of infectious disease. In using them, one important fact must be borne in mind, namely, that it is not possible to charge the air of any ward or room with any medicinal substance or disinfectant while the doors and windows are open, as of course any amount of the substances which may be eliminated will be too rapidly dispersed to be of any service. It is not practicable, therefore, to combine either effective inhalation or disinfection with free ventilation.

The Globe Oro-nasal Inhaler.

It occurred to me, that some at all events of the advantages gained by increasing the surface of exposure, might be obtained by applying the principle of extension to the construction of oro-nasal and oral inhalers, and with this object I made a variety of experiments. Without giving in detail the results of these, I will now describe two inhalers I have devised, the 'Globe Oro-nasal Inhaler' and the 'Globe Oral Inhaler.'

68   THE APPARATUS EMPLOYED IN INHALATION.

That the oro-nasal inhalers at present employed, are defective in construction in many ways, and are all but useless in practice, particularly in affections of the deeper bronchi and air cells, has, I think, been sufficiently demonstrated. One defect is that, as a rule, they are not accurately adapted to the contour of the face of the wearer, so that air enters freely through other than the proper channels. Another is that the amount of the substances used is generally much too small, and this is also the case with the materials on which these are placed; then the nature as well as the size of the materials selected, such as sponge or cotton-wool, is but ill-suited to the object in view, since they hold but a few drops of liquid, and even these are sufficient so to clog them as to render them more or less impermeable to the air, which should pass in all directions freely through them.

FIG. 16.—GLOBE ORO-NASAL INHALER.

The globe oro-nasal inhaler should consist of a covering for the nose and mouth of a material such as celluloid, which, becoming flexible in hot water, may be accurately adapted to the face of the wearer, or it should be furnished with a border of indiarubber tubing. To the oro-nasal covering is attached a light glass globe, filled with a suitable material, packed so lightly as to allow of its ready permeation by the air drawn through it in inspiration; this material is then charged with the medicines desired to be employed.

The most suitable material, in consequence of its retentive properties, for filling the globe is cotton-wool lint, not flax lint, or sheep's wool, since these allow the liquid to pass through too freely. The lint is prepared as follows : it is torn into very narrow strips ; these are cut across into pieces about a quarter of an inch in size, which are then pulled into fine shreds with the fingers. A quantity of the material thus prepared should be kept in readiness for use.

The globe must be most carefully charged in the following manner—it being remembered that the object is to divide and disperse the medicaments employed as much as possible, and to allow the air to pass through the lint at all points freely. On no account must clogging be allowed, as when this occurs, evaporation is greatly impeded. 40 to 50 grains of the lint is about the quantity which should be taken. The globe is charged with the lint by means of the forceps, it being added little by little and without squeezing, as follows : the requisite quantity of the solution of carbolic acid or other medicaments used, generally 1 to $1\frac{1}{2}$ drachms, is measured off with the pipette and allowed to flow into a wine-glass. A small quantity of the lint, unmoistened, is placed at the bottom of the globe, and afterwards the rest of the lint, little by little, the liquid being sprinkled drop by drop over the lint in the globe as each successive quantity of the lint is added. By the time all the lint is used and the liquid sprinkled over it, the globe will be found to be full. On no account must the lint be pressed or squeezed in any way, as it should be kept as light and open as possible, nor must any clogging be allowed to take place.

If the liquid be poured on the lint too rapidly, a shrinking occurs, so that the globe is imperfectly filled and the air during inspiration, in place of passing through the lint, makes its way along the space between it and the globe.

Another and better method of filling the globe is the following : put a small quantity of the cotton lint unmoistened in the globe as before ; place the whole of the remainder, which is of course the chief part, in a small saucer, and proceed to sprinkle every part of it with the liquid by means of the pipette drop by drop ; lastly, transfer the moistened cotton lint to the globe in very small pieces, until it is lightly and equally charged. It is easier to avoid clogging by this method than by that first described.

The quantity of carbolic acid used for each charge, supposing the water bath not to be employed, may vary from 30 to 40 grains, usually the latter; of creasote, from 20 to 30 grains; and of thymol, from 10 to 15 grains. The last two substances not being completely soluble in the small quantity of water allowed, must be dissolved with the aid of a little spirit, no more being used than is absolutely necessary. When the water bath is used, as it should be in most cases, about half the above quantities will be sufficient.

The inhaler, if used daily without the water bath for one or two hours only, will not require refilling for two days.

For the quantities of other medicaments used with this inhaler, the medical attendant should be consulted. The same quantity of liquid should, as a rule, be used, no matter what the medicines prescribed may be. If the weather be cold, evaporation is greatly checked, and hence the action of the inhaler will be promoted by placing the hand round the globe or by allowing the globe to rest on the mouth of a metal jug filled with hot water, or better still upon *the small water bath which has been specially constructed for the purpose*. This last proceeding is recommended particularly, when the inhalation of a large quantity of carbolic acid or other medicament is desired.

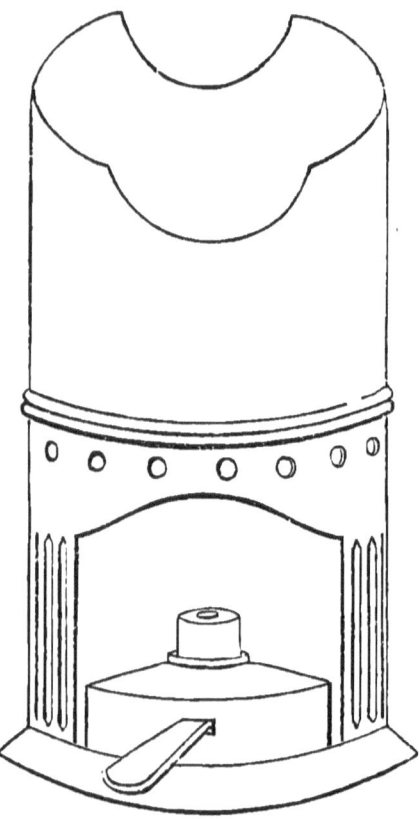

FIG. 17.—WATER BATH FOR GLOBE INHALERS.

The quantity of carbolic acid inhaled varies according to the more or less perfect way in which the inhaler is charged, the temperature, the depth and force of the inspirations taken, and the length of time during which the inhalation is continued. If for two hours and the temperature be warm, the amount of carbolic acid inhaled will vary, but will usually be about $\frac{1}{2}$ gramme $=7\cdot7$ grains; but if the inhaler be placed over a jug of hot water, renewing the water at the end of the first hour, or on the water bath specially designed for the purpose (fig. 17), the loss may amount to a gramme, $=15\cdot4$ grains. With some of the other antiseptics employed the loss will be more considerable.

It is necessary, that persons using any inhaler, the medicines employed in which are intended to pass into the lungs, should inspire as deeply as they can. If an inspiration be taken only with the mouth and cheeks and not by expanding the chest, the medicaments will not pass beyond the mouth and throat. This fact is well exemplified in the case of tobacco smokers. As most persons use cigars and pipes, the smoke does not enter the lungs at all, and for it to do so it is neccesary that a deep and continuous inspiration should be taken. All persons do not use the oro-nasal inhalers alike; some inhale by both the nose and mouth, many by the nose only, the mouth being kept shut. The most effective plan is to inhale as much as possible by the mouth, since in this case the medicaments pass directly to the throat, whereas the course by the nares is longer, more indirect, and there is a greater liability of the substances becoming absorbed in their passage over the mucous membrane covering the turbinated bones of the nose.

Although the author has repeatedly used his 'globe oro-nasal' inhaler with a charge of $2\frac{1}{2}$ grammes of carbolic acid, and has inhaled for two hours at a time, he has never experienced any ill effects; still he would recommend that the quantity used to begin with should not exceed $1\frac{1}{2}$ gramme. The Pharmacopœial dose of carbolic acid for internal administration is one grain only. What portion of the antiseptic inhaled actually makes its way into the lungs, has yet to be determined, but there is no question that, with the 'globe inhaler,' a medicinal quantity does really enter the system, part of it being absorbed by the mucous membrane of the mouth and fauces. While using this inhaler, the carbolic acid may be strongly smelt and tasted, and in a short time the skin of the lips becomes whitened and a tingling sensation may even be experienced; these last effects are produced wholly by evaporation, as, although the quantity of water or other menstruum used is larger than that employed in ordinary inhalers, not a drop ever escapes from the inhaler into the mouth, as so often occurs with most other oro-nasal inhalers.

The Globe Oral Inhaler.

This inhaler consists of a glass globe with, on the distal side, an aperture and cribriform plate for the entrance of the air during inspiration, and on the near side a flattened tube for the mouth, guarded by a valve, which prevents the air being discharged into the globe during expiration. The nose is uncovered; the flattened tube only enters the mouth, and the air is drawn through the globe in the same manner as through a cigar or pipe. By the use of this inhaler, the objection so often urged against oro-nasal inhalers is removed—namely, that by covering the nose and mouth they restrict the entrance of the air and greatly impede respiration. Moreover, its employment is far less irksome and fatiguing;

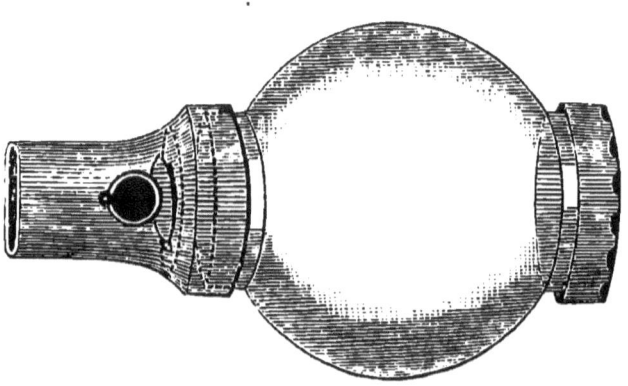

FIG. 18.—GLOBE ORAL INHALER.

it may be taken up or put down in a moment, and when once charged it may be used for one or two days in most cases, without requiring to be refilled. The ordinary quantities of carbolic acid for this inhaler are the same as those for the 'globe oro-nasal inhaler.' The 'globe oral inhaler' has a distinct advantage over the 'globe oro-nasal inhaler,' in that the whole of the medicaments pass at once into the mouth, none being lost on the skin of the face and lips, on the moustache or beard, or in the convolutions of the nares.

The valves of this inhaler are too small, and should be larger. It is most effective when used with the water-bath.

Inhaler for Dry or Concentrated Vapours.

The author has recently devised the following apparatus for the vaporization and inhalation, in a concentrated form, of the vapours of carbolic acid, creasote, and other medicaments. The description given below is mainly taken from the 'Lancet' of August 16, 1884, in which the account first appeared.

In many cases the substances intended for inhalation in diseases of the organs of respiration are prescribed in a form far too dilute to be effective. It is often desirable that the vapours should be inhaled in a comparatively concentrated state. Even when the medicaments used are undiluted, they usually become largely admixed with atmospheric air before reaching the lungs.

The means at present employed for the volatiliza-

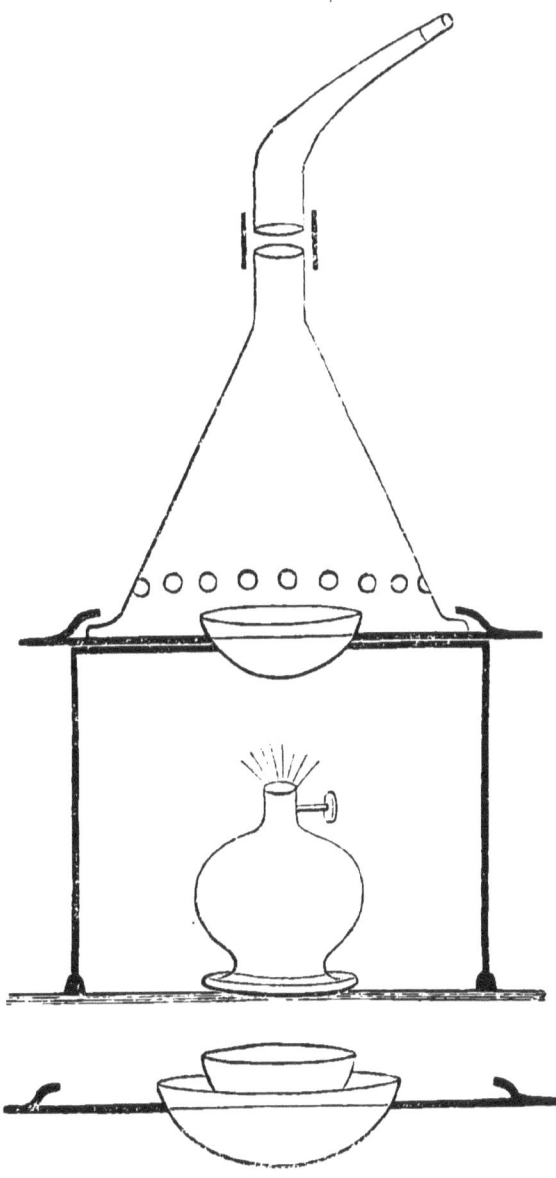

Fig. 19.—Inhaler for Dry or Concentrated Vapours.

tion, and inhalation in a concentrated form, of the vapours of certain substances are often very defective. Thus iodine and carbolic acid are sometimes vaporized from a porcelain or metal plate placed near the patient; in this way the greater part of the fumes are lost in the air of the apartment, and the small portion really inhaled is diluted many thousand times. The author has therefore been led to devise the very simple apparatus for the accomplishment of the object desired, as shown in fig. 19. It is suited alike for solids and liquids; and when an increase of temperature above that of the air is required, this is obtained by applying the lamp either directly to the capsule containing the substance, or to the small water bath. One great recommendation of the apparatus is its absolute cleanliness. There is no probability, unless it is used in the most careless way, of its becoming contaminated with bacilli, and hence there is no risk of infection, such as may happen when flexible tubes and mouthpieces enter into the structure of the apparatus employed, especially when more than one person makes use of such an inhaler. This contingency constitutes a real danger, and one not usually recognized.

The drawing almost explains itself, yet it may be well to give a brief general description. The apparatus consists of a small spirit lamp, furnished with a fine rack, so that the flame can be nicely adjusted; a quadrangular stand; two pieces of copper sheeting, with a central aperture in each; a small copper water bath; two small porcelain dishes; a funnel of glass or other suitable material, furnished with openings below for the admission of air; and a glass or metal mouthpiece, which may be used or not, as may be required. The funnel should turn up at the bottom, so as to allow of its being fastened to the copper plates by two or three simple springs. This apparatus is suited for the vaporization of carbolic acid, creasote, iodine, thymol, eucalyptol, pine oil, and many other substances. The porcelain capsules should have a diameter of about $1\frac{3}{4}$ inch, the water bath $2\frac{3}{4}$ inches, the base of the funnel $3\frac{1}{2}$ inches, and the length of the mouthpiece should be about 4 inches.

Apparatus for the Inhalation of Tar Vapours.

A very convenient apparatus for this purpose is a modification of that for the inhalation of concentrated vapours; only it must be made on a very much larger scale. The porcelain dish should have a diameter of 8 or 9 inches, and the water bath should be of corresponding size; in lieu of a glass or metal funnel one of paper may be extemporized for temporary use. The porcelain dish may be used either with or without the water bath.

Mr. L. Kay-Shuttleworth's Inhaler.

In this apparatus, which was described in the 'Lancet' of January 5, 1884, the outer air is made to pass through a coiled pipe placed in a receiver containing hot water, the temperature of which is maintained by means of a spirit lamp. So far as the coiled pipe and the receiver are concerned it therefore bears a resemblance to the apparatus which has been already described for cooling air, and this in its turn reminds one of the worm and tub ordinarily employed in distillation (see fig. 1). The tube or pipe is 9 feet long, the lower end being open to the air and the upper end 'expanded into a small chamber, closed by a hollow plug of metal, in the interior of which tow could be placed to receive the carbolic or other disinfectant, and which was connected with an ordinary oronasal mouthpiece properly padded by a large indiarubber tube. I then placed 13 grains of carbolic acid on 8 grains of dry tow, making a total weight of 21 grains, and placed this in the hollow plug. After 10 minutes' inhalation the total weight was reduced to 13 grains; after 20 minutes, including the time occupied with the first weighing, to 8 grains +, after 30 minutes to 8 grains —, showing, as I believe, that the total amount of carbolic acid was completely vaporized in about 20 minutes, and from my sensations of tingling lips, sore nose and throat, pain in the chest, and slight vertigo, with exhilaration of spirits, I believe faster than desirable for ordinary inhalation. The temperature of the water in the boiler ranges from 170° F. at the commencement to 198° F. at the termination; a slightly lower temperature causes the vaporization of the carbolic acid to be more gradual.'—See the 'Lancet.'

CHAPTER IV.

ON INHALATION CHAMBERS.

THERE may be said to be three kinds of inhalation chambers: the pneumatic chamber already described, used chiefly for the inhalation of compressed and rarefied air; the chamber in which water, holding in solution the medicaments, is discharged into the air in a finely divided or atomized state by means of various kinds of spray producers; and, thirdly, the chamber, the whole air of which is uniformly charged with the substances required to be inhaled in a gaseous or aëriform condition. The construction of this last description of inhalation chamber, so far as I am aware, has not until recently been accomplished in a satisfactory and scientific manner.

The following are the conditions and principles of the construction of such a chamber, as first described by me in an article communicated to the British Medical Association at the meeting in Liverpool in August 1883, and which was afterwards published in the journal of the Association of January 12, 1884, under the title of 'The Principles of the Construction of Chambers for Inhalation in Diseases of the Lungs.'

First, the dimensions of the chamber should be moderate, and in some cases even small; and this on account of the great quantity of medicaments which would be required for charging a very large space. Secondly, the fireplace to the chamber should be closed, and the inlets and outlets should be only those of ordinary and moderately well-fitting doors and windows; and this because, if the air of the room were rapidly changed, of the difficulty in keeping the air charged to any given extent and of the great waste of material which would take place. Every chamber may be provided with one or two small outlets, to enable those inhaling to get rid of the used air emitted during expiration, the air being thereby maintained in a state of greater purity. For the purpose nothing more would be necessary than a piece of elastic tubing, furnished with a covering for the nose and mouth. Thirdly, the temperature of the room should be kept at about 17° C., = 62° F. : this regulation of the temperature is requisite in order that the volatilization of the substances used should proceed at one fairly uniform rate. Fourthly, the air of the room should be somewhat humid, the degree of humidity being ascertained by means of wet and dry bulb thermometers. A moderate, but not too great an amount of moisture in the room is desirable.

The means whereby the air of the chamber is to be charged with suitable medicaments have already been referred to, particularly in Chapter II., but not in sufficient detail. The two great principles relied upon for accomplishing the object in view are *extent of surface* and *augmentation of temperature*.

It occurred to me, that if any substance possessing even a feeble volatility at ordinary temperatures were dissolved and spread over a very large surface, the amount of the substance which would become volatilized would be in proportion to the quantity taken, the extent of the surface exposed, the temperature of the chamber, and the amount of the moisture therein contained.

To test this point, 50 grammes of carbolic acid (= 771 grains) were dissolved in water, two Turkish towels being saturated with the solution and exposed to the air of a room at a temperature of 22° C., = 72° F., the room also containing rather much aqueous vapour.

At the end of 2 hours a portion of the air of the room was, by means of a large syringe, drawn through a solution of alcohol, a few drops of which, on evaporation, gave with bromine water decided evidences of the presence of carbolic acid ; 13 cubic centimetres of the vapour or moisture which had settled on the windows (= 200 grains) were collected, and were found to contain 0·029 gramme of the acid, = 0·447, or nearly half a grain.

After the lapse of 48 hours the towels had become quite dry, and I was not a little surprised, on examining them, to find that they did not possess the slightest smell or taste of the carbolic acid. One of the towels was then tested, and found to contain only 0·07 gramme of acid, = 1·08 grain ; that is to say, the whole of the carbolic acid employed, namely, 771 grains, had disappeared, with the exception of the small quantity above referred to.

This result appeared to me so surprising, although I was prepared for a considerable reduction in the amount of the acid taken, that I repeated the experiment in a more precise manner. I took one of the Turkish towels only. It measured 48 by 18 inches, thus presenting a surface of 1728 square inches, and it was saturated with an aqueous solution containing 25 grammes, or 385 grains, of carbolic acid. This experiment was commenced at 10 A.M., the towel being moistened in the course of the day once or twice with a little water. At 10 A.M. the next day the towel was still somewhat wet, but at 4 P.M. it was only just damp. It was then tested for the acid, 0·09 gramme only being obtained, = 1·39 grain. Thus, in the space of 30 hours the whole of the acid used, 385 grains, had practically disappeared.

These results are of the highest importance, as they place in our hands the means of charging inhalation chambers to almost any desired extent with all those medicinal substances which possess even a feeble volatility at the ordinary temperature of the air.

The following experiments were also made with carbolic acid :— Turkish towels were cut into pieces, measuring $18\frac{1}{2}$ inches in one direction and $9\frac{1}{2}$ inches in the other. Each of these was charged with 5 grammes (77 grains) of carbolic acid dissolved in about 100 cubic centimetres of water, each piece being tested at intervals of *four hours*.

There were recovered out of the 5 grammes, or 77 grains, taken—

After 4 hours 2·77 grammes = 42·66 grains
 ,,   8   ,,   1·82   ,,   = 28·03   ,,
 ,,   12   ,,   —   ,,   = —   ,,
 ,,   16   ,,   0·49   ,,   = 7·55   ,,
 ,,   20   ,,   0·20   ,,   = 3·08   ,,

To show that the principle holds good in other cases I will now state the results of a few further experiments.

Portions of Turkish towels were taken, $9\frac{1}{2}$ by $4\frac{1}{2}$ inches, and were each charged with 1·25 gramme of creasote, = 19 grains. In one case the creasote was placed on the cloth without any addition; in another water was used; and in a third case the creasote was dissolved in alcohol. After 4 hours the cloths were examined. From that with creasote only 0·91 gramme was recovered, from that with creasote and water 0·41 gramme, and from that with the alcohol 0·39 gramme: amounts which correspond with 14·01, 6·31, and 6·01 grains respectively. It will thus be seen, taking equal areas of the Turkish towels, that the loss of creasote at the end of the 4 hours was even greater than that of carbolic acid, both having been treated with water; the proportion recovered in the one case being 2·77 grammes, and in the other 1·65 gramme, = 42·7 and 25·4 grains. The experiments also showed that the loss, as might have been anticipated, is much less when the creasote is simply sprinkled on the cloth, because in this case it is less diffused and covers a smaller surface. When alcohol is employed the loss appears to be about the same as when water is used, although of course the towels, when moistened with alcohol, become dry in a much shorter time than with water.

In other corresponding experiments, in which the time allowed was 24 hours in place of 4, the amounts of creasote recovered, approximated to the quantities obtained after 4 hours; and this would appear to show that the loss is proportionately greater during the first few hours of its exposure; and that this should be so will be readily understood when it is stated that creasote, unlike carbolic acid, is not a definite substance, but is made up of different compounds, having different degrees of volatility.

The last experiments made were with thymol. Thymol possesses only very slight volatility at ordinary temperatures, and is nearly insoluble in water; so that, in the case of this antiseptic, it is necessary to use a solvent. This in the experiments about to be described was alcohol. 5 grammes of thymol, = 77 grains, dissolved in 80 cubic centimetres of spirits of wine, were spread over a piece of Turkish towel, 18½ by 9½ inches; and after 4 hours 3·44 grammes, = 52 grains, of the thymol were recovered. From another portion of Turkish towel of the same size, and containing the same amount of thymol, 1·70 gramme, = 26·27 grains, was obtained at the end of 50 hours.

These experiments show that the loss of thymol, so little volatile at ordinary temperatures, is very considerable; although the proportionate loss is much less than in the case of carbolic acid and creasote, still it is sufficiently great for practical purposes.

It is thus seen that the idea or principle upon which the foregoing experiments were founded was correctly conceived.

But there is another principle which may be brought to bear on the subject of inhalation, and which, combined with the principle of evaporation over a large surface, is capable of furnishing results equally striking with those already narrated; and that is the principle of increasing the volatility of but little volatile substances by means of heat; that is to say, by a regulated temperature, varying according to the volatility and properties of the substances employed.

The best and most practical method of applying this increased temperature, is by means of hot water spread over a comparatively large surface. Owing, however, to the increase of volatility obtained by the augmentation of temperature, the extent of surface required will be very many times less than that which is necessary when Turkish towels or other suitable fabrics are used, while the time occupied will be proportionately reduced—two great advantages.

The effect of the combination of a somewhat large surface with increase of temperature is well illustrated by what takes place in the case of thymol. 40 grammes of thymol, = 616 grains, were melted in hot water in a dish having a superficies of 64 inches, the temperature being maintained at $72°$ C., = $161·6°$ F. At the end of 12 hours it was found that the whole of the thymol had disappeared, a substance which, as already remarked, possesses but little volatility at ordinary temperatures, and which, even when spread over the large surface of a Turkish towel, is far less volatile than carbolic acid, creasote, and a variety of other substances.

In another communication, 'Investigations relative to Inhalation and Disinfection, with Descriptions of New Forms of Apparatus,' inserted in the 'Lancet' of October 6, 1883, I gave some further data illustrating the effects of evaporation on extended surfaces, exposed either to the air or in water, and also of augmentation of temperature in increasing the rate of evaporation.

A piece of Turkish towel, 16 by 11 inches, giving a superficies on the two surfaces of 352 inches, was moistened with 100 cubic centimetres of water, containing 100 grains of carbolic acid. Six of such pieces were similarly treated, suspended in the air on glass rods, and the amount of carbolic acid contained in them determined at the end of each successive hour; the quantities recovered were 82·79, 75·49, 63·89, 57·02, 48·73, and 41·39 per cent.

These figures show a progressive loss each hour of 17·21, 24·51, 37·11, 42·98, 51·27, and 58·61 per cent. Thus if the 6 pieces had all remained suspended for the 6 hours, of the 600 grains of carbolic acid used in charging them, no less than 350 grains would have disappeared, or more than one-half.

Six cloths of the same dimensions were next taken, each being moistened with a mixture of spirit and water in equal parts, containing 120 grains of creasote; the entire quantity in this case for the 6 cloths would therefore be 720 grains. There were recovered at the end of each successive hour, the following amounts: 103·5, 86·8, 83·7, 80·8, 72·0, and 66·0, showing a loss of 16·5, 33·2, 36·3, 49·2, 48·0, and 54·0. Thus of the 720 grains taken, no less than 324 grains would have disappeared had all the cloths been suspended for 6 hours.

Lastly, 5 cloths were each moistened with a mixture of spirits of wine and water, in equal parts, containing 5 grammes of thymol, = 77 grains; these cloths therefore contained 25 grammes of thymol, or 385 grains. The following amounts were recovered each hour: 4·74, 4·44, 4·08, 3·88, 3·12 grammes, = 73·0, 68·3, 62·8, 59·7, and 48·0 grains respectively; the loss for each cloth in grains standing thus: 4·15, 8·78, 14·10, 17·4, and 29·15 respectively. Here again the loss is very considerable, although not so great as in the two previous cases. Had all the cloths been exposed the whole time, the loss would have amounted to 145 grains out of the 385 originally taken. A considerable loss of course also takes place in the case of other but little volatile substances, exposed in the same manner on an extended surface, and dissolved, if necessary, in suitable menstrua; the extent of the loss being proportionate to the natural volatility of the substance and its solubility. If the loss is so great resulting from the exposure of such a limited superficies, as is afforded by a suitable fabric of 16 by 11 inches, it is obvious that by extending the surface we possess a most powerful means for charging the air of a chamber with any medicament, which possesses only a feeble volatility at ordinary temperatures, and this to any extent required.

The further experiments instituted for the purpose of showing the effects of the exposure of the substances on a comparatively extended surface of water, or even by dissolving them in it, and then raising and maintaining the temperature of the water at certain fixed degrees, furnished the following results :—

Two and a half grammes of carbolic acid were dissolved in 150 c.c. of water and placed in a small glass dish, having a superficies of 8 inches; 5 other dishes were similarly charged, and all were floated on a water bath, the temperature of the liquid in the dishes being maintained at about 74° C., or 165° F. And here I would remark, that the temperature of the liquid should always be taken in the dishes themselves, and not in the water in the outer bath, as there is often a difference of 6° to 8° F. between the two, according to the nature of the vessels employed. The quantity of carbolic acid, therefore, contained in the 6 dishes amounted to 15 grammes, or 231 grains. The amounts recovered were 1·838, 1·470, 1·116, 0·868, 0·661, and 0·511 grammes, = 28·3, 22·6, 17·2, 13·3, 10·1, and 7·8 grains respectively. It thus appears that of the 38·5 grains originally present only 7·8 grains remained at the end of 6 hours. It follows therefore that of the 231 grains originally taken all but 47 grains would have been dissipated, equivalent to three-fourths, had all the dishes been exposed for the same length of time.

With similar quantities of creasote, namely, 2·5 grammes for each dish, the results stand as follows :—In the first experiments the creasote was allowed to remain at rest, the greater part sinking to the bottom, it being only partially soluble in water, and part floating on the surface. There were recovered successively 1·245, 0·894, 1·10, 0·51, 0·21, and 0·37 grammes, =19·17, 13·76, 16·94, 7·85, 3·23, and 5·69 grains. Thus, in this case, had the temperature of all the dishes been maintained for the 6 hours, no less than 197 grains would have disappeared out of the 231 grains originally taken.

When, however, the contents of the dishes were stirred occasionally with a glass rod, whereby the heavier portions of the creasote, which had fallen to the bottom of the dishes, became broken up and dispersed, the loss was still more considerable, as indicated by the following figures : 0·76, 0·62, 0·43, 0·13, 0·035, 0·00 grammes, = 11·70, 9·54, 6·62, 2·00, 0·54 grains, while from the last dish the whole of the creasote had disappeared.

The experiments with thymol furnished results almost equally striking. This substance is soluble only to a slight extent in water, and when exposed to the air at ordinary temperatures, although it gives off its pleasant smell freely, it scarcely loses weight at all; thus 15 grammes exposed to the air were found to have lost only the most trifling amount at the end of a fortnight. But thymol melts rapidly in hot water, and, being lighter than water, floats upon and spreads over its surface in a thin scum, and in this condition it becomes rapidly volatilized, as will be apparent on an examination of the annexed figures. 2·5 grammes being added to each of the 6 dishes containing the usual amount of water, the temperature of which was maintained at 74° C., the following amounts were recovered: 1·82, 1·23, 0·874, 0·64, 0·36, 0·096 grammes, = 28·02, 18·94, 13·46, 9·85, 5·54, and 1·47 grains. Thus, had the experiments been continued with all the dishes for 6 hours, the loss would have amounted to 123 out of the 131 grains taken at the commencement.

Armed with these principles and data, I was led to devise the several forms of apparatus, which have already been described in Chapter III.

In the article communicated to the British Medical Association already referred to, 'On the Principles of the Construction of Inhalation Chambers,' I stated that, with a view to carry out these principles, I was about to construct an inhalation chamber at San Remo. This was completed and has been in operation for more than a year; the following are some details respecting it, together with the results of further experiments, and which first appeared in the 'Lancet' of January 19, 1884. The chamber has a cubic capacity of 1170 feet; it has only one door and one window, but no fireplace or chimney. The walls are washable and the floor tiled. It is warmed by the admission of a little hot air through a grating in the centre of the floor, from a calorifère placed in the basement of the house, the temperature of the chamber itself being raised usually to about 64° F. The chief apparatus contained in the chamber consists of one of my No. 1 Chamber Inhalers, which was described in the 'Lancet' of October 6, 1883, and consists of a very convenient and portable arrangement of a cotton fabric disposed in layers. Each of the 5 cloths or layers of the apparatus employed by me has a superficies on one side of 2500 inches, and on both sides, of course, of 5000 inches. It is suspended from the ceiling about four feet above the grating in the floor, through which the warmed air enters, and it can be raised or lowered at will. The quantity of carbolic acid used to charge the whole of this apparatus is 2000 grains; but it is better to charge part of it only, and to begin with 800 grains for the first day or two, raising the amount to 1000 or 1200 grains in the course of a few days. The latter quantity is that which I have hitherto chiefly employed, and so far no inconvenience has been experienced. The time of inhalation has varied from one to two hours daily, but it is best at first to be content with half an hour only.

There are several circumstances which affect the rate of evaporation of carbolic acid and of other more or less volatile substances, in addition to those to which I have elsewhere adverted, namely, complete solution and dispersion over a great extent of surface. The chief of these additional circumstances are the state of the air as to moisture, motion, and above all temperature; the dryer the air the more rapid its movement, and the higher the temperature the quicker the volatilization. On the other hand, the reverse conditions of a moist, still air and comparatively low temperature, greatly retard and may even stop volatilization.

The following experiments may be cited in illustration of the above statements :—100 grains of carbolic acid dissolved in 100 c.c. of water, =1,540 grains, were distributed over a piece of cotton fabric having a superficies on one side of 153 inches. This was exposed at a temperature of 58° F. for 10 hours of the night in my inhalation chamber, at the end of which time 58·2 grains of the acid were recovered. A portion of the same fabric, of the same dimensions and similarly charged, was exposed to the rays of the sun at a temperature of 90°, there being a light breeze blowing at the time. At the end of 4 hours 4·74 grains only were recovered ; that is to say, nearly the whole of the acid had disappeared. From a third portion of the fabric charged in the same manner and exposed in the chamber during the day for 6 hours at a temperature of 60° F., there were recovered 54·2 grains, so that nearly one-half of the acid had disappeared. The above experiments clearly show that the still, sunless air of a chamber having a temperature of only 60° F. is far less favourable to evaporation than free exposure to an actively moving and warmer air. Still, the rate of evaporation, even at so low a temperature, is amply sufficient for all practical purposes. When the temperature of my inhalation chamber is raised to 64° F. the results, as will now be shown, are still more favourable. In an experiment in my chamber, the temperature of which was maintained at about

64° F., using 1200 grains of carbolic acid, there were recovered at the end of 6 hours 500 grains, showing a loss of no less than 700 grains. This result is very satisfactory. From the foregoing experiments it is also apparent that, in order to arrive at fixed results, it is necessary that the chamber should be worked as nearly as possible under the same conditions, especially of temperature. It has been stated that the apparatus employed for promoting the vaporization of the carbolic acid is suspended from the middle of the ceiling, and can be lowered or raised at will, but usually it is kept at a distance of about four feet from the floor; also that the warmed air enters at the centre of the floor. The result of this arrangement, not at first foreseen, is that while the temperature of the room is usually not more than 64° F., the air as it passes through the grating is much warmer, and when it reaches the lower part of the apparatus it has a temperature of about 80° F. This causes the evaporation of the acid in the lower folds of cloth to proceed at a much greater rate than in the upper folds. Another effect observed is that part of the carbolic acid vaporized becomes gradually deposited on the walls of the room.

But there is still another important circumstance which has to be taken into consideration—namely, *the composition and quality of the fabric over which the medicaments employed are dispersed.* I find that a cotton fabric answers the purpose much better than one of linen, and also that, whatever the composition, it should not be too thick and rigid, as in this case it retains the moisture for too long a period, the vaporization being thereby much retarded. The following experiments show this in a striking manner. From a light cotton fabric, with a superficies on one side of 153 inches, and weighing 483 grains, there were recovered at the end of 6 hours' exposure in the chamber, at a temperature of 60° F., 54·2 per cent. of the acid. From a thick coarse brown flax fabric, with a superficies of 157 inches, and weighing 1027 grains, there were recovered under the like circumstances no less than 77 per cent. Lastly, from a thick close cotton fabric with a surface of 153 inches, having a weight when dry of 662 grains, 77 per cent. of the acid were likewise recovered. These experiments are also very instructive. I may here mention, that my first experiments on evaporation by means of cloths, namely, those recorded in the 'Lancet' of October 6, 1883, were, I have since ascertained, made with a somewhat dense flax material, and hence they were not the most favourable to evaporation; yet the results arrived at were very striking.

It has been stated that my inhalation chamber has a capacity of 1170 cubic feet. A cubic foot contains 1728 inches; a man 5 feet 7 inches in height has a respiratory capacity of 222 inches = nearly 8 inspirations per foot, or 9368 per 1170 feet. Taking normal respiration at 18 per minute, the consumption of air would be 3888 inches per minute and 233,280 inches per hour =134 cubic feet.

Now, supposing the apparatus be charged with 1200 grains of carbolic acid, about 600 grains, or one-half, would at a temperature of 64° F. be vaporized at the end of six hours, this being at the rate of 100 grains per hour. Supposing, further, that the cloths have been charged for one hour before the patient enters the chamber, and that he remains there for one hour, the quantity of carbolic acid which would be vaporized in the two hours would be 200 grains, and if the patient remained in the chamber for two hours the amount would be raised to 300 grains. The quantity of acid, therefore, which would be contained in the 134 cubic feet of air at the end of the first hour of inhalation, supposing none of it were inhaled, would be 22·9 grains, and at the end of the second hour 34·3 grains. But it must not be supposed that the whole of the above quantities would actually be inhaled by any patient who occupied the inhalation chamber for the periods above named; the quantities would in reality be much less, owing to several causes. First, the capacity of but few *poitrinaires* is as great as 222 cubic inches; secondly, a loss of material is constantly going on to some extent by escape through the crevices around the window and the door; and then, thirdly, there is a sensible deposition of carbolic acid on the walls of the chamber itself: so, formidable as is the amount of carbolic acid originally taken, the quantity actually inhaled is surprisingly diminished. Still great caution and watchfulness should be observed, lest in any case the chamber should become overcharged; but I have never once observed any ill effects to ensue from the use of the chamber in the manner I have described.

The loss of about 100 grains per hour during the first 6 hours at a temperature of 64° F. is readily effected by means of my No. 1 chamber inhaler, 11 by 16 inches. It is preferable however, to employ an inhaler double or treble the size, as in this case it will not be necessary to charge the whole of the cloths at one time. Should a larger elimination of the acid be desired, in consequence of the greater size of the chamber, it is only necessary that a proportionately larger apparatus should be employed. Amongst the advantages of an inhalation chamber is the avoidance of fatigue, often a consideration of great moment, since many patients are unable from weakness to make an efficient use of oro-nasal or oral inhalers.

Again, it is not necessary that all patients should resort to a specially prepared chamber, since, under proper medical advice, the method of inhalation with either of my chamber inhalers, No. 1 or No. 2, may be carried out, though less perfectly, in the patient's own room.

That inhalation chambers, constructed in the manner and on the principles I have described, furnish a powerful and valuable means of treating many diseases of the organs of respiration, including some cases of consumption, is unquestionable, and no hospital or institution for the treatment of this class of affections should be without one or more of such chambers. But it is not only in lung diseases that they are required; with certain modifications they would doubtless prove of great service when used as chambers for disinfection.

In carrying out a system of inhalation, it is not well to rely exclusively on any one means, but as far as practicable to combine them all. Thus, while the air of the chamber occupied by the patient should be medicated, he should also use a spray apparatus or a really efficient oro-nasal inhaler, while in some cases antiseptic or other appropriate remedies should also be administered by the mouth. Although I have referred in this article chiefly to carbolic acid, yet it must be understood that a variety of disinfectants, stimulants, antispasmodics, and even sedatives may also be vaporized in the manner I have described.

In this place may now be noticed one or two other forms of apparatus, which have been devised for charging the air of rooms with medicaments.

Mr. Robson's Eucalyptus Machine.

Mr. A. W. M. Robson, F.R.C.S., of Leeds, has devised a machine for impregnating the air of a room with the vapour of oil of eucalyptus. The machine was shown at the meeting of the

British Medical Association at Worcester, and was described in the journal of September 2, 1882. The apparatus is worked by a bellows, and is made by Messrs. Mayer, Meltzer, and Co., who let it out on hire.

Mr. Robson, writing in the 'Lancet,' states, 'I direct the nurse or other attendant to work the bellows for a few minutes at occasional intervals, so as to keep the air of the room odorous of eucalyptol. By converting the air of a sick chamber into a pure and antiseptic atmosphere it refreshes and soothes the patient; and in rooms occupied by consumptive cases the antiseptic treatment may be effectually carried out without muzzling the invalid by a respirator.'

Mr. H. Langley Browne's Apparatus.

Mr. Browne described in the 'British Medical Journal' of February 2, 1884, an apparatus on a different principle for the diffusion of oil of eucalyptus or turpentine through the air of a room, the dispersion being effected by means of a fan. This apparatus was intended to take the place of the spray in operating, or to fill the operating room with antiseptic air, and so do away with the necessity of using the spray. It consists of a box with two chambers; a revolving fan, worked by turning a handle, sucks the air through a filter in one chamber and then drives the air through the second chamber, which contains oil of eucalyptus on stick charcoal, and then delivers the air through a wide-mouthed box, at the rate of 200 cubic feet per minute, thoroughly charged with eucalyptus vapour. Turpentine answers as well and is much cheaper. A room can speedily be filled with antiseptic air, and by means of a water current the supply can be made continuous. This apparatus is made by Messrs. Cuxon and Co., of Wednesbury.

Dr. Corrigan's Vaporizer.

The late Dr. Corrigan, of Dublin, several of whose courses of lectures I attended in my student days, devised many years ago an inhaler which will still be found very convenient for the continuous inhalation of many medicaments in the vapour of hot water. It consisted of a lamp, lamp stand, porcelain dish of about 6 inches in diameter containing water, and an inverted flask, supported by a retort stand, and closed by a perforated cork, through which a few

ON INHALATION CHAMBERS. 89

threads of cotton were passed, so as to allow of the regular and constant dropping of the contained solution into the hot water beneath it. There must be a pin hole in the flask or other contrivance, to allow of the entrance of air. This apparatus is adapted for partially charging the air of that part of a room in which the patient is either sitting or sleeping.

Messrs. Savory and Moore's Vaporizer.

An apparatus similar in principle to the preceding, for the continuous volatilization of carbolic acid and some other medicaments is the well-known vaporizer of Messrs. Savory and Moore. This consists of a stand, small lamp, a curved metal plate, and 'a reversible dropper, capable of holding a considerable amount of carbolic acid or other suitable ingredient. When the vaporization is to commence the lamp must be lighted and the dropper turned over, when the acid will escape in successive drops, and falling on the heated dish will become speedily dissipated.' The vaporizer of the ordinary size will continue in action for three hours.

Dr. Neale's Chemical Lung.

Of course, after the inhalation chamber has been in use for some hours, and since its ordinary outlets are closed, the air becomes contaminated, to some extent, by an excess of carbonic acid, ammonia, and organic matter. Supposing the chamber to have been used by only one person for an hour or so, the extent of the contamination would be but slight, seeing that an adult man ordinarily gives out per hour only about 6 cubic feet of carbonic acid gas, so that even at the end of two hours the deterioration of the atmosphere would not be considerable. Still, as far as possible in practice, I allow an interval to elapse after each use of the chamber; the window and door are opened for a time, and the air then recharged. When this is not practicable, recourse may be had to Dr. Neale's chemical lung and punkah.

The purifying agent in this apparatus consists of a strong solution of caustic soda or lime; these alkalies have a remarkable affinity for carbonic acid and in a less degree for ammonia and sulphuric acid, these constituting the chief impurities in the air of insufficiently ventilated rooms, and they speedily remove the excess of those substances when the air is brought into contact with the alkalies.

The apparatus consists of an endless sheet passing over two rollers, the lower of which is sunk into a trough, which holds about three gallons of a saturated solution of a caustic alkali. The apparatus is suspended from a bar attached to the ceiling of the room, and as the punkah swings to and fro a wheel and a ratchet cause the sheeting to revolve, one complete revolution of the sheet being accomplished every four minutes. There is a water gauge to indicate the amount of fluid contained in the trough, a pocket, whence the water and pieces of caustic alkali are supplied to the trough, and a tap for drawing the solution from the trough.

In the absence of any special means of purification, the door and window should then always be thrown open after the inhalation chamber has been used once or twice, and at the end of each day it is better to allow them to remain open until the following morning. In this way not only will the air in the room become renewed, but any carbolic acid or other medicament which may have settled on the walls will have had time to become entirely dissipated.

But a very effectual means of ensuring the purification of the air of the chamber is to cause the patient to expire through a tube communicating with the outer air. This, however, imposes on the patient a proceeding, which is somewhat irksome, and at the same time, with due precautions entirely unnecessary.

Dr. Braxton Hicks' Apparatus for Dry Air Disinfection.

Dr. Braxton Hicks showed, at the meeting of the Medical Society of London, held October 27, 1884, an apparatus for medicated dry air disinfection. It consists of a cylindrical vessel of pewter, the apparatus being worked by a foot-bellows, the conducting pipe of which enters the vessel near the bottom. The interior is filled with sponge clippings saturated with the medicament to be employed, while a discharge tube proceeds from the upper part of the receptacle. Should it be desired to increase the volatility of the substances used, this object may be effected by placing the apparatus in hot water. The apparatus exhibited was 10 inches high and 4 inches in diameter.

CHAPTER V.

ON THE QUANTITIES OF THE MEDICAMENTS AND ON THE MANNER, FREQUENCY, AND DURATION OF THE INHALATIONS.

SEVERAL important practical questions have now to be considered in relation to the subject of inhalation in diseases of the organs of respiration, as the *doses* of the medicaments, the frequency of their repetition, and the duration of the inhalations.

First, let us enquire how far there is reason to believe that the medicaments, when introduced into any part of the respiratory track, are really effective.

Formerly it was the general belief, that the only natural and effective channel for the administration of medicines was the stomach, because it was in this organ and in the duodenum, it was conceived, that the lacteal and lymphatic systems were centred, in consequence of which, it was reasoned, that the remedies must necessarily be the more readily absorbed and thus spread over the system. But in this view of the matter certain facts were for a long time overlooked; one was that the acids in the stomach, altered in some cases the constitution of sundry of the medicines used, impairing their efficiency, and the other that the presence of food in the stomach often hindered and delayed their absorption.

Soon, however, from reasoning and experience, it was ascertained that powerful local effects were obtainable by the injection of medicines into the rectum. At first this method of administration was considered to be of somewhat limited application, but as the proceeding became more general, it was found that medicines might often be usefully introduced into the rectum in cases in which remote constitutional effects were required, and that even some kinds of food might be thus administered, with the result of prolonging life.

Another great step in advance was the subcutaneous injection of remedies. The success of this plan must have astonished those who first resorted to it, although now it admits of easy explanation. In the cellular tissue beneath the skin there are no acids to alter the composition of the remedies, and no bulky food to dilute them and impede their absorption by the lymph spaces and vessels which there abound. The consequence of these circumstances is that medicines injected beneath the skin are more quickly and completely absorbed, and that much smaller quantities are needed to produce the required effects. Once introduced into the absorbents, the remedies are conveyed to the blood in most cases in an incredibly short time, but in others a somewhat longer period is necessary. When the medicament is injected at once into the blood, without its having to pass through the absorbents, this brief interval is still further shortened, but still there must always be some interval between the administration of a remedy and the effects resulting from it.

It has been already remarked that the time occupied in the absorption of remedies subcutaneously injected, although very short, varies. This variation depends mainly on the diffusive power of the substances used, in which there is a great difference. Most of those which are employed for hypodermic injection have a considerable power of diffusion, but some of those given by the stomach or bowels possess but little diffusive capability. The success of the administration of remedies by the rectum, and still more by injection beneath the skin, has proved a great encouragement to the practice of inhalation in affections of the lungs. The mucous membranes of these organs and the submucous cellular tissue abound in lymphatic absorbents, whereby the medicaments are rapidly taken up, absorbed, and conveyed away, and thus remote effects are produced in a manner even more effective and prompt in most cases than when given by the stomach.

But the principal advantage of inhalation in diseases of the organs of respiration is, that the remedies are brought into actual contact with the parts affected, so that their local effects are obtained, just as though the parts were fully exposed, as on the surface of the body. The above views and statements are fully borne out and confirmed by the results of actual trial and experience.

In reference to this subject Dr. Cohen has expressed himself to the following effect :—

'As has again and again been proven by actual experiment on the lower animals and on man, the mucous membrane of respiratory organs has a much greater capacity for absorption than that of the stomach, than which it is much more delicate; and for articles not desirable to be exposed to the solvent principle of the gastric juice, inhalation is at least as advantageous a mode in many instances for the administration of appropriate remedies as the skin or connective tissue. The material inhaled comes directly in close juxtaposition to the blood while in its most vital state, and is thus more promptly and more thoroughly absorbed into the tide of the circulation than when it traverses part of the venous circuit before exposure to the inspiratory effort. It is often advantageous, too, that nothing shall interfere directly with the digestive functions.

'The promptness with which the respiratory mucous membrane absorbs is well shown in the action of general anæsthetics, and he who is sceptical as regards the facility with which other articles of the materia medica are absorbed can perform upon himself the experiment of taking a certain amount of a narcotic by the stomach, and at another time, under similar conditions of body, as well as may be, inhaling a similar amount of narcotic in solution, by some of the methods described in the pages of this volume. Again, in a memoir presented by Sales Girons to the Paris Academy of Medicine on Respiratory Therapeutics it is claimed for these inhalations of nebulized liquids that they may be advantageously resorted to for the purpose of producing constitutional effects in a variety of other diseases besides those of the organs of respiration, since the mucous lining of the bronchi, and particularly of the air cells and lungs, exceeds all other mucous membranes, even that of the stomach and small intestines, in its powers of absorption, in consequence of the rapid passage of the entire mass of the blood through the lungs, exposing within the space of half a minute, almost every globule of that fluid to the action of any remedy through the endosmotic action of an extremely attenuated membrane of great absorptive power. In proof of this capability of absorption, it is stated that 25 litres of water may be injected into the bronchi of a horse within 6 hours, and become at once absorbed without occasioning any sensible injury to the animal.'

The practice of medicated inhalation in affections of the organs of respiration rests, then, in this respect on a sure and scientific basis.

94 ON THE QUANTITY OF MEDICAMENTS EMPLOYED.

With respect to the actual quantities of medicaments which should be used, it is evident, from what has been stated, that as a general principle, subject to modification from several causes, the doses should be less than would be given by the stomach, and should approximate to those employed in hypodermic injections, especially when the more remote effects are desired.

When the remedies are inhaled for their local effects, the case is different. We must, then, consider and weigh well the physical action and properties of the medicaments used and the sensitive nature of the surfaces and parts to which they are to be applied.

Now one of the principal circumstances which modify in a highly important manner the amounts of the medicaments to be employed, is the quantity or proportion which is actually lost during the act of inhalation, I will not say lost in the lungs themselves, but in the throat, stomach, and lungs, the quantity, in fact, which passes into the system through all the parts named.

On this point, until my investigations were undertaken and made known, but few precise or definite observations and experiments existed. It was erroneously assumed, that a large and in some cases the larger proportion of what was used was really utilized; and this assumption, as I have shown, was a fundamental error of great magnitude, and hence it came about that the doses generally employed were in most cases far too small.

Thus it has been conclusively proved, when oro-nasal inhalers are used for such substances as carbolic acid, creasote, thymol, &c., so little volatile at ordinary temperatures, that about four-fifths of the quantity taken are recoverable, after one or even two hours' inhalation, from the sponge or cotton-wool of the inhaler; so that, using oro-nasal inhalers in the manner generally adopted, it would be necessary to charge them with five times the quantity of carbolic acid or other similar substance which it is desired should actually enter the system; I do not say the lungs, for we have no precise information as to how much really reaches the air passages; only of this we are sure, that the quantity is extremely small when oro-nasal inhalers are used.

In the case of air and steam sprays, although the loss is less, it is still very considerable, amounting to fully two-thirds of the quantity taken, and possibly it is even greater. With air sprays of the most recent construction, that it is to say, with one tube only, so curved as to actually enter the mouth, the loss, if the apparatus be very carefully employed, may be less considerable.

With fumes and vapours as ordinarily inhaled, especially in the vapour of hot water, the loss is even very much greater than with oro-nasal inhalers; in fact, only a fractional portion of the amount with which the apparatus is charged actually becomes inhaled.

Several attempts have been made to determine the proportion of the sprayed liquids inhaled, or rather of that which really enters the system, and which is in part either absorbed by the mucous membrane of the mouth and throat, swallowed, or inhaled. Owing to the inherent difficulties of the enquiry the results are by no means satisfactory or precise; indeed, the determination of the exact quantity which really passes into the respiratory track would appear to be impossible; but an approximation may be arrived at of the quantity which enters the mouth and is retained in the system. In some experiments which have already been referred to, I showed that the loss of the sprayed liquid used amounted to not less than two-thirds, and was probably greater.

Beigel, with Siegle's apparatus, found that with 1 oz. of steam $\frac{1}{2}$ oz. of the medicated liquid passed over as spray, together $1\frac{1}{2}$ oz. Of this he affirmed that about one-half recoiled from the intercepting screen into the air, while of the other half, a moiety was either swallowed or spit away, and the remaining 3 drachms of mixed steam and medicated fluid passed the glottis and penetrated into the respiratory track. The above conclusions were, however, not founded upon precise experiments, but were chiefly conjectural.

With respect to the composition of the spray, Beigel states that the conclusion may be fairly drawn that it contains a far larger quantity of medicated fluid than vapour, because the lighter steam flies off and becomes readily mingled with the atmosphere, so that the strength of the medicated fluid is altered but very little through its mixture with the steam. This conclusion he confirmed by an analysis of the condensed liquid itself.

This statement seemed to be one of so much importance that I determined to subject it to the test of certain experiments.

Three separate experiments were made, Siegle's steam spray apparatus being employed for the purpose. In each case 1500 grains of distilled water were placed in the boiler, and in the bottle for the medicaments, 500 grains of water holding in solution 10 grains of chloride of sodium or salt. The spray was intercepted by a small glass disk having a diameter of 1¾ inch at 4, 8, and 12 inches distant from the points of the tubes from which the spray issued. The results are set forth in the fo'lowing table, and, as will be seen, they are of a very instructive character :—

IN GRAINS.

	Inches	Temperature of Spray	Left in Boiler	Drippings	Containing Chloride of Sodium	Left in Bottle
		°F.				
1	4	88	1027	310	4·23	46
2	8	78	1201	154	2·19	37
3	12	72	1232	69	0·92	37

The temperature of the room at the time was 64° F. The proportion of salt which should have been contained in the 310 grains of drippings in the first experiment is about 6 grains, whereas little more than 4 were recovered, showing that the medicament does undergo dilution by the steam, although not perhaps to the extent which might be anticipated. The diminution of the quantity of the drippings in the second and third experiments, and the great reduction in the amount of chloride of sodium, are remarkable, and show that if you desire to obtain the full benefit of the spray it must not be held at too great a distance from the mouth. At 8 inches, only one-fifth part of the salt originally taken was found in the drippings, and at 12 inches, one-tenth part only. Had a volatile substance been selected for the experiments the loss would doubtless have been much greater.

Some attempts were also made by Mr. James Collins, of Philadelphia, to determine the amount of certain medicaments received into the system when a steam spray apparatus was used. There were recovered from the drippings of 2 grammes of each of the medicaments employed, the following amounts: of ferro-ferricyanide of potassium 1·445, of sulphate of copper 1·181, of bicarbonate of potash 1·545, of nitrate of potash 1·285, and of chloride of ammonium 1·845 gramme. The amounts lost, therefore, would stand in the following order: 0·555, 0·819, 0·455, 0·715, 0·155 ; but it does not follow that these quantities were actually received into the system, as there was no doubt a further loss in the steam which passed into the air and from other causes.

THE MANNER OF INHALING.

As already pointed out, the fact of the entrance of certain medicaments into the respiratory track may in some cases be established by the sensations and effects experienced by the patient, and a general idea may be formed therefrom of the quantities really inhaled.

When an atomized spray, or, as Dr. Cohen calls it, 'a nebulized fluid,' really enters the respiratory track the patient is conscious of certain feelings and sensations; if the spray be cold, he will be sensible of some degree of coldness in the trachea and upper part of the chest behind the sternum, and in proportion as the spray penetrates deeply, so will the area of this sensation become extended; if, on the other hand, the spray be warm, there will be a corresponding sense of warmth or heat. When strong solutions are inhaled they may give rise to a sense of constriction, or even soreness; and if too long continued they can occasion wheezing 'râles,' and even hæmorrhage. It is necessary, therefore, to make careful enquiries of the patient as to the nature of the sensations experienced. Sometimes the inhalation is provocative of cough; this may occur at first, but it soon ceases and may not subsequently be reproduced. As a rule there is some amount of coughing at the end of each inhalation, excited by the presence of increased fluid or secretion.

The manner of inhaling, also, as before shown, is a matter of some importance in determining the quantities which enter the lungs. The patient should lean a little forward, with the elbow resting on the table. The object of this position is to incline the opening of the larynx towards the spray; if the head be held quite erect, the spray is apt to strike against the throat, and is deflected from its course only by a greater effort. He should inspire continuously and deeply; the mouth should be well open and the tongue depressed. If the liquid accumulate in the mouth and be likely to produce unpleasant constitutional effects, it may be expectorated; but if, on the other hand, the effects are desirable and promote the object in view, then it may be swallowed. The spray producer should be raised to the level of the patient's mouth, and this may be done either by means of a stand, or the apparatus may be furnished with a handle and so held at the required height. When the object is that the spray should reach the pharynx only, deep inspiration should be avoided, lest the medicaments be carried into the air passages.

The distance at which the mouth is held from the apparatus makes a considerable difference in the result of the inhalation. If too close and the spray contain steam it may be too hot; then, again, the force may be so great that a strong effort is required to deflect it into the air passages. The mouth should be at a distance of some 4 or 6 inches from the apparatus, but some recommend as much as from 1 to 2 feet, and consider that, as at the longer distances much of the projecting force of the spray has been expended, the vapour is more readily deflected and inhaled. This is a mistake, however, as proved by the experiments on page 96, which show that, at a distance of one foot, the amount of the medicament contained in the spray and which enters the mouth, is reduced to about one-tenth part of that originally taken. It should be remembered, that the temperature of the spray rapidly lessens with the distance from the apparatus.

Patients who are too weak to sit up may lie on their sides and inhale, while with children it is often a good plan to take them on the lap and encourage them by showing them the action of the spray, in which they may soon be made to become interested.

The medical attendant should instruct the patient at first as to the right method of employing the inhaler which he uses, and should be careful to make strict enquiries as to the local effects produced.

In view of the preceding facts and observations, it is obvious, that the quantities of medicaments which have hitherto been relied on in lung inhalation, and which were based upon the assumption that a large portion of them was inhaled, were much too small, and in proof of this position I will now adduce further evidence.

In this country, inhalation in diseases of the lungs has been chiefly carried on by means of oro-nasal inhalers. I will now give some evidence as to the quantities of medicaments with which these are usually charged, as of course the amounts volatilized depend greatly on the quantities placed in the inhalers.

Dr. J. Carrick Murray, physician to the Northern Counties Hospital for Diseases of the Chest, stated in the 'British Medical Journal' of July 23, 1881, that he employed a mixture, which has probably been more used than any other, and to which surprising results have been attributed. Its composition was as follows: tinct. iodi etherea ʒj., acidi carbolici ʒij., creasoti aut thymolis ʒj., sp. vini rect. ad ʒj. Of this he found that 20 minims

QUANTITIES OF MEDICAMENTS EMPLOYED. 99

dropped at once on the cotton-wool of an oral inhaler were enough for two days, and that more were too irritating to the glottis; also that half an hour's inhalation twice a day was sufficient.

Now the number of 20-minim doses in 1 ounce of the above mixture amounts to just 24. Each 20 minims, therefore, would contain 5 grains of carbolic acid and $2\frac{1}{2}$ grains of creasote. Further, since only about one-fifth part of the carbolic acid and creasote taken disappears during an inhalation lasting from 1 to 2 hours, the loss of those antiseptics in the time specified would be equal to only 1 and $\frac{1}{2}$ grain respectively; but since the inhalations in the trials made by Dr. Murray were only of half an hour's duration, the actual loss would be but little more than $\frac{1}{4}$ and $\frac{1}{3}$ grain respectively.

Mr. J. Brindley James, writing in the 'British Medical Journal,' December 8, 1883, states that he employed a mixture resembling the above, only that thymol was substituted for the creasote. Of this, 20 drops were added night and morning to the sponge of Dr. Burney Yeo's inhaler, which allows much of the very small amount of the carbolic acid which really becomes evaporated, to be dissipated in the surrounding air.

The quantity of the medicaments employed when sprays are used is, as a rule, even less than in the case of the oral or oronasal inhalers, but then a larger proportion is conveyed into the mouth and inhaled.

For the Vapor Coniæ of the British Pharmacopœia 20 minims of a mixture containing 1 part of extract of hemlock in 12 parts of a weak solution of potash, equal to 1 grain of extract in 12 minims, are to be placed on a sponge in a suitable apparatus, so that the vapour of hot water passing over it may be inhaled.

For the Vapor Creasoti 12 minims of creasote are to be mixed with 8 ounces of boiling water in an apparatus so arranged that air may be made to pass through the solution for inhalation.

In the Throat Hospital Pharmacopœia 1 grain of thymol is directed to be inhaled in the vapour of 20 ounces of hot water by means of the eclectic inhaler.

I have elsewhere demonstrated how utterly useless are the three inhalations above referred to. In the case of the vapor coniæ, scarcely a trace of the alkaloid is to be discovered by the most delicate tests in the vapours which come over even when these are collected and concentrated. In that of the vapor creasoti about $\frac{1}{2}$

grain at the outside is vaporized. In the vapor thymolis the effect produced by the vaporization of a single grain in 8750 grains of water must be nil.

The quantities or doses contained in some of the medicated cigarettes are equally inefficient. Dr. Macnaughton Jones gives the formula for a cigarette of powdered eucalyptus leaves saturated with oil of eucalyptus, with $\frac{1}{2}$ grain of carbolic acid in each cigarette. If this latter constituent is to be of any service, the quantity is far too small.

These few examples, which might be indefinitely multiplied, are enough to show that the quantities of the medicaments used in inhalation are generally so small that the remedies cannot possibly be sufficiently effective.

We have now to deal with the question what should be the proper doses of the medicaments employed? To this question, from the nature of the case, only a very general answer can be given.

First, it must be determined whether the remote and constitutional, or the direct and local effects are required, and if the latter, on what part they are to be mainly exerted. Then again if the local effect is needed, not only must the quantity of the remedy taken be considered, but *the degree of concentration.* 5 grains of alum or tannin dissolved in $\frac{1}{2}$ ounce of water will be twice as strong and perhaps as effective, as the same quantity in 1 ounce of water. This consideration has doubtless had much to do with the general employment on the Continent of percentage solutions, but these still leave the practitioner in doubt, unless the quantities of the different solutions which should be used in each case are specially indicated, and this can only be done exceptionally. On the whole, then, I have come to the conclusion that the best plan is to define the doses approximately, which should be used for each inhalation, leaving the quantity of the menstruum to be determined by the knowledge and judgment of the prescriber.

Again, in determining the dose, regard must be had to the apparatus employed, and we must recall to mind the extent of the loss experienced. With oro-nasal inhalers, when carbolic acid, creasote, or thymol is used, I have shown that about four-fifths of the quantities taken do not become evaporated at all during the period of an ordinary inhalation, but are still to be found in the sponge or cotton-wool of the inhaler. With spray producers, as formerly used, the loss is also very considerable, reaching at least to two-thirds of the original amount, while when medicaments are inhaled in the fumes of hot water the loss is greatest of all.

Then, again, *the degree of volatility* of the medicaments must be noted. In the case of substances but little volatile at ordinary temperatures, as carbolic acid and creasote, it is all but useless to employ them in any doses, particularly in cold weather, unless the volatilization be aided by heat, the extent of the increase of temperature being regulated by the relative volatility of the several medicaments.

The above remark applies also to the oils of eucalyptol, fir-wood, and juniper; but with the more volatile substances, as chloroform and ether, the case is different, and these may probably be used with advantage in oro-nasal inhalers without any increase of temperature. (See page 23.)

If, notwithstanding what has been stated, ordinary oro-nasal inhalers be used without artificial heat for the inhalation of carbolic acid, creasote, &c., then at least five times the quantity of these must be taken above that which it is desired should be inhaled—in a general way from 25 to 40 grains.

When a spray apparatus is employed, then the quantity should at least be three times as great. And when the medicaments are to be inhaled in the vapour of hot water, from ten to twenty times as much must be added.

Again, when oro-nasal inhalers are employed without artificial heat, not only is the amount lost small, but the evaporation is spread over a long time, and the local effects produced are but weak, indeed insignificant.

When air or steam spray machines are resorted to, the inhalation occupies only a few minutes, but the effect produced is much more rapid and considerable.

The same remark is applicable to the inhalation of the medicated vapour of hot water.

It is, however, still more applicable to the inhalation of dry or highly concentrated fumes and vapours, and it is probably to the use of these that we must look for the greatest amount of benefit derivable from inhalation in certain diseases of the organs of respiration.

The larynx, bronchial tubes, and lungs will bear the inhalation of many vapours in a considerable degree of concentration, provided that their application be not too long continued, and also that they be not mixed with a variety of other more or less irritating but non-essential substances, as particularly preparations containing any considerable amount of alcohol.

Alcohol, ether, and chloroform are frequently added under the notion that they increase the volatility of carbolic acid, creasote, or thymol; but the observations which I have elsewhere recorded show that this idea, except in the case of ether, is illusory, and even then the effect is but small. The practice, then, of adding such substances to aid volatilization is one which should be for the most part discarded.

For precise and definite information as to the doses of the different medicaments to be employed, the reader is referred to the next chapter.

That remedies may be introduced, if the right means are adopted, into the lungs in quantity sufficient to produce local and constitutional effects is undoubted; but the question may now be briefly considered whether it is possible to inhale them in such amounts and in such a degree of concentration as to kill the tubercle bacillus, which so many, following the teaching of Koch, regard as the cause of consumption and of other tubercular diseases. Koch considers this to be impossible with the cholera bacillus, owing to the large quantities of the most powerful antiseptics which are required to effect its destruction.

But the circumstances in the case of the cholera and tubercle bacilli differ greatly: the former are far removed from any outlet of the body; they are inaccessible to the air, and cannot be reached, or only with great difficulty, by the direct application of any medicaments. But the reverse is the case with the tubercle bacilli: here the parts are in close proximity to a main inlet of the body; the air enters into them at each inspiration, and topical applications of considerable power can be freely applied.

Nevertheless, even with these advantages it is at present doubtful whether any remedies can be so employed as to kill the bacillus outright. But it is quite conceivable, and from analogy even probable, that the bacilli may be destroyed in other less direct ways; the medicaments may be strong enough to gradually impair their vitality and power of multiplication, and to so alter the surrounding conditions that their prolonged existence becomes impossible. That tubercle bacilli, like other organisms, have their vulnerable points, particularly at certain stages of their development, is most probable.

Dr. Solomon C. Smith, in a very suggestive paper on 'Antiseptic Inhalations,' read in the Section of Medicine at the meeting of the British Medical Association in Liverpool in August 1883, referred to some experiments of Professor Tyndall relative to the preparation of sterile solutions. It appears that the Professor was much troubled by the prolonged boiling which was necessary to destroy the germs and organisms contained in them; 'but he found that if, instead of giving his solutions one long boil, he heated them several times for quite short periods, leaving them at the ordinary temperatures at the intervals, he could readily devitalize them. This method he described as sterilization by discontinuous heating. He discovered, in fact, that however hard and resisting the germ may be, there is in the life history of every bacterium a period when it is very soft, tender, and easily destroyed.' And this it can hardly be doubted must be true in the case of the tubercle bacillus.

Again, Koch's experiments and observations on the cholera bacillus were made with the solutions of certain disinfectants, but the effect of vapours does not appear to have been ascertained. I believe that the mucous membrane of the air passages will bear the application for short periods of many vapours in considerable concentration, and it is on the employment of these, perhaps, that the most reliance is to be placed when the object is the destruction of the bacillus.

Even if the efforts to kill the bacillus of phthisis should fail, yet that many beneficial effects may be brought about by judicious and persistent inhalation, is undoubted. There is, therefore, much reason for perseverance in the practice of inhalation even in this disease.

Dr. Fischer and Dr. Schüll, of the Imperial Health Office in Berlin, have made some important observations relative to the vitality of the tubercle bacillus, and they find that if a 5 per cent. solution of carbolic acid in water be mixed with an equal bulk of sputum the bacillus was rendered powerless. This conclusion was arrived at by the results (on animals) of injections of the uncarbolized and carbolized sputa; the effect in the former case was invariably to give rise to the formation of tubercles in the lungs, while in the latter no such result followed. It will be perceived that these experiments have only an indirect bearing on the subject of the destruction of the bacillus when within the lungs.

Dr. J. Sormani, Professor of Hygiene at the University of

Pavia, has also made known the results of some observations on the vitality of the tubercle bacillus. He states in his communication to the Hygienic Congress at the Hague at the last meeting, that its vitality is not impaired by drying, and that it is only killed by 5 minutes' ebullition in water; that the principal disinfectants which in small quantities destroy it, are bichloride of mercury, camphor, creasote, iodide of propyle, carbolic acid, and turpentine. The substances were kept in contact with the sputa for about two hours at a temperature of 95° to 104° F., and were afterwards injected under the skin or into the peritoneum of a number of guinea-pigs.

It appears from a communication by M. Schnetzler, addressed to the Académie des Sciences, that water containing one-hundredth part of formic acid, added to an equal quantity of liquid teeming with *Bacterium subtile*, which is most difficult to kill, is sufficient to sterilize the solution. Its effects would probably be not less destructive to the bacillus of tubercle.

The next questions for consideration are, How frequently should the inhalations be repeated, and for how long should they be continued?

The answers to these questions must be general.

As to *frequency*, this of course will depend very much on the objects aimed at and the nature of the malady.

If the affections be chronic, twice a day will usually be sufficient; if acute, then they must be more frequent, every 1, 2, or 4 hours, according to the urgency of the case.

The *duration* must vary not merely according to the malady and the condition of the patient, but also in accordance with the properties of the substance inhaled and the nature of the apparatus employed.

If an ordinary oro-nasal inhaler be used for such substances as carbolic acid, creasote, and some others, the inhalations must be continued for one or two hours, and then unless artificial heat be employed, but little benefit is to be anticipated; but if with these inhalers more volatile substances be used, as chloroform and ether, much shorter times will be sufficient, varying, however, with the relative volatility of the substances inhaled.

When either an air or steam spray apparatus is used, an inhalation of from 10 to 15 minutes will suffice. The inhalation of medicaments in the vapour of hot water must also be of brief duration, and a similar remark is applicable to the fumes arising from the combustion of medicated pastilles, cigars, &c., as also to those fumes which are inhaled in a concentrated form without the use of hot water.

The practice of medical men as to duration differs greatly, as will be apparent from the following brief references.

Dr. J. Carrick Murray, using only 20 drops of the mixture, the formula for which has already been given, states that he found half an hour twice a day sufficient in a case of phthisis.

Mr. J. Brindley James states that in somewhat similar cases, and using 20 drops of a mixture closely resembling that above referred to, some of the patients after a time slept with the inhalers on all night.

Dr. Burney Yeo sometimes advises his patients to wear their inhalers day and night.

Dr. E. Blake recommends his patients to wear them all day if possible.

The ordinary practice, however, is for an oro-nasal inhaler to be worn for one or two hours night and morning.

On the other hand, there are many medical men who forbid their patients to wear their respirators either all day or all night, and who consider that any possible advantage is more than counterbalanced by the impediment to respiration and the due aëration of the blood which they occasion; to say nothing of the interference with sleep which such a covering over the mouth and nose must inevitably entail.

For myself I entirely agree with those who forbid the continuous wearing of oro-nasal inhalers.

It is not to be inferred from these remarks that I am opposed to prolonged and continuous inhalations in such a malady as phthisis: on the contrary, I consider they are most desirable; but this purpose should be effected by charging the air of the room occupied by the patient with disinfectants, such as carbolic acid, creasote, or eucalyptol, an object which can be effectually carried out, especially at night, by the chamber inhaler No. 1 or No. 2. I do not, however, rely on this method alone; it must be combined with inhalation twice daily, by means either of the globe inhaler and water bath, or better still by the use of the apparatus for concentrated vapours.

But the employment of inhalation sometimes presents practical difficulties arising from the age or condition of the patient. In most chronic cases the patient can carry out the instructions given him unaided, and this is always very desirable where possible, as it occupies the patient, interests him in the treatment, and saves expense.

In acute and very advanced cases, the necessary proceedings must be entrusted to an attendant or nurse, and in critical cases must be carried out by the medical man.

In bringing this chapter to a conclusion, I would remind the reader that the medicaments to be effective in diseases of the organs of respiration must be made to enter the lungs in quantities sufficient to produce decided and appreciable effects; a result only to be arrived at by the observance of certain precautions as to the doses and the manner of inhaling.

This remark applies especially to inhalations of fumes, vapours, sprays, and powders. In a true inhalation chamber, such as that I have already described, in which the air is charged with the medicament in a more or less completely gaseous form, the case is different; under these circumstances the patient cannot avoid drawing into his lungs whatever medicaments are present in the air, and this without effort or fatigue, a point of the greatest importance with most invalids.

CHAPTER VI.

THE MEDICAMENTS EMPLOYED IN INHALATION IN DISEASES OF THE ORGANS OF RESPIRATION.

GASES.

Atmospheric Air.

THE simplest form of inhalation is that practised in ordinary respiration, namely, the inhalation of atmospheric air. The air is not a chemical compound, but a uniform mixture in certain proportions of two substances in the gaseous state, nitrogen and oxygen, the latter being the most important and essential constituent, and the former playing the part chiefly of a dilutor of the oxygen, which would otherwise be too powerful and stimulating. The amount of oxygen contained in the air varies, but only to a very limited extent, with the locality; thus, according to Dr. Angus Smith, while ordinary air contains 20·96 per cent. of oxygen by measure, mountain air contains 20·98, and the air of towns 20·90, or even as low as 20·87 per cent. According to other authorities the figures are somewhat different, and the average amount of oxygen in the air is stated to be 20·81 per cent., but the figures quoted are sufficiently accurate for the purpose in view. It is certain that these variations in the oxygen are far too small to account for the different effects of the air of plains, mountains, and cities.

But atmospheric air contains other constituents, some of which are essential, while others must be regarded either as accidental, or even as impurities. Among the other essential constituents are the active form of oxygen, ozone, and the vapour of water.

The proportion of watery vapour in the air varies greatly, from almost nothing to saturation, that is to say, to the full amount it is capable of holding in suspension.

A very important gas present in all air, but which yet must be regarded as an impurity, is carbonic acid. The amount of this in ordinary air varies from 0·2 to 0·5 per 1000, = 2 to 5 volumes in 10,000 volumes, or it may reach to 0·6 or 0·7 per 1000 volumes.

Another constant impurity is organic matter; the presence of this is always revealed by the odour detectable in a room in which the carbonic acid reaches 0·7, and is very marked indeed when it amounts to 1 per 1000 volumes.

Besides the substances just mentioned, a host of others are sometimes found in variable amount in the air, gases, vapours, and salts, especially chlorides and sulphates, and even bodies in the solid state. To all these it is not necessary to allude in detail, but a few of the more important, as bearing on the subject of inhalation, may here be referred to, as the sporules of fungi, different kinds of bacilli, including particularly those of phthisis and some other diseases, possibly also of cholera and the pollen of innumerable plants, notably of those which give rise to hay fever.

The chief source of the bacilli of phthisis is the expectoration, which, when dried, is very light, pulverulent, and easily diffused. It is in this way, that the presence of the bacilli in the air of chambers and houses occupied by the phthisical, especially when the needful precautions for the collection, disinfection, and destruction of the sputa are not strictly carried out, is to be explained.

Now, the air we breathe at the level of the sea exerts a pressure on the body of 14·696 lbs. to the square inch. This pressure is general; that is to say, it acts equally not only over the whole of the exterior of the body, but also on the internal surfaces which are in direct communication with the outer air, chiefly the lungs. Now this pressure is increased if one descends below the level of the sea, as in mines, and the air then becomes more or less condensed, according to the depth. On the other hand the pressure is lessened if heights, such as mountains, are ascended, the air becoming proportionately thinned or rarefied.

But temperature exerts a marked effect on the density of the air, as well as relative altitude. Reduction of temperature renders the air more condensed, while increase of temperature causes it t become more rarefied.

Now these conditions of condensation and rarefaction of the air produce marked effects on the functions and structure of the lungs, and advantage has been taken of them in the treatment of pulmonary affections, as will be hereafter explained.

The systematic and forcible inhalation of ordinary atmospheric air often proves of considerable therapeutical value, not by reason of any quality in the air itself, but because of the increased efforts made to fill the chest, whereby the muscles of inspiration are strengthened, the capacity of the pulmonary organs augmented, and the blood more thoroughly ventilated. This mode of treatment was in much favour with the late Dr. Ramadge, in the early stages of consumption, he causing his patients to inhale through a long tube three times a day.

Cold Air.

The effect of cold is to occasion some degree of contraction or condensation of the air, the extent of the increase of density varying with the degree of cold. Since cold air occupies less space than warm air, it is richer in nitrogen and oxygen, but has a less capacity for moisture. Its expansive force is in proportion to its condensation, and is of course greater than that of air of the ordinary density.

Properties.—Cold air is tonic, sedative, refrigerant, and antipyretic, and is therefore an important agent in the treatment of many diseases of the organs of respiration, and particularly of most of those attended with elevation of temperature.

Applied to the air passages and lungs, cold air exerts several important and marked effects : the vessels become contracted, caloric is abstracted, abnormal heat reduced, and congestion and inflammation checked, and in this way, as well as by its anæsthetic and benumbing effects on the extremities of the nerves, it acts soothingly and relieves pain. If the application be long continued, it will produce in feverish conditions a general as well as a local reduction of temperature.

Again, with the contraction of the vessels and tissues, the supply of blood to the part is lessened, and, as a consequence, transudation and secretion are diminished.

Following the application of cold, reaction sets in ; in strong and healthy persons this is proportionate to the degree and duration of the cold.

However cold the air, it quickly takes up caloric, and its temperature is raised when brought into contact with a warm body; thus by the time the inspired air reaches the deeper bronchi and air cells its temperature has become much increased.

When cold air, which is rich in oxygen, is habitually inhaled, oxydation is increased.

Lastly, since cold air has a diminished capacity for moisture, its tendency is to check evaporation from the lungs.

Administration.—The air of a room may be cooled by the admission of the outer air when this is cold, and in many cases such air may be allowed to enter freely at night and may be breathed with advantage, provided the inlet be properly screened. This cooled air is much better borne than is usually supposed, and ill effects but seldom follow. The air must enter the room indirectly, so that no current can reach the bed, and the patient must be covered sufficiently to keep him quite warm. The patient may have his doubts and fears at first, but these will quickly subside. Of course on foggy and damp nights, the window must be closed, but on such occasions a partially opened and screened door may be substituted. The warmer the air is kept of a room occupied by an invalid, the more sensitive his skin becomes, till at last he morbidly shrinks from the least degree of cold.

It is desirable in many cases, that the patient should breathe a specially and greatly refrigerated air, the apparatus for the inhalation of which has already been described in Chapter III.

Indications.—Cold air is indicated in catarrhal and inflammatory affections of the throat, larynx, and bronchi. Being condensed, it is grateful to many asthmatics and affords relief to some who suffer from emphysema, acting in this case by promoting the contractility of the fibrous tissue of the air cells.

The inhalation of cold air is also indicated in most hæmorrhages of the air passages and lungs, as in epistaxis, hæmoptysis, and pneumonia; in phthisis and other maladies of the organs of respiration accompanied by fever. It does not appear to have proved so efficacious in the treatment of the smaller hæmorrhages from the lungs, as might have been anticipated, and this partly because the air becomes to some extent warmed before it reaches the source of the bleeding. Dry, cold air is particularly well borne by many consumptives, lessening the fever and the consequent exhaustion, and thus helping to produce a general amelioration; and its beneficial effects are well exemplified by the favourable results which often ensue from a prolonged stay in certain mountain and Alpine regions.

The inhalation of cold or artificially cooled air in pulmonary diseases, has been advised by several medical men, and has been more or less put into practice. Dr. C. Drake, of New York, caused the patient to inhale cooled air at temperatures ranging from 32° F. to three or four degrees below zero. He employed these inhalations in chronic catarrhs, and found them to relieve the cough and to render the pulse slower and fuller. Dr. Ramadge employed artificially cooled air for the purpose of exciting a catarrhal inflammation of the mucous membrane of the air passages in cases of consumption, with, he states, curative results in some cases. Ice-cold air has been used by Professor Langenbeck in pulmonary consumption, in hæmoptysis, and in fevers, it effecting in the latter case a very considerable reduction in temperature.

Warm Air.

Warm air is in many respects the opposite of cold air; it is more rarefied, hence occupies more space, and its expansive force is less; it is poorer in nitrogen and oxygen, but has a greater capacity for the vapour of water.

Properties.—It is stimulant to the nerves, relaxant to the tissues, and sudorific.

Administration.—In winter and in certain conditions and diseases, the body becomes too much cooled down, so that it is necessary to restore its warmth. This may be done by artificially warming the air of the rooms occupied by the patient, by additional clothing, and by the direct inhalation of warmed air. The methods whereby this latter object is accomplished have already been pointed out.

Indications.—The direct inhalation of specially warmed air is particularly indicated in the case of those whose organs of respiration are either very susceptible, weak, or even diseased. Many such persons are apt to suffer when called upon to pass suddenly from a warm to a colder air. The inhalation of warm air is sometimes simply precautionary against taking cold, or it may be to avert the setting up of cough or the onset of difficulty of breathing, or even an attack of asthma.

RESPIRATION OF COMPRESSED AND RAREFIED AIR BY MEANS OF TRANSPORTABLE APPLIANCES.

Inhalation of Compressed Air.

It has been already stated, that at the level of the sea, the air presses on the surface of the body with a weight but little short of 15 lbs. to the square inch; but this pressure may be, and is, either increased or lessened in accordance with elevation and temperature. If the air pressure be increased, as in mines, then the air becomes thickened or condensed, and if it be diminished, as in high altitudes, thinned or rarefied. Similar effects are produced n a less degree by variations of temperature either above or below 32° F.; if the temperature be lower, condensation of the air ensues, and if higher, rarefaction.

As there are no convenient depths to which we can descend, where the air would be sufficiently condensed to produce therapeutical effects, so we have been led to devise ingenious forms of apparatus for producing compression of the air and for facilitating its inhalation. The more useful and practical of these have already been described.

This compressed air, then, when received into the air cells of the lungs, the pressure being in part removed, expands and in its turn presses on the walls of the cells, and so helps to distend or enlarge them, an effect heightened by the increased temperature of the air which ensues on its being brought into contact with the blood.

The compressed air inspired may be expired either into the ordinary outer air, or into an apparatus containing either compressed air or rarefied air. The effects of such expiration will of course differ in each case. Ordinarily, only compressed air is inspired, with expiration into the outer or into rarefied air.

Now, when either compressed or rarefied air is received into the lungs only, and is not applied to the general surface of the body, it is technically described as 'one-sided pressure,' in contradistinction to the complete pressure which takes place in the pneumatic chamber, on mountain heights and in ordinary air.

The effects of one-sided inhalation are chiefly mechanical on the lungs themselves and on the circulation, the chemical changes being mainly consequent on the mechanical effects.

The extent to which the air for practical purposes is compressed or rarefied is by no means considerable. While the ordinary pressure of the atmosphere is represented by the figure 1, the one-sided or transportable appliances for compressed air are used for the most part with a pressure of $\frac{1}{100}$ to $\frac{1}{40}$ and even to $\frac{1}{30}$ of an atmosphere.

The inspiration of compressed air, then, causes expansion of the air cells of the lungs, and expiration into rarefied air allows of a more complete emptying of the cells than could take place by expiration into ordinary or, still more, into compressed air, this greater facility being due to the diminished resistance offered by the rarefied air. Thus it follows that the lungs being more completely emptied the inspirations made are deeper and fuller, so that on the whole a larger quantity of air is made to enter the lungs.

In some cases the compressed air is medicated with other gases or respirable substances.

The effect of the continued inspiration of moderately compressed air, in favourable cases, is expansion of the lungs and thorax; in other words, there is increased capacity of both, which after some time may become permanent.

By moderate compression, the elastic tissue of the air cells of the lungs is increased, whereby the force of respiration, both inspiration and expiration, is improved; and Oertel particularly points out, that it is an unfavourable sign if the expiration power be not increased, and he advises, in such a case, that the inhalation of the compressed air should be discontinued.

When the air is too condensed, the air cells are rendered liable to be overstretched, and their contractility and expulsive power consequently impaired; the lungs become incapable of emptying themselves completely, difficulty of breathing arises, and even emphysema may be brought about.

Speck has shown, that with moderate pressure of the air, the respired air and oxygen are increased, as well as the expired carbonic acid.

Application.—Some of the more useful and practical transportable contrivances for the inhalation of compressed air will be found described in Chapter III.

Indications.—The cases best suited for the employment of this method of treatment are chiefly those in which the air cells of the lungs have become more or less closed by external pressure, as by fluid in cases of pleurisy, by the contraction of the elastic tissues surrounding the air cells in interstitial pneumonia and fibroid phthisis, or by deposition in, or consolidation of the air cells.

It is in the young and middle-aged, that the inhalation of compressed air promises the best results, as in them the lungs and thorax still retain their resiliency. In the aged, the cartilages of the ribs have become ossified, when of course such inhalation would be useless, as under these circumstances any permanent expansion of the walls of the thorax could not take place; in the emphysematous it would prove in the highest degree mischievous, unless the compression were slight and followed by expiration into rarefied air.

In carrying out the treatment with compressed air, careful measurements of the chest should be made from time to time; in favourable cases the capacity is increased to several hundred, and Oertel states to even 1000 c.c., = 61 cubic inches, the normal capacity being about 3500 c.c., = 213 inches.

Expiration into Compressed Air.

The inhalation of compressed air gives rise to a feeling of distension and fulness of the chest; and since, in expiration, the lungs are never completely emptied, the feeling of distension continues for some time afterwards, owing to the pressure of the residue of compressed air.

This result is still more likely to occur when, in addition to the inspiration of compressed air, expiration is made to take place also into compressed air, since this offers greater resistance, whereby the tension in the air cells is increased and expiration rendered more difficult. The method of expiration into condensed air would appear to be of little, if any, therapeutical value, and it might prove in many cases even dangerous.

Inspiration of Rarefied Air.

Rarefied or thin air, as has been already more than once stated, occupies a greater space than compressed or condensed air, and when it enters the lungs, in consequence of the increased temperature it there meets with, it undergoes a still further rarefaction or expansion. At the same time inspiration is rendered more difficult, and greater strength is required to overcome the difficulty. It is only possible to draw in or inspire rarefied air when the air in the lungs is more rarefied than that in the apparatus, because the heavier air always passes to the lighter. If the air in the apparatus and in the lungs is of the same density, no interchange can take place, and if that in the apparatus is lighter than in the lungs it will press out from them and pass into the apparatus, the movements of the thorax become arrested, and breathlessness or apnœa takes place.

The extent of rarefaction of the air which can be borne is limited, usually, to − 0·30 of an atmosphere. Nevertheless the inspiration of thin or rarefied air, in the same way as expiration into compressed air, by the increased energy of respiration evoked, causes an increase in the amount of air respired and of gas exchange. Speck found that the inspiration of rarefied air brought about an increase of respired air, of absorbed oxygen, and of expired carbonic acid. Therapeutically, therefore, by the inhalation of rarefied air the muscles of respiration become strengthened, because they have greater efforts to make to overcome the increased difficulty of inspiration.

According to Waldenburg, the lung capacity is increased by the methodical inhalation of rarefied air in the pneumatic apparatus, as much as by the inspiration of the rarefied air of mountain heights. Oertel considers it is still uncertain whether this result is obtained by the artificial inhalation of rarefied air; and if so, the increase must always be less than is the case by the inhalation of compressed air. The treatment by the inhalation of artificially rarefied air has to this date been but little practised.

Expiration into Rarefied Air.

From what has already been stated, it will be evident, that expiration into rarefied air is accomplished with great ease, although of course the rarefaction of the air in the gasometer is lessened by the air added at each expiration; but this result is, in a measure, obviated by the further rarefaction of the air brought about by a reduction of the gasometer pressure.

Not only is the act of expiration performed with greater ease, but it is also more complete, so that less air remains in the lungs at the termination of the act; this allows of course of the more complete filling of the lungs at the next and all succeeding inspirations.

During expiration into rarefied air, the person feels as though the thorax were contracted and the diaphragm thrown up at the same time. With people of limited lung capacity, this feeling is experienced on a comparatively slight rarefaction of the air.

It has been stated that the lungs become more completely emptied on expiration into rarefied air, and consequently the quantity of expired air is greater; but the actual quantity varies considerably, being dependent of course on the amount of air originally received into the lungs, and this again in part is determined by the elasticity of the lungs themselves. The increase may amount to several hundred and even to 1000 c.c., = 61 cubic inches. This effect is well seen in emphysematous affections when these have become far advanced and the resiliency of the lungs is almost lost; in such cases the increase may amount, according to Waldenburg, to 2000 c.c., when the air is rarefied $\frac{1}{60}$ to $\frac{1}{40}$ of an atmosphere, although the lung capacity was only 2000 to 3000 c.c.

Indications.—Expiration into rarefied air is therefore very important, when the object is to strengthen the elasticity of the lungs, an object which is not satisfactorily achieved even by residence in high mountain resorts. By expiration into rarefied air, the air remaining in the lungs is not only lessened, but is also rarefied: thus inspiration of the ordinary air is so complete that it powerfully draws the air even into the alveoles of the lungs. Again, since the residual air which is so easily expired contains, according to Oertel, 4 per cent. of carbonic acid, and the ordinary air respired hardly 0·1 per cent., so by continued expiration into rarefied air the gas exchange is increased and the lungs are better ventilated. Further, by this pumping out of the residual air, and by the streaming in of the oxygen-charged air, the blood is decarbonised and dyspnœa is relieved.

' Actual estimations of the effects of expiration into rarefied air on the gas exchange in the lungs, are given by Speck, and show even by moderate rarefaction, by means of 7·8 c.c. water, a noticeable increase of the respiration air, from 1 to 1·62; of the absorbed oxygen 1 to 1·14, and 1 to 1·30 of the carbonic acid ' (Oertel).

The improvement in the strength of the elastic tissue of the air cells of the lungs and of the muscles of respiration is a permanent one, and therefore it is in emphysematous cases, not too far advanced, that the method of treatment by expiration into rarefied air is proved to be specially serviceable; the power of both expiration and inspiration is increased, and the lung capacity much augmented. In some cases of emphysema, in a few weeks it has been increased 500 to 1000 or even 1200 c.c.

'Let us once again,' Oertel writes, 'sum up the different effects of expiration into rarefied air: on the one hand, the removal of larger quantities of air from the lungs than in ordinary expiration, the increase of gas exchange by the pumping out of a considerable part of the residual air charged with carbonic acid, and the absorption of greater volumes of atmospheric air charged with oxygen; on the other hand, the lessening of the circumference of the thorax, the retraction of the lungs, the increase in the inspiration and expiration power, and finally the increase of the vital lung capacity. Therefore, from simple theoretical considerations, this expiration into rarefied air is capable of removing dyspnœa caused by insufficiency of expiration, and of preventing its return. In this opinion the experience of practical persons perfectly accords.'

On the Respiration of Compressed and Rarefied Air in the Pneumatic Chamber.

It has already been stated that each sitting in a pneumatic chamber should be of about two hours' duration; that during the first half-hour the pressure should be gradually increased; during the next hour it should be maintained at one even rate, and during the last half-hour the pressure should be still more gradually and carefully removed.

The increase of pressure ordinarily employed, ranges between $\frac{2}{5}$ and $\frac{3}{7}$ of an atmosphere, although in some cases as much as $\frac{1}{2}$ an atmosphere, or about 22 lbs., is used.

For persons who are weak, the pressure should be much less, namely, about $\frac{1}{5}$ of an atmosphere, = about 3 lbs.

The duration of the treatment varies greatly, from a few days to many weeks and even several months, according to the nature of the case. If the treatment be very long continued, it should be suspended from time to time for a few days.

Although this is not the place to point out the indications and contra-indications for the use of the pneumatic chamber, yet it may be here stated that, should the patient become thin, it will generally be proper to leave off the treatment, although before doing so the effect of a diminution of pressure may be tried.

The Inhalation of Compressed Air.

It has already been stated that the action of compressed air in the pneumatic chamber differs in several important particulars from that of the air in the transportable apparatus. In the latter case the air presses only on the interior of the organs of respiration and on the parts lying near to or associated therewith; while in the former the pressure is exerted not only on the interior surface of the lungs, but on the exterior surface of the whole body as well, and hence the different results produced in the two cases.

The effects produced by the inhalation of compressed air in the chamber may be described under several heads: the effects on the air itself, on respiration, on the lungs, and on the vascular system.

Compressed air, as before pointed out, is richer than ordinary air in its several constituents, especially in oxygen; its power of expansion is greater, and it has less capacity for moisture and for caloric; hence when ordinary air is compressed its temperature rises almost in the ratio of the pressure exerted. This elevation of temperature, however, enables it to retain a larger proportion of moisture than it otherwise would be capable of doing, as is clearly shown by the fact that, as the condensed air becomes cooled down, the moisture, which was before invisible, makes its appearance in the form of a mist or cloud.

The compressed air, entering the lungs with great expansive force, opens out the passages and air cells, and so enables the patient to breathe more fully; the inspirations taken are deeper and the lungs more completely filled. These effects are at first temporary, but after a period of treatment varying in length according to the nature of the case they become more or less permanent. At the same time that inspiration is easier, expiration is rendered more difficult, is slower, and there is a diminution in the number of respirations taken.

The effect on the lungs of the continued inhalation of compressed air, is an increase in the lung capacity, a strengthening of the elastic tissue of the air cells, increased mobility and circumference of the thorax. Vivenot found an increase of 400 c.c. of capacity at the end of the first month's trial, of another 200 c.c. at the end of the second month, and an additional 100 c.c. at the termination of the third month, making in all an addition of no less than 700 c.c., = 42·6 cubic inches, or one-fifth of the natural lung capacity, as fixed by Hutchinson.

The more important physical effects on the circulation and heart, are a partial emptying of the superficial capillaries and vessels, including those of the surface of the body and of the lining membranes of the air passages and lungs; corresponding with this is an increase of blood in the deeper vessels, a diminution of the frequency of the heart's beats, a contraction of the vessels, and according to some observers a decrease of blood pressure. It is supposed that the pressure produced on the heart by the inflation of the lungs hinders its expansion, slows its action, and diminishes its force. The pulse becomes slower and smaller, and under a pressure of two or three atmospheres so reduced in frequency and force as to be small, thready, and hardly to be felt.

According to Panum, Lange, and Jacobson, there is an increase of arterial tension, as might reasonably be inferred would be the case from the blood being driven from the surface into the deeper vessels. Here it will be well to quote some remarks by Dr. Burdon Sanderson from the paper already noticed, based upon his own observations at Reichenhall as well as on the experiments of Dr. von Vivenot. 'As has been already stated, the principal physiological effect of increased pressure is to retard the action of the heart. For the purpose of ascertaining the cause and signification of this diminution of frequency, as well as its relation to the other physical conditions of the circulation, a large series of observations have been made by Dr. von Vivenot, which, though not made with those precautions which we now regard as necessary in order to render the tracings available for the direct determination of changes of arterial pressure, were so carefully verified in other respects that they afford very valuable material. Dr. von Vivenot's observations were made partly on healthy persons, partly on persons affected with chronic bronchitis and emphysema. The effects observed in every instance may be summed up in a few words: diminished amplitude of the oscillations of the lever, hence diminished expansive movement of the arterial wall; increased obliquity of the ascending limb, hence more gradual filling out of the artery and postponement of the acme or maximum of expansion. Thus, as von Vivenot rightly observes, the pulse, besides being retarded, assumes the character of the *pulsus lentus*, that form in short which characterises *increased arterial tension*. By way of counter-experiment, von Vivenot has recorded a number of equally important observations on persons under *diminished* pressure, not, however, by the aid of the sphygmograph. He found that in a chamber in which the density of the air was reduced from 30 to 20 inches the circulation and respiration were effected even more markedly than in compressed air. While in five individuals simultaneously experimented on, including the author, the mean frequency of the heart's action increased from 72 to 86, the radial pulse became manifestly more voluminous and the extent of the expansive movements of the artery became greater.' . . . 'From all these facts it is to be inferred that the fundamental physiological effect of compressed air consists in its altering the distribution of the blood, i.e. diminishing the quantity or volume of the blood contained in the veins and auricles, and consequently increasing the quantity contained in the ventricles and arteries.' . . . 'The effect of diminished fulness of the venous system is to retard the filling of the ventricles during the period of their relaxation, and consequently to lengthen the diastolic interval and diminish the frequency of the pulse.'

Quincke and Pfeiffer have shown, that by expansion of the thorax in inspiration, all vessels of the lungs are widened and more filled with blood, while in expiration, on account of the greater elasticity of the lungs, their emptying is promoted.

Such, then, are the mechanical effects of the inspiration of compressed air in the pneumatic chamber. We have now to consider the chemical and physiological results under the heads of gas exchange, oxidation, and metamorphosis of tissue.

Various observers, as Vivenot, Sandahl, Panum, and Lange, have established the fact that there is a not inconsiderable increase in the quantity of carbonic acid thrown out. Dr. G. von Liebig, on the strength of certain analyses, on the other hand seeks to prove that there is even a diminution in the carbonic acid thrown out in respiration in compressed air, and he considers that the quantity actually evolved depends less on the depth than on the number of respirations taken. Liebig, however, does not think that less carbonic acid is actually formed.

Again, P. Bert is of opinion that there is no change in the amount of carbonic acid expired.

This evidence is somewhat contradictory, but on the whole it goes to prove that an increased quantity of carbonic acid is evolved.

With regard to the taking up of oxygen, the evidence of Vivenot, Sandahl, Panum, and Lange is not positive or founded on investigations; they only assume from certain indirect considerations that there is an increase in the oxygen absorbed, partly only by the lungs and mechanically by the skin. Liebig, however, found a decided increase in the oxygen absorbed, the quantity depending on the depth of the breathings, and not the frequency; the amount therefore is related to the more or less completeness or depth of the act of inspiration.

Bert examined the blood and found only a slight increase of the oxygen.

Here again the direct evidence as to the increased absorption of oxygen is by no means as conclusive as might have been expected.

Nearly the same difficulties are experienced in determining by experiment an increase of oxidation as one of the results of the respiration of compressed air in the pneumatic chamber. One alleged effect is, however, an increase of temperature. Vivenot has made some experiments on this point, the temperature being determined by the thermometer placed under the armpit. He found as the result of numerous observations, that at the commencement of the period of constant pressure the temperature was 0·503° C., at the end of this period 0·344°, and at the termination of the sitting 0·212° more than before the commencement of the same.

Vivenot endeavoured to explain the increase of temperature in two ways: by augmented oxidation, and by diminution of the loss of warmth, in consequence of diminished radiation of heat and evaporation of watery vapour occasioned by the pressure of the compressed air on the blood-vessels and their consequent diminished supply of blood.

Stembo again, on the other hand, found a reduction of temperature all through the sittings and for some time after; in his experiments the temperature was taken in the rectum, a loss of warmth by the cooling of the room being guarded against. He therefore arrived at the opinion that the constant decrease of temperature was due to a diminished formation of warmth in the body. Lastly, Stembo considered that Vivenot in taking his observations, had used his thermometer for too short a time.

Many other observers have arrived at the conclusion that there is increased oxidation, but exact observations are still needed to clear up the question.

Although an increase of oxidation has not been satisfactorily proved by direct experiment, it yet appears to be clearly established in other ways; one of these is by the thinning effect produced by the continued respiration of compressed air in the pneumatic chamber; by its influence on divers and by the experiments of Simonoff and Sandahl; these are only to be explained by increased oxidation, especially of the tissues themselves.

But this thinning effect is by no means the general result; if an increased amount of nourishment be consumed and digested, there will be an increase of weight, and with the increased destruction of tissue there will be augmented nutrition, which will more than make up for the loss. The observations of Sandahl, Lange, Lewinstein, Vivenot, and Simonoff all show a considerable increase of weight, with at the same time a decrease in the formation of and a disappearance of fat.

The results of experiments on the urine are contradictory, and so far do not appear to throw any light on the subject. Hadra was of opinion that the solids of the urine were increased, but Fränkel, on the other hand, did not find any increase and combats Hadra's views.

Such, then, are the effects, mechanical, chemical, and physiological, of compressed air in the pneumatic chamber.

The mechanical effects are very great, so much so indeed as to necessitate much care in the removal of the pressure, which must be slowly and gradually effected, or a variety of evil consequences may, and sometimes do, ensue, as congestion and itching of the skin, pains in the muscles, redness of the face, a hard, full, and frequent pulse, difficult and noisy respiration, dyspnœa, and a sense of suffocation, while in extreme cases hæmorrhage, from the nose, ears, or lungs, may occur.

Taking into consideration the preceding facts and data, the therapeutical effects of the inhalation of compressed air in the pneumatic chamber may thus be summarised :—The direct mechanical effects of the pressure on the lungs themselves are a partial emptying of the superficial vessels of their lining mucous membranes, a lessening of their secretion, a promotion of the absorption of the effused deposits either in or upon those membranes, augmented power of the tissues, and increased capacity of the lungs. The more remote effects are, 1st, an increased supply of blood to those parts which are not under the influence of the pressure, as especially the organs and contents of the ventral cavity, liver, kidneys, intestines, &c., their nourishment being thereby increased and their secretions augmented; and, 2nd, lowering of the frequency and energy of the action of the heart and blood vessels.

The chemical and physiological effects are an improvement in the condition of the blood, increased oxidation, augmented tissue change and nutrition, followed by a general increase of strength.

The above results form the groundwork on which the indications for treatment, as will appear more clearly hereafter, are based.

Indications.—The special affections in which this treatment is indicated are therefore hyperæmic conditions of the mucous membrane of the air passages, without or with effusion, deposition, and structural changes in the membrane itself, as in acute, subacute, and chronic catarrhal or inflammatory affections; in bronchitis, in which the mucous membrane is not merely unduly vascular but is also swollen and thickened; in maladies of the lungs with restricted capacity, arising from consolidation of any of the air vesicles or from their temporary obliteration by external pressure, as in catarrh of the apex of the lung; in bronchopneumonia and lobular pneumonia; in chronic parenchymatous lung inflammation and in chronic desquamative pneumonia; in emphysema; in bronchitic asthma; in deficient lung capacity; in chlorosis and anæmia, often the precursors of lung mischief; in hereditary predisposition to phthisis; in phthisis in its earlier stages; in hæmorrhage and in certain forms of heart disease often associated with lung mischief. The greater or less benefit accruing in the above cases is mainly due directly or indirectly to the mechanical effects of the pressure. The chemical and physiological effects render the treatment suited particularly for anæmia, chlorosis, and obesity.

Contra-indications.—The use of compressed air in the pneumatic chamber is contra-indicated principally in the following cases: in great weakness and emaciation; in high fever and wasting; in blenorrhœa, putrid bronchitis, and bronchiectasis where there is danger of the resorption of the abundant decomposing and sometimes putrid secretions; in advanced phthisis, especially with extensive softening, suppuration, and cavities, where there is not only a danger of resorption, but the possibility of hæmorrhage occurring from some portion of the ulcerated and weakened surfaces.

Maladies suited for Treatment by the Compressed Air of the Pneumatic Chamber.

We may now indicate in more detail the special diseases of the organs of respiration for which the inhalation of compressed air in the pneumatic chamber is best adapted.

Oertel considers, that acute and subacute inflammations of the respiratory mucous membrane are more effectually treated by compressed air in the pneumatic chamber than by any other known means. In these affections the principal condition is one of hyperæmia of the mucous surfaces, with but little, if any, structural change. He states, that a cure is generally effected in three or four sittings. If there be any alteration of the membrane or obstruction of the air vesicles, a longer treatment will be required.

In the chronic form of the above maladies the indications are not so favourable, owing to pathological changes in the mucous membrane, and a longer period of treatment will be necessary, especially in affections of the throat and larynx, with persistent deviations from the normal structure.

Chronic bronchitis is more amenable to the pneumatic treatment than affections of the larynx, because the mucous membrane is more delicate and less liable to become thickened by structural changes than that of the throat and larynx. By this treatment the hyperæmia and tumefaction are lessened, the blood becomes better oxygenated and decarbonised, and the difficulty of breathing relieved or removed.

If the bronchitis be complicated with bronchiectasis and emphysema, the amelioration will be slower and less decisive, and this remark will apply still more to those cases in which there is considerable dilatation of the right ventricle of the heart, although even in these cases, relief of the attacks of dyspnœa is usually speedily obtained.

If the emphysema has not progressed to such an extent as to give rise to a change in the form of the thorax, then the best results are obtained.

Emphysema has long been regarded as one of the most suitable diseases for the pneumatic treatment, and this especially when accompanied by bronchitis, which being relieved leads to a great improvement in the breathing. The earlier observers were very much convinced of this, but more recently doubts have arisen as to whether the treatment does more than relieve some of the symptoms without ever curing the malady; while even it becomes aggravated in some cases, by increase of the dilatation of the air cells, particularly if the treatment has not been carried out with great care and judgment. Ordinarily the breathing becomes easier, is rendered slower and deeper, in consequence of the descent of the diaphragm, and more oxygen is absorbed, with relief of the dyspnœa. If the inhalations be long continued the power of inspiration and expiration is increased, the tissues become more elastic, and finally there is an augmentation of lung capacity.

The inspiration of compressed air brings about an increased flow of blood to the lungs, whereby not only are the functions of those organs better performed, but the lungs themselves become strengthened and better nourished.

It must be understood, however, that the enlargement of the air cells is not removed, and that the lungs do not become smaller by the inhalation of compressed air, unless this be followed by expiration into ordinary or rarefied air.

In asthma with bronchitis, benefit may be expected from treatment in the pneumatic chamber, since if the bronchitis be relieved, the asthma, so far as it depends on the bronchial complication, will be ameliorated, the attacks will be milder, and the intervals between them longer. In so called nervous and spasmodic asthma the results are not so favourable.

Oertel regards consumption, and maladies related thereto, as specially suited for the pneumatic treatment, and he goes so far as to state, that it is even more important than climate; this statement being borne out, he considers, by experience, provided the treatment be commenced early.

' The mechanical and chemico-physiological effects, the unfolding of the lungs, the raising of the inspiration and expiration pressure, the increase in the depth of the inspiration, the augmentation of the vital lung capacity, the increased amount of oxygen absorbed, the widening of the bed of the stream of blood, and the improved nourishment of the lungs, dependent on this, all make it of value' (Oertel). Not only is the condition of the lungs themselves improved, but the general health of course as well. To ensure beneficial results it is necessary that the treatment should be begun early; in fact, it may be resorted to as a preventive, and it is necessary in most cases that it should be long continued.

The conditions of the lungs, most favourable to the treatment, are not only those attended with a general deficiency of lung development or capacity, but those in which the capacity is more or less impaired or restricted by deposition in and consolidation of lung tissue. Such a condition exists in the apices of the lungs at the commencement of very many cases of consumption, in the *Spitzenkatarrh*, or apex catarrh, of the Germans. The hyperæmia is lessened, absorption promoted, and the air vesicles of the part become expanded.

In chronic parenchymatous inflammation of the lungs and in chronic desquamative pneumonia the results are less favourable, and any amelioration which takes place is due partly to the mechanical effects of the pressure in improving the local conditions, and partly to the increased absorption of oxygen, whereby tissue change is augmented and the appetite and nutrition are promoted.

Lobular inflammation with caseous formations, and bronchopneumonia with coexistent ulceration, are still less hopeful, although benefit may sometimes be obtained in those cases in which the extent of the mischief is but limited and the attendant fever not very high.

Cases of pleurisy with effusion, after the subsidence of active symptoms are, on the other hand, well adapted for the treatment, more particularly in those cases in which the fluid thrown out is of a serous character. The effects of the treatment are to promote expansion of the compressed air cells of the lungs and to hasten resorption. Simonoff is of opinion, that the pressure may in some cases have a tendency to promote the escape of the fluid through the lungs. It does not appear, although fever sometimes ensues in the course of the treatment, that any apprehension need be entertained of blood-poisoning from the absorption of the effused matter in cases of pleurisy.

The treatment of hæmorrhage from the lungs by the pressure exerted on the bleeding vessels by the compressed air of the pneumatic chamber, is one which would naturally suggest itself, and it no doubt does exert a beneficial effect in some cases: those are best suited for the treatment, which are of a slight and chronic character. It does not appear that there is any danger to be apprehended from the pressure simply, but that great caution is required in the reduction of the pressure down to normal. If this be effected otherwise than very slowly, there is a risk of a recurrence of the hæmorrhage. It may be remarked, that the ordinary remedies for the treatment of hæmorrhage from the lungs are so effective that it is not likely recourse would often be had to this mode of treatment.

Certain affections of the heart, especially of its right side, are so often associated with or dependent on diseases of the lungs that it will be well now to consider briefly the effects of the pneumatic treatment in such cases. This treatment, however, must be regarded as being to a considerable extent palliative rather than curative. One of the principal effects of the inhalation of compressed air consists in its altering the distribution of the blood, diminishing the quantity in the veins and auricles, and increasing it in the ventricles and arteries. In consequence of the lessened fulness of the venous system, the filling of the ventricles is retarded and the freqnency of the pulse diminished.

'The bearing of these considerations,' writes Dr. Sanderson in the communication already quoted, 'on the therapeutical action of compressed air as a means of relieving dyspnœa, is not difficult to explain. The cases in which it is useful are precisely those in which the dyspnœa is dependent on dilatation of the right side of the heart, fulness of the venous system, impairment of the pulmonary circulation, and consequent emptiness of the arteries, a state of things which exists in a vast number of cases of chronic bronchitis with emphysema. In other words, the dyspnœa, which it relieves, is that which arises when the feebleness of the contraction of the left ventricle and the diminution of arterial tension are due, not to the defective vigour of the heart itself, but to the interference with its supply of blood *a tergo*. In all such cases immediate relief may be confidently expected by an agency which tends to facilitate the filling of the left ventricle, so that at the end of each period of relaxation, the mass of blood which it contains is sufficiently large for it to grasp vigorously in its systole.'

Not only is the breathing relieved and the pulse rendered less frequent, but there will be a lessening of the congestion and œdema of the lungs.

The results of the treatment are occasionally very striking: patients who were compelled to pass the night in the erect position have been enabled to lie down and sleep comfortably even after a single sitting.

An essential condition for improvement is, that the heart muscles should not be too weak, and especially that this weakness should not be caused by fatty degeneration.

The Inhalation of Rarefied Air.

Thus far, we have been treating of the pneumatic effects of compressed air, and we now pass on to the subject of the inhalation of rarefied air, the effects of which are, in most particulars, the very reverse of those of compressed air.

The effects of the respiration of rarefied air in the chamber, are somewhat the same as when mountain air is breathed; but when expiration takes place from the compressed air of the chamber, the conditions are very different. In the transportable apparatus, the practice usually is to inspire compressed air, and to expire into rarefied air, the surface of the body being subjected to the ordinary pressure of the atmosphere; on the other hand, in the chamber the whole body is included, and there is the same relative amount of pressure both on the surface of the body and in the interior of the lungs, no matter what may be the degree of rarefaction or of compression of the air. If the patient be made to breathe out of the chamber into more or less highly rarefied air, as is occasionally done, then again the conditions will be different from those in the transportable apparatus, because the surface of the body is still subjected to the pressure of the compressed air in the chamber. These differences cannot fail to have their effect in the treatment of many lung affections, although they have not yet been very clearly worked out.

The effects on the respiration of the inhalation of rarefied air in the pneumatic chamber are, according to von Vivenot, as follow :—

When the air is rarefied, say to the extent of $-\frac{2}{7}$ of an atmosphere, about equal to 14,000 feet above the level of the sea, the breathing, according to Vivenot, becomes at first more shallow and frequent, but this soon gives place to a fuller and deeper respiration; the inspiration takes place with effort, while the expiration is easier, and the desire for more air is shown in involuntary deeper breathings. The lung capacity at the same time is less, according to Vivenot and Bert, and this is explained by the expansion of gas in the intestines under the lessened air pressure and the throwing up of the diaphragm.

There are some variations in the statements of different experimenters as to the amount of carbonic acid eliminated, the quantity of oxygen absorbed and present in the blood, and the extent of oxidation and tissue change, but the evidence for the most part goes to show that there is a diminution under each of the above heads, at all events at first and until a compensatory action of the lungs is set up and their capacity augmented.

The effects on the circulation of the inspiration of rarefied air in the chamber, may be thus described :—in consequence of the partial removal of the pressure, the heart acts more strongly, as do also the arteries, causing the pulse to be fuller and more frequent; more blood is conveyed to the lungs and to the vessels on the surface of the body generally, causing the skin to become more or less congested and red, with an increased feeling of warmth. The augmented flow of blood to the lungs causes an increase of secretion, while, in consequence of the acceleration of the flow of blood to the lungs, the deficiency of oxygen in the rarefied air is compensated for, if the rarefaction be not too great. 'The absolute blood pressure is lowered by diminished pressure, while the relative blood pressure is increased' (Oertel).

Although at first there is a feeling of increased warmth of the body, yet the continued inhalation of rarefied air tends to a lowering of temperature : the blood receives less oxygen, and hence arise breathlessness and great prostration.

Indications.—It may be stated at the outset, that the inhalation of rarefied air is contra-indicated in nearly all those cases for which compressed air is suited.

'Most maladies of the organs of respiration, of the heart and circulation, as well as those cases of deranged nutrition and excitement of the nervous system, in which raised atmospheric pressure and greater oxygen tension exert a beneficial effect, are excluded from the treatment of rarefied air' (Oertel).

To the above statement, the cases of those whose respiratory muscles are weak. or whose lung capacity is below the normal, the lungs themselves being free from disease, may, however, be regarded

K

as forming exceptions, since respiration in either compressed or rarefied air is, in such cases, calculated to prove beneficial, by strengthening the muscles, by increasing the lung capacity and the power of the elastic tissue of the air cells of the lungs.

By the effect of the inhalation of rarefied air in expanding the alveoles of the lungs, and increasing the lung elasticity, the lungs are enabled the more completely to empty themselves of the residual air, and so the method proves of value in some cases of emphysema and asthma; but this result is more likely to occur when expiration takes place into rarefied air rather than when the same air is both inspired and expired, in which case the pressure is equalised during both acts.

But little practical experience has hitherto been obtained of the use of rarefied air in the pneumatic chamber; it is not, in fact, very frequently employed. It has been considered that the breathing of mountain air produced nearly the same physical effects, with certain advantages in its favour, as the inhalation of rarefied air in the pneumatic chamber; but this is true only to a certain extent, since the conditions in the chamber and on mountain heights differ in many respects. Thus in the chamber as ordinarily employed, namely, by the inhalation of air compressed to the extent of usually $\frac{1}{40}$ and afterwards rarefied sometimes to $-\frac{3}{7}$ of an atmosphere, the effect produced even as regards the rarefaction of the air is very much greater, than it is on most mountain heights, the range of the air pressure being so much more considerable in the chamber. The majority of mountain resorts may be said to have an elevation of only about 3500 to 5500 feet, which corresponds to a diminution of about 2 or 3 lbs. of atmospheric pressure to the square inch.

The beneficial effects, therefore, of a prolonged stay in high altitudes are due less to the rarefaction of the air, than to other causes.

Expiration into Rarefied Air.

By expiration from the compressed air of the chamber into rarefied air, not only is expiration made easier, and the emptying of the air vesicles of the lungs more complete, but the alternate contraction and expansion of the elastic tissue of the lungs is greater, whereby its tone and strength are, in favourable cases, increased.

Another effect of the alternating pressure is the consecutive emptying and filling of the capillaries of the lungs, which, however, if carried to too great an extent, may give rise to irritation or even inflammation.

Indications.—Expiration from compressed air into either ordinary or rarefied air, has been found serviceable in cases of emphysema and of asthma complicated therewith and attended with an inability to empty the air cells of the lungs sufficiently.

The pneumatic chamber doubtless places in the hands of the physician a very important means of treating certain affections of the organs of respiration, some of which are of a very intractable nature; yet this method, especially in this country, is but little available. There is, I believe, only one such chamber in England, namely, at the Brompton Hospital for Consumption, and this is of comparatively recent construction.

On the Inhalation of Naturally Rarefied Air.

It has already been stated, that the air at the sea level presses on the surface of the body with a weight equal to 14·69 lbs. to the square inch. An ascent of about 900 feet reduces the pressure on the surface of the body by about $\frac{1}{2}$ lb., the decrease continuing at an unequal ratio in proportion to the heights attained.

At the same time that the pressure is diminished, there will be as a rule, at an elevation of about 4000 feet, a reduction of temperature, lessened humidity, greater movement of the air, stronger light, and more sun radiation.

The air is also freer from impurities, especially from organic germs, and it is to this greater freedom that the retardation of the putrefaction of meat and other organic substances is to be attributed.

Owing to the rarefaction of the air and its greater freedom from moisture, the air is more transparent, the soil becomes heated more quickly; but it also parts with the heat thus acquired more readily, the radiation not being restrained by the vapour in the air. Thus at night, the earth becomes much cooled, and the air near it proportionately warmed.

Under such circumstances, as will be readily supposed, the inhalation of highly rarefied air produces very marked effects on some of the principal functions of the body. These effects at an elevation of 3000 or 4000 feet are, an accelerated pulse and quickened respiration; the lung capacity is at the same time diminished temporarily, and there is augmented elimination by the skin and lungs. At great altitudes the superficial vessels become swollen, and there is a proneness to hæmorrhage from the nose and lungs. At heights not exceeding 7000 feet there is marked improvement in digestion, sanguification, and nutrition, with increase of nervous and muscular power.

With the expansion of the air, there is also a diminution in the amount of oxygen in a given space of air, the extent of this being determined by the amount of the pressure exerted and by the temperature. A foot of dry air at the level of the sea at 32° F. contains 130·4 grains of oxygen, but at an altitude of 5000 feet the quantity would be reduced to 108·6 grains, and so on for still higher altitudes. The increased frequency of respiration, however, compensates fully, and perhaps does even more than this for the deficiency, so that at least as much of that gas actually enters the lungs as at lower levels.

Indications.—The preceding particulars furnish the key to the curative power of naturally rarefied or mountain air. Thus by its effects on digestion, sanguification, and nutrition it is doubtless of great efficacy in simple atonic dyspepsia, in anæmia from whatever cause, in gout, in rheumatism, and, though last not least, in the early stages of phthisis. It is certain, however, that the beneficial effects of mountain air are not entirely due to rarefaction, but there are many other causes at work which contribute to those results, and notably the greater purity of the air, its greater dryness, and its lower temperature, particularly at night.

On the Inhalation of Oxygen.

Oxygen constitutes rather more than one-fifth part of the atmosphere; it plays a very important part, not only in transforming carbohydrates into the carbonic acid which is so freely evolved by the lungs, but in other oxidizing processes. There is, therefore, no question but that the addition of oxygen to the air inhaled must produce valuable therapeutic results; but the exact manner in which these are brought about has not yet been clearly established.

When atmospheric air is brought into contact with the blood, as in the lungs during inspiration, part of the oxygen becomes absorbed, but the chief portion unites with the hæmoglobin of the blood corpuscles; the quantity of oxygen in the blood therefore bears a relation to the amount of hæmoglobin present.

Speck has found, that the amount of oxygen absorbed by the blood varies with the amount of oxygen in the air and with the pressure of the air; the greater the quantity of oxygen and the higher the pressure the more oxygen is absorbed. With this increased absorption, however, he was unable to establish the fact either of an increased oxidation, a raised temperature, or an increased formation and expiration of carbonic acid. Neither in the experiments on himself was he conscious of any alteration in his condition, and in particular the breathing was not affected, no matter whether the air inhaled was rich or poor in oxygen. Bert has proved, what indeed could scarcely be doubted, that blood shaken up with oxygen under increased air pressure does take up much larger quantities of oxygen.

Properties.—Oxygen promotes the formation of red blood corpuscles, increases oxidation, improves appetite and digestion, and augments strength. Demarquay states, that with some exceptions, the first few inhalations of oxygen produce a slight sensation of warmth in the mouth, larynx, and interior of the chest; this feeling is also quickly experienced in the hypogastric region, but disappears very shortly after the inhalations have ceased. The pulse is usually accelerated and becomes hard, but this condition soon ceases. Some persons also experience a sensation of heat in the skin, while in some instances a disposition for increased muscular activity shows itself. After the continuance of the inhalations for a few days, increase of appetite is generally manifested.

Administration.—As it is not easy to generate oxygen in sufficient quantity, to materially increase the proportion in the air of an inhalation chamber or room, the simplest and most practical method of administration is by means of the apparatus of M. Limousin, already described and figured. The oxygen may be inhaled pure, but is usually mixed with variable proportions of atmospheric air a quarter, third, or even half; as much as from 20 to 30 litres of the gas should be inhaled twice daily, the inhalation being continued often for weeks, according to the nature of the case. Oxygen gas is usually prepared on a large scale by the action of heat on peroxide of manganese, which when pure yields about one-ninth of its weight in oxygen. Chlorate of potash may also be used, but this salt is much more expensive.

Indications.—Sometimes employed in the non-febrile forms of phthisis, and is said to be of special service in the stomach derangements of phthisical patients. Useful in several forms of anæmia, as in that of insipient phthisis and in that from loss of blood or from suppuration; also in cases of dyspnœa, in which there is excess of carbonaceous products in the blood. It is affirmed by Beddoes and Demarquay to be useful in asthma.

On Ozone.

While ordinary oxygen is sometimes regarded as the negative, ozone is described as the active form of oxygen; thus oxygen does not oxidize the organic matter of the air directly, but ozone, which is said to consist of three atoms of oxygen united in one, attacks it energetically. In this way it acts as a purifier of the air. Owing to this oxidizing property, ozone if too concentrated when inhaled cannot, it is stated, enter the blood, because it immediately attacks the mucous membrane or other organic surfaces with which it is brought into contact. When mixed with decomposing blood or other animal liquids it destroys their offensive odour.

Properties.—Messrs. Dewar and MacKendrick find from their investigations that the inhalation of an atmosphere highly charged with ozone reduces the number of respirations and beats of the heart, is very irritating to the mucous membrane of the air passages, and renders the blood venous.

Administration.—The air of a room may to a certain extent be charged with ozone. One method is by very slowly mixing three parts of sulphuric acid with two parts of permanganate or bichromate of potash. This method has been employed by Dr. Fox. Another plan is to half immerse a stick of phosphorus in tepid water contained in a wide-mouthed bottle, while a third way is to heat a platinum wire by a Bunsen cell. Water charged with it may be used as a spray.

As is commonly known, the amount of ozone in the air can be loosely determined by means of starch papers containing iodide of potassium; the iodine is set free, and this, uniting with the starch, turns the paper more or less blue, according to the amount of the ozone in the atmosphere.

Indications.—There does not appear to be much field for the use of ozone, except as an air purifier in maladies of the organs of respiration; when ozone is abundant in the air it is believed to give rise to colds in the head and to laryngeal and bronchial catarrhs. From the energy of its action on organic matter it might possibly prove serviceable when applied to the false membranes of croup and diphtheria.

On Peroxide of Hydrogen, or Hydroxyl.

This is usually prepared by adding pure hydrated peroxide of barium to a mixture of 1 part of concentrated sulphuric acid and 5 parts of water in a vessel surrounded by ice; the liquid, freed from sulphate of baryta and any free acid, is slowly evaporated over sulphuric acid *in vacuo*, until a transparent oily fluid of 1·452 specific gravity is obtained. It has no smell, but an astringent, bitter taste, like that of tartar emetic. Peroxide of hydrogen is also produced by the oxidation of the oils of turpentine, eucalyptus, and some other essential oils, and it forms the active constituent of the well-known disinfectant 'Sanitas.' It does not redden, but gradually bleaches litmus paper; it is an unstable compound, decomposing with facility into oxygen and water. Oxygen is given off very slowly when peroxide of hydrogen is kept at a low temperature, but at $20°$ C., $= 68°$ F., the evolution of gas becomes plainly visible, while at $100°$ C., $= 212°$ F., a concentrated solution evolves oxygen so rapidly as occasionally to give rise to an explosion.

The liability to decomposition forms an obstacle to the medicinal employment of hydroxyl; but it has been found that ether restrains the decomposition, and this has led to the production of what is known as ozonic ether, which contains as much as 30 volumes of hydroxyl and is miscible in water.

Hydrogen peroxide, by reason of its oxidizing properties, is a powerful bleaching agent, acting like chlorine, but more slowly. It oxidizes arsenious into arsenic, and sulphurous into sulphuric acids. In some cases it acts as a deoxidizer or reducing agent, as with potassium permanganate. In this and other similar cases oxygen is thrown off not only from the hydrogen peroxide, but from the reagents with which it is brought into contact.

Properties.—It is a powerful oxidizer, stimulant, and disinfectant, both when applied externally and administered internally. Applied to the skin or mucous membranes, it produces after a time great irritation, and on abraded surfaces it causes the formation of a white coating of coagulated albumen. Added to venous blood, effervescence takes place; in a few minutes the blood is rendered colourless, and a white flocculent coagulum separates. With pus, effervescence also takes place and flocculi are precipitated, while the pus corpuscles become shrunken and even destroyed.

Administration.—This reagent is far too costly to allow of its being employed for the generation of oxygen, for which there are other cheaper and more suitable sources. It is usually sold in aqueous solutions, more or less strong, the best makers guaranteeing the number of volumes of peroxide present in them. The solution for medical purposes usually contains about ten volumes of hydroxyl. In lung inhalation it is most conveniently used in the form of spray, the application being repeated two or three times, according to circumstances.

Doses.—From $\frac{1}{2}$ to 2 drachms of a 10 per cent. aqueous solution in $\frac{1}{2}$ to 1 ounce of water as a spray. Of ozonic ether, $\frac{1}{2}$ to 1 drachm in the same quantities of water.

Indications.—From its composition and properties it might be presumed, that the inhalation of hydrogen peroxide would be indicated in the same cases as those for which oxygen is suited, namely, those in which there was a deficient supply of oxygen to the lungs and in the blood, and consequent excess of retained carbonaceous matters.

From its effects in promoting the healing of unhealthy sores and ulcers the injection of a weak aqueous solution in the form of spray into the throat, larynx, or bronchi would probably prove serviceable. May be used as a spray or gargle in bronchitis and whooping-cough, and has been applied to the throat in cases of scarlet fever and diphtheria. It has also been given internally in bronchitis and whooping-cough.

I may here remark, that one cannot help feeling some disappointment at the comparatively subordinate part which oxygen and its compounds appear to play in respiratory therapeutics. Perhaps when the subject is more thoroughly investigated and understood there will not be the same ground for disappointment; perhaps also it is not reasonable to look for great curative effects from the somewhat augmented consumption of a gas, which is received into the lungs in considerable amount some eighteen times per minute during the whole course of life.

The Inhalation of Nitrogen.

This gas constitutes nearly four-fifths of the atmosphere we breathe. One of its chief functions seems to be to act as a diluent of the oxygen of the air; it being of itself very unirritating, it renders the oxygen with which it is mechanically mixed much less exciting and irritative than it would be otherwise. Whether the nitrogen is ever directly appropriated for the purpose of nourishment, or for any other purpose, by the human body there is no present evidence to prove, but the belief is generally entertained, that whatever therapeutical power it possesses is due to its lessening the amount and effects of the oxygen inhaled.

Properties.—Sedative by diminishing irritation. According to Treutler, the inhalation of nitrogen occasions paleness and coldness of the skin, and causes the pulse to become small and more frequent. When a large amount of nitrogen is inhaled giddiness ensues, which quickly subsides, followed sometimes by headache, but never by fainting or asphyxia. Immediately after the inhalation, a feeling of comfort and greater freedom of breathing are experienced, as also sometimes lassitude.

Administration.—Pure nitrogen is very conveniently obtained by deoxidizing atmospheric air. This object is accomplished by allowing the air to pass through iron filings which have been moistened with a solution of sulphate of iron. The nitrogen thus obtained may be collected in a receiver of indiarubber similar to that employed for the storage of oxygen, or into a gasometer, or other suitable receptacle.

A very good method of obtaining pure nitrogen suitable for inhalation, is to subject the crystals of nitrite of ammonium to ignition : the salt is decomposed with the evolution of nitrogen gas and a little steam, which may be readily separated by condensation.

On an average, the patient should inhale about 120 litres, according to Kohlschütter, of a mixture usually containing about 90 per cent. of nitrogen, the oxygen thus being reduced to about 11 per cent. Most patients bear 96 per cent. of nitrogen, but sometimes a difficulty of breathing is experienced. The nitrogen should be inhaled for half an hour daily for four weeks, and after an interval of six weeks, the inhalation should, if necessary, be recommenced.

There are certain mineral springs which contain free nitrogen ;

some of the most notable of these are at Lippspringe, near Hanover. They contain no less than 44 per cent. of nitrogen, while of the gases evolved upwards of 80 per cent. consist of nitrogen.

Indications.—The employment of nitrogen is indicated in cases in which there is too great a waste of tissue, as in the following affections: in bronchial and lung catarrhs accompanied with fever, especially, according to Oertel, in apex catarrh of the lungs; in pneumonia and other inflammations of the organs of respiration, in the hectic fever and the suppuration of advanced phthisis. Treutler observed in cases of phthisis, after inhalations of nitrogen continued from eight to fourteen days, quiet sleep, increased appetite, diminution of the night perspiration and diarrhœa, increase of lung capacity and weight of the body.

Kohlschütter has employed inhalations of nitrogen in certain cases in which, from inflammation and other causes, there has been diminished lung capacity, and he states with good effects, as expansion of the compressed alveoli and absorption of effused material; these effects he attributes partly to the forced inspirations necessitated by the construction of Treutler's apparatus, and partly to the small quantity of oxygen contained in the inhaled air.

Nitrogen inhalation has also been recommended in spasmodic and bronchial asthma, and in emphysema with much irritation.

On Nitrous Oxide Gas.

This gas has hitherto been but little employed in the treatment of affections of the respiratory organs. It is always used in a diluted state. Sometimes the air of a room is more or less charged with the gas, but usually it is diluted either with atmospheric air or with oxygen: if with the former, some 20 or 30 per cent. of the nitrous oxide gas may be used; and if with the latter, about the same proportion as that which is present in the atmosphere, namely, 20 per cent.

A case of whooping-cough is recorded in Dr. Cohen's work on 'Inhalation,' which was treated with this gas apparently with very beneficial results. $\frac{1}{2}$ ounce of sulphuric acid was poured into a tea-cup placed on a sand bath, and to this was gradually added $\frac{1}{2}$ ounce of pulverized nitrate of potassa, the gas being disengaged for an hour each night in the patient's bedroom.

S. Klikowitsch found in a case of bronchial asthma treated with inhalations of a mixture of 1 part of oxygen with 4 of nitrous oxide, that the respirations were made less frequent and deeper. In two phthisical cases, the cough was lessened and sleep obtained. In several instances where inhalations of oxygen did little or no good, nitrous oxide has produced favourable results.

The Inhalation of Carbonic Acid.

When atmospheric air is inhaled, the oxygen contained in it combines with the carbonaceous matters with which it is brought into contact in the lungs, carbonic acid being thereby formed and freely evolved during expiration.

It has already been stated, that the amount of carbonic acid in atmospheric air of fair purity and in a fit state for respiration should not exceed 0·4 per 1000 volumes, and that air containing 0·7 per 1000 is perceptibly tainted, while if it contains 1 per 1000 volumes disagreeable effects are produced. The observations of Helfft have shown that from 2 to 4 per cent. of carbonic acid in the air gives rise to the following effects: the respiration is accelerated, the act of inspiration is shortened and that of expiration is made longer and stronger; there is quickening of the pulse, lessening of the secretion of the air passages, and dryness of the throat. Gradually, increased perspiration breaks out, finally redness of the face and cerebral disturbance ensue, these symptoms being regarded by Helfft as evidence of the irritating effects of the gas on the organs of respiration.

Properties.—Stimulant to the skin, produces tingling, redness, and perspiration, but after a time it lessens the sensibility and relieves pain.

Administration.—This gas may be readily obtained in any quantity in a state of purity, and may be inhaled in solution in ordinary water and in many mineral waters, or in the dry state mixed with atmospheric air. Cold water is capable of taking up large quantities of this gas, especially under pressure. The ordinary aërated beverages in use contain it in large quantities, and these may be freely partaken of without any marked effects being produced.

If it be desired, that the gas should be inhaled in the dry state, it may be supplied by the chemist in an elastic bag or receptacle similar to that used for storing oxygen.

Indications.—This gas has been employed in inhalation in several diseases, in the belief that the effects produced were due to its acting as a diluent rather than to any properties possessed by the gas itself. It was under this idea that it was used by Dodson, Beddoes, and others in confirmed phthisis, and they have reported even anæsthetic and antiseptic results from these inhalations.

According to Helfft, who has studied the action of this gas carefully, its inhalation is very serviceable in cases of breathlessness produced by masses of secretion in the alveoli of the lungs, because the gas stimulates the mucous membrane and so promotes expectoration.

Carbonic acid, either in the free state or in solution, is inhaled at some health resorts, and has been used in the treatment of chronic coryza, chronic inflammations of the air passages, phthisis, and asthma. Willemin is of opinion that it ought to be avoided in phthisical cases. At Vichy it is employed in cases of dyspnœa and asthma; at Ems it is used by Sprengler in the form of spray in granular pharyngitis, but Oertel considers it to be unsuited for this complaint, lest it should induce irritation and even inflammation. Sir James Simpson was of opinion that the inhalation of carbonic acid was beneficial in irritable coughs, in chronic bronchitis, and in asthma. It is very useful in irritable conditions of the stomach.

The Inhalation of Hydrogen.

Hydrogen, like nitrogen gas, is simply a diluent, and it might therefore be employed for the same purposes as the latter, but it has not yet been subjected to any prolonged trial.

This gas mixed with air, was employed by Beddoes in the treatment of certain chronic affections of the respiratory organs, under the belief that by diminishing the amount of oxygen in the air it would produce soothing effects. It relieved pain and promoted sleep in cases of phthisis and in acute bronchitis. Other cases have been reported of the soothing effect of the gas in phthisis.

Beddoes employed 1 part of hydrogen to 7 of atmospheric air, the inhalation being continued for 15 minutes. Demarguay used a mixture of 4 parts of hydrogen with 1 of oxygen.

On Carburetted Hydrogen.

In Dr. Cohen's work on inhalation, some interesting particulars are given relative to the employment of carburetted hydrogen or coal gas in whooping-cough. 'Early in 1864 at Amsterdam, children with this affection were sent to the place of manufacture of the illuminating gas, and allowed to breathe the gaseous atmosphere for a certain period with very satisfactory results, many cases having been considered to be cured in this manner and without injurious consequences in any one instance. The same method was employed at Calais in the winter of 1864 during an epidemic of pertussis, a number of children thus affected being sent to the gas house to breathe the vapour at the moment of escape after subjection to the purifying process. As soon as the children breathed this air they began to improve, and thorough cures resulted; two or three visits sufficed to put an end to the paroxysms of cough.' The testimony of M. Commerge is adduced to the same effect.

So far back as 1864, the methods of purification of coal gas were by no means perfect, and the gas after the operation still contained a considerable quantity of compounds, some of which are of known value in inhalation as benzene, sulphur, and ammonia, and it is probably to the presence of these that the beneficial effects experienced were due, and not to the pure carburetted hydrogen itself.

On Sulphuretted Hydrogen.

The pure gas is never inhaled alone; it is nearly always in combination with a base forming a sulphide, although the sulphuretted hydrogen in these compounds is only feebly retained. In mineral waters the gas is for the most part, but not wholly, thus combined, and even when sulphur itself is administered a minute portion of it becomes converted into a sulphide; and it is in this way that the beneficial action of the administration of from 5 to 10 grain doses of sulphur in severe chronic bronchitis with profuse expectoration, as recommended by Graves, is explained.

In sulphuretted mineral waters the gas, whether free or combined, is mixed up with a variety of other constituents, gaseous and saline, as nitrogen and carbonic acid, the carbonates and chlorides of the alkalies, &c.

Properties.—Sulphuretted hydrogen in small quantities is stimulant, reduces the frequency of the pulse, increases the secretions of the mucous membrane and skin, and promotes expectoration; its inhalation gives rise to a feeling of comfort and enables the patient to inspire more freely. In large quantities depresses powerfully and is dangerous.

Administration.—Any of the sulphuretted mineral waters may be freely inhaled in the form of an atomized spray. It is of course best to inhale the waters at the springs themselves, but where this is not practicable the imported waters may be employed. They should be inhaled not less than two or three times a day. Ringer is in the habit of employing a mixture consisting of one grain of sulphide of calcium in half a pint of water, which is of about the strength of the Harrogate waters. In connection with some bath establishments there are special rooms for the inhalation of sulphuretted hydrogen, as at Aix-la-Chapelle, Eaux-Bonnes, Bagnères de Luchon, and many other places.

Indications.—In catarrhs, whether of the throat, larynx, or bronchi; in chronic bronchitis with profuse secretion; in non-inflammatory phthisis with much secretion, and some forms of bronchial asthma and emphysema with bronchitis.

On Sulphurous Acid Gas.

Sulphurous acid gas is a dioxide of sulphur. The solution of the British Pharmacopœia consists of the gas dissolved in water, and contains 9·2 per cent. of the gas by weight, equal to about 30 times its volume of sulphurous acid gas. It is pungent and possesses the odour of sulphur, and by long keeping or exposure to the air the sulphurous acid becomes gradually converted into sulphuric acid. The gas is usually generated from sulphite or hyposulphite of soda.

Properties.—Deoxidizing; antiseptic, probably disinfectant, and destructive of many of the lower forms of vegetable life. Although by arresting putrefaction it prevents the formation of offensive odours, it has but little power of destroying them when once formed.

Administration.—Either the gas in the form of vapour may be inhaled, or the solution in the atomized condition. For the inhalation of the vapour 1 to 2 drachms of the pharmacopœial solution should be added to 10 ounces of boiling water, and the vapour inhaled by means of an eclectic inhaler or any other suitable apparatus. For the spray, $\frac{1}{2}$ to 1 drachm of the solution in 2, 3, or 4 times the quantity of water, or with a mixture of glycerine and water. Thus diluted, the spray excites but little irritation. Dr. Dewar, of Kirkaldy, who has employed sulphurous acid gas in the form of spray for a variety of diseases, makes use of an apparatus with vulcanite tubes, and recommends the

THE INHALATION OF CHLORINE. 143

assistants 'to hold the nozzle of the instrument about 6 inches from the patient's mouth, and administer 3 or 4 whiffs to begin with; then after a corresponding interval, during which a cough or two is given, the process is repeated, about 20 squeezes in all, which represents the injection of from 40 to 60 minims of acid. The acid should be pure.' In the case of children, the apparatus should be held at a greater distance.

Indications.—Dr. Dewar has successfully employed this remedy in many different affections of the mucous membrane of the throat and lungs; in colds in the head, in tonsillitis, in sore throat, including that incidental to scarlet fever and diphtheria; in hoarseness, in croup, in chronic laryngitis, bronchitis, phthisis, and asthma. The inhalation should be repeated at short intervals in most of the above-named affections.

On Chlorine.

This is a very powerful gas, so much so indeed as to be almost irrespirable in the concentrated state.

The British Pharmacopœia contains several preparations which may be employed in inhalation. One of these is liquor chlori, which consists of a solution of chlorine gas in water: 1 ounce of this contains 2·66 grains of the gas; another is calx. chlorata, prepared by exposing slaked lime to the action of chlorine gas until it ceases to take up more of the gas. It is not a very definite chemical compound, but good chloride of lime should contain about 30 per cent. of chlorine when liberated by hydrochloric acid. Calx. chlorata is partially soluble in water. A soluble preparation is liquor calcis chloratæ P. B.: this consists of one part of chlorinated lime to 10 of water, and 1 ounce should contain about 13 grains of chlorine.

Vapor chlori P. B. is made by moistening 2 ounces of chlorinated lime with water and inhaling the vapour from a suitable apparatus.

Liquor sodæ chloratæ P. B. is a mixture of hypochlorite of soda, chloride of sodium, and bicarbonate of soda in water. This preparation is in most cases preferable to liquor calcis chloratæ, it being less caustic.

Properties.—Chlorine gas possesses powerful chemical affinities, and by combining with the hydrogen of certain organic and inorganic substances it alters their composition. The lime and soda compounds give off hypochlorous acid, and this is an oxidizing agent, chlorine at the same time being set free; they are therefore chemical deodorizers and disintegrators, breaking up ammonia compounds and destroying the odour of sulphuretted hydrogen and many other malodorous substances. Not only do they remove bad odours, but they prevent their development, since they are antiseptic and prevent fermentation by their action on the minute organisms which are essential to that process, hindering their growth and probably even destroying them. How far chlorine is destructive of the active principles of infectious diseases, has not yet been sufficiently determined.

Chlorine and the compounds above referred to may therefore be described as stimulant, antifermentive, antiputrescent, deodorant, and in some cases disinfectant.

Administration.—Chlorine may be inhaled in the gaseous state, in the vapour of hot water, or in solution as an atomized spray.

The air of a room or inhalation chamber may be more or less charged with the gas in the manner to be presently described. The vapour may be inhaled by the employment of the vapor chlori P. B., while the solutions of chlorine which have been already mentioned and described may all be used in the form of atomized sprays. Of the liquor chlori from 10 to 20 minims in ℥ss. to 1 ounce of distilled water may be used for each inhalation. Of the liquor calcis chloratæ, double the quantity may be taken, and of the liquor sodæ chloratæ the same as of liquor chlori.

For the inhalation of chlorine in the vapour of hot water, larger quantities of the several preparations must be employed, from 1 to 2 drachms of liquor chlori in 2 to 4 ounces of hot water, the temperature of the water being maintained by a hot-water bath or the flame of a spirit lamp. It would be well in some cases to assist the liberation by a little dilute hydrochloric acid. Dr. Corrigan's apparatus will be found useful in some cases, for keeping up a continuous evolution of chlorine. The arrangement Dr. W. G. Walford has suggested, is similar to the above in principle. It consists of a funnel supported on a retort stand, furnished with a perforated ground stopper, so as to allow the liquid in the funnel to pass through in drops continuously. This liquid should consist of a solution of hydrochloric acid. Beneath the funnel, a dish should be placed containing chloride of lime in water, or a solution of chlorinated soda; chlorine gas will become disengaged in proportion to the quantity of acid which falls from the funnel.

Indications.—Chlorine has been found very useful in coryza attended with fœtor, and for the correction of the offensive odour of other secretions and discharges, as of bronchitis and pulmonary gangrene; in ill-conditioned ulcers of the fauces and tonsils, either with or without fœtor; in scarlatina anginosa, and in diphtheria.

Opinions differ as to the value of inhalations of chlorine in phthisis. Several authorities have stated, that the workmen in bleaching establishments where much chloride of lime is used are particularly exempt from phthisis, and that those who are suffering under the disease become visibly improved. Other observers again have testified to the beneficial effects of inhalations of chlorine in phthisis; Louis, however, has reported that he did not obtain any satisfactory results from the inhalation in 50 cases of phthisis, although he found it to be of great benefit in certain cases of chronic bronchitis, in which the symptoms resemble those of phthisis. Toulmouche also obtained most satisfactory results in chronic bronchitis. He employed a simple Woulfe's bottle inhaler, one-fourth of which was filled with hot water, to which 10 to 40 drops of chlorine water were added, the inhalations being continued from 10 to 15 minutes and repeated several times a day. There can be no question, that the inhalation of chlorine, unless practised with the greatest care, is far too irritating in phthisis, and this is in the opinion of Laennec, Stokes, and others.

Euchlorine.—If strong hydrochloric acid and chlorate of potassa be gently heated in a saucer placed in warm water, a compound called euchlorine is formed, which consists of a mixture of free chlorine and chlorous acid. The preparation of this requires great care, as there is the danger of explosion if too much chlorate of potash be used. Euchlorine has been particularly recommended by Professor Stone, of Manchester, as greatly preferable to chlorine alone, on account of its pleasanter smell and less irritating properties. If prepared, by adding a few grains of the chlorate from time to time to fuming nitric acid contained in a wine-glass, there is in this case no danger of explosion.

The following experiments are instructive, as bearing on the use of chlorine as a disinfectant.

Various methods may be pursued for charging the air with chlorine for disinfecting purposes.

The substance usually employed, mainly on account of its cheapness, is chloride of lime: sometimes the dry powder is used; at others this is placed in saucers or shallow pans containing water. Neither of these plans, though so commonly practised, is effective; the dry chloride, though it smells so strongly, gives off scarcely any of the gas, while, as the following figures will show, the quantity evolved is but little greater when the lime is placed in water, whereby the caustic alkali contained in it becomes slaked, with of course increase of temperature.

Three different quantities of 5 grammes, = 77·1 grains of chloride of lime, or, speaking more correctly, of chlorinated lime, containing 32·66 per cent. of chlorine, were each placed in 100 c.c. of water, = 1540 grains, and exposed to the air for 1, 2, and 3 hours, at the end of which times the following amounts of the gas were recovered: 1·647, 1·576, and 1·605 grammes, = 32·94, 31·52, and 32·09 per cent. These figures show that practically none of the chlorine is evolved, even after an exposure of 3 hours.

A similar result was obtained with a solution of chlorinated soda. Six different quantities of 50 c.c. of the solution, = 770 grains, and each containing 0·1295 gramme of chlorine, were exposed to the air in 6 different glass dishes for 1, 2, 3, 4, 5, and 6 hours, the following amounts of chlorine being recovered at the end of the several hours: 0·1260, 0·1260, 0·1260, 0·1267, 0·1260, and 0·1280 gramme. These figures prove that none of the chlorine had been evolved even after 6 hours' exposure, and this although it was in solution, whereas in the chlorinated lime experiment it was in an insoluble condition.

The preceding figures are of practical importance, because they show that the method of fumigation and disinfection so commonly practised by chloride of lime, either in the dry or moist state, is utterly valueless.

I next treated the chloride of lime and the chlorinated soda solution with hydrochloric acid, 1 part of the strong acid to 3 of water. Three separate experiments were made with the lime: 5 grammes of chloride of lime in 100 c.c. of water and 12 c.c. of the acid were taken in each case, and were exposed for 1, 2, and 3 hours. On the addition of the acid, a considerable effervescence ensued and continued for some time. The solutions were at first acid, but soon ceased to be so and became alkaline. It will be remembered that the 5 grammes contained 1·633 gramme of chlorine, = 32·66 per cent. There were recovered after 1, 2, and 3 hours 1·065, 1·107, and 1·022 per cent. respectively. It thus appears, that while the acid greatly promoted the elimination of the chlorine nearly two-thirds of that gas still remained in the solutions at the end of the experiments.

In the next trials a larger quantity of the acid was used, namely, 20 c.c., and there were recovered at the end of 1, 2, and 3 hours the following amounts of chlorine : 0·966, 0·770, and 0·639 gramme,=19·31, 15·34, and 12·78 per cent. Thus with the larger quantity of acid, equal to one-sixth of the whole solution, a greater loss of chlorine ensued, and this was to some extent progressive ; but still, even in these three experiments taken together, it did not amount to more than one-half of that originally contained in the chloride of lime used.

With a solution of chlorinated soda, 50 c.c. being taken, to which 14 c.c. of the acid were added, the results stand thus : There were recovered at the end of 1, 2, and 3 hours 0·426, 0·440, and 0 468 gramme, or less than one-half that originally contained in the solution, namely, 0·937 gramme. The figures further show that there was no progressive loss of chlorine, but that the whole of what was lost was evolved during the first hour; indeed, I may say that the chief part of the chlorine was liberated and thrown off during the first few minutes of exposure, it escaping in a burst on the addition of the acid.

In the next series of experiments, 5 grammes of chloride of lime and 100 c.c. of water, to which were added 14 c.c. of hydrochloric acid of the same strength as before, were placed in 6 shallow round glass dishes, each having a diameter of 3 inches. These dishes were floated on a water bath, the temperature of which was raised to 62° C.,=143·6° F. The water was kept at this temperature for 2½ hours, when of the 1·633 gramme of chlorine originally contained in each 5 grammes of chloride of lime only ·0284,=·0568 per cent., were recovered; that is to say, practically the whole of the chlorine had become evolved.

The results of these experiments, it is evident, are of great practical importance, as they show the marked effect on the elimination of the chlorine produced by increase of temperature. For charging the air of a room, therefore, with chlorine the chamber inhaler No. 2, described and figured in Chapter III., will be found a most effective apparatus.

The results with 50 c.c. of the solution of chlorinated soda, mixed with the same quantity of water, acidulated with the same amount of acid as in the previous case, and treated in precisely the same way as to temperature and time, furnished results nearly equally favourable. Of the 0·937 gramme originally present 0·309, = 1·62 per cent., only was recovered.

The next and last experiments were as follow :—

Three pieces of Turkish towel, 18 inches by 9 inches, were each moistened with 5 grammes of chloride of lime in 100 c.c. of water, and were exposed to the air for 1, 2, and 3 hours. Of the 1·633 gramme of chlorine originally present, =32·66 per cent., there remained at the end of the several hours the following amounts: 0·142 = 2·84, 0·142 = 2·84, and 0·085 gramme = 1·70 per cent. Thus practically nearly the whole of the chlorine was liberated, and this in the first hour, but really, as was subsequently ascertained, in the first few minutes of exposure.

The effect of the exposure of chloride of lime in water on such a fabric as that of a Turkish towel is most remarkable; a fine froth of chlorine gas gathers almost immediately on the cloth, from which it quickly passes into the atmosphere.

With 50 c.c. of the solution of chlorinated soda in an equal quantity of water the results were much less favourable. The 50 c.c. of the solution in this case contained only 0·61 gramme of chlorine; 5 cloths were similarly charged and exposed for 1, 2, 3, 4, and 5 hours. The quantities of chlorine obtained at the end of each successive hour were 0·340, 0·340, 0·336, 0·335, and 0·330 gramme. In this case, therefore, half the amount of the chlorine originally present was lost, it also escaping not gradually, but, as in the previous case, almost immediately on exposure. This sudden liberation enables one to charge the air of a chamber with chlorine with great ease and rapidity; but to keep it fully charged fresh cloths should be suspended from time to time, care being taken of course to close the doors and windows.

The facts elicited by the experiments now recorded are either entirely new or but little known or practically acted on, but they are of considerable importance.

Other methods of obtaining chlorine for disinfecting purposes may be adopted, but when it is desired that the gas should be inhaled, there are no better or more suitable plans than the two methods which I have just indicated—namely, that by exposure of the chlorinated lime on Turkish towels or any similar fabric, or by placing it in water the temperature of which is raised to about 144·6° F.

Parkes gives the following formulæ for the liberation of chlorine in large amounts: 4 parts by weight of strong hydrochloric acid are poured on 1 part of powdered binoxide of manganese; or 4 parts of common salt and 1 part of binoxide of manganese are mixed with 2 parts by weight of sulphuric acid and 2 of water, and heated gently; or 2 tablespoonfuls of common salt, 2 teaspoonfuls of red lead, half a wineglassful of sulphuric acid, and a quart of water are taken; mix the lead and salt with the water, stir well, and add the sulphuric acid gradually. Chlorine is evolved, and is absorbed by the water, from which it is slowly driven out.

DILUENTS AND REFRIGERANTS.
Cold Water.

These remedies consist chiefly of cold air, already treated of, cold water, or ice water.

The effects of cold water resemble in some respects those of cold air; it lowers temperature and contracts the vessels; but in addition it moistens the frequently dry mucous membranes of the air passages, and helps to dilute and detach any irritating mucus or other secretion which may rest upon them. It takes up caloric, however, much more rapidly than does cold air, so that when it reaches the deeper bronchi and air cells it is scarcely cold at all.

Administration.—The water may be rendered cold by the addition of pieces of ice, or the still colder water of melted ice may be used.

Indications.—This remedy is therefore indicated in dry and hyperæmic conditions of the mucous membrane of the throat; in catarrhs, either acute or chronic, where there is heat and dryness; in inflammation of the tonsils, and in pharyngitis.

DILUENT, WARMING AND SOOTHING REMEDIES.
Hot Water.

The principal remedy of this class consists of hot water in the form of vapour, or of an atomized liquid or spray. The water may be used alone or it may be medicated in a variety of ways, as by the addition of demulcent, soothing, or anodyne substances, including the following: gum arabic, glycerine, or infusions containing gummy or oily substances, as of marsh mallow, linseed or sweet almonds, or infusion of the flowers of camomile or decoction of poppy-heads.

Hot, in the same way as cold water, acts as a diluent of any mucous or other secretion which may be upon the surface of the respiratory mucous membrane, and helps to loosen and thin the same; it thus becomes a soothing and indirectly an expectorant remedy, and hence it often greatly relieves irritation and cough.

But hot water, in the form either of vapour or spray, possesses other properties, dependent upon the increased amount of caloric it contains; thus, in place of contracting the vessels as cold water does, and so lessening the supply of blood to the parts and diminishing secretion, the vessels are relaxed, their calibre and the quantity of contained blood in them are increased, and as further consequences the secretion of mucus and the formation of the cell elements of mucus, epithelium, and pus are all considerably augmented. By this increase, the vessels and surrounding tissues become greatly relieved, and the entire effect of the remedy is to soothe, to relieve irritation and pain, and to promote secretion.

When simple gum, or gummy vegetable extractive matters, or glycerine, are added to the hot water the moistening effect of the spray is rendered more continuous and it is more emollient, so that any dryness which may be present is more quickly removed. If the anodyne principles of the hop or poppy be also added, then the effect produced is still more soothing, and greater relief is afforded to irritation, pain, and cough.

Indications.—The vapour or spray of hot water therefore, either alone or still more when medicated, is indicated in throat affections, and to a less extent in those of the larynx, trachea, and lungs; in acute and chronic catarrhs, where there is great irritation, and pain, redness, congestion and swelling, with deficiency of the normal secretion; in cases in which it is desirable to promote suppuration or the separation of false membranes, as in croup, and diphtheria.

Glycerinum.

Glycerine is much employed in respiratory therapeutics for several purposes : as a solvent of certain medicaments, as gallic acid ; as promoting the adhesion of others to the mucous surfaces with which it is brought into contact; and, lastly, on account of its own particular action.

The following are the principal glycerides which have been employed in respiratory therapeutics : namely, those of carbolic, gallic, and tannic acids, of borax, and subacetate of lead.

Administration.—Usually employed as a spray, the glycerine or glycerides being more or less diluted with water; or as a vapour, the glycerine being volatilized by a spirit lamp.

Indications.—It has been employed in the following cases: in aphonia and hoarseness, in dry conditions of the mucous membrane of the pharynx, larynx, and bronchi, as in chronic laryngitis and bronchitis, and in croup and whooping-cough. It has been recommended in œdematous conditions of the epiglottis and larynx, with a view to promote exosmosis and so relieve the congestion and effusion.

Oleum Olivæ, Ol. Amygdalæ.

A remedy not sufficiently employed in respiratory therapeutics is oil, either olive or almond oil.

Administration.—It may be used either undiluted or mixed with gum as an emulsion, or it may be medicated in various ways according to the indications it is intended to fulfil. As a spray; since neither oil nor glycerine will pass through very fine apertures, the tubes of the spray apparatus used should have somewhat larger openings.

Properties.—It acts partly as a lubricant, but by its exosmotic effects it produces very striking results when applied to epithelia and granular structures and formations, as mucous membranes, mucus and pus and some other secretions.

Indications.—Its employment is therefore indicated in dry conditions of the mucous membrane of the throat, larynx, and bronchi, with a view to moisten the surfaces, to promote expectoration, and to destroy by exosmosis the vitality of the granular corpuscles which form so large a part of the pulmonary secretions ; in whooping-cough and croup. Above all, it seems to me that the remedy is one which might possibly prove of value in the treatment of phthisis by its exosmotic effect on the tubercle bacillus.

Verbascum Thapsus.

The leaves of this plant, the Great Mullein, have lately come into use in Ireland, mainly through the recommendation of Dr. Quinlan, in cases of incipient phthisis, and they are extolled for their soothing, demulcent, nourishing, and weight-increasing effects. A decoction may be prepared either from the fresh or dried leaves, the former, of course, being preferable, or from the root, and made either with water or milk; for the fresh leaves 1 part in 5, and for the dried leaves, 1 in 30 of the menstruum. For the relief of dryness of the throat and of cough, the decoction may be liberally employed in the form of spray. It is said, that smoking the dried leaves affords relief in the irritative cough of phthisis. A tincture has been prepared, the dose of which ranges from 20 to 60 minims, but whether it possesses the properties of the fresh leaf is questionable. When given in milk, much, and according to some the whole, of the fattening effects must be attributed to this. To settle this point, experiments are needed with mullein alone, without the milk.

CHEMICAL SOLVENTS.

While water acts partly mechanically as a diluent, solvent, and soothing agent, the remedies, chiefly alkaline and belonging principally to the ammonia, potash, and soda groups, now to be noticed, exert in addition certain chemical effects or properties; thus they form, with the albumen of mucus and pus, soluble albuminates, and the cells of epithelium and the corpuscles of mucus and pus swell up and become either partially or wholly dissolved, according to the strength of the solutions.

Liquor Ammoniæ, Ammoniæ Carbonas, Ammonii Chloridum.

Ammonia and its carbonate are alkaline, very volatile, very soluble, highly diffusive, and are readily eliminated through the breath, skin, and urine; they exert a solvent action on the animal secretions, mucus, and pus, and on the tissues themselves when the strong solution of ammonia is applied to them. The solvent or destructive action of caustic ammonia on the tissues, owing to its less affinity for water, is not so great as either potash or soda, and hence its effects are more limited. Externally applied, owing to its diffusive power it readily penetrates the skin, giving rise to redness, inflammation, vesication, and even sloughing.

Chloride of ammonium differs considerably from the solution and the carbonate, both in its chemical nature and medicinal properties. It is far less stimulant, is a neutral salt, and is readily volatilized by heat. It is formed when the vapours of strong hydrochloric acid and ammonia are brought into contact, as in the apparatus which has been already described, white dense fumes of the salt being freely generated; these, however, quickly condense and form a crystalline deposit.

Properties.—Ammonia is powerfully stimulant, increasing the force and frequency of the pulse, antispasmodic, alkaline, antacid, solvent, augmenting the secretion of the skin and mucous membrane, especially of the bronchi; expectorant. Externally applied, it is irritant, rubefacient, and may even be escharotic.

The carbonate possesses similar properties, but is less stimulant, and its action on the skin is much weaker.

Chloride of ammonium is non-alkaline, slightly stimulant only, is less solvent, but in most other respects acts in the same way as the carbonate. It is diaphoretic, diuretic, and it also increases the secretion of the pulmonary mucous membrane; is expectorant, alterative, deobstruent, and by its effects on the nervous system, sometimes acts as an anodyne. It also increases considerably the elimination of urea, from which it is inferred that tissue metamorphosis is augmented.

Administration.—The fumes of the solution of ammonia or its carbonate may be directly inhaled through the nose, or a solution of the carbonate may be inhaled by the mouth in the form of an atomized spray, and so be applied to the mucous membrane of the throat, and even in a less degree to the respiratory passages. More frequently, however, the fumes of chloride of ammonium are inhaled in the manner described in a previous chapter.

Doses.—Of the liquor ammoniæ P. B., 10 to 20 minims, and of the carbonate of ammonia 3 to 6 grains, in ½ ounce to 1 ounce of water, both as sprays; of chloride of ammonium, if inhaled as a spray, 10 to 20 grains in 1 ounce of water. The fumes should be inspired for a few minutes only. Cigarettes, each containing 5 grains of chloride of ammonium, have been prepared by Messrs. Corbyn an Co. at the suggestion of Dr. Macnaughten Jones.

Indications.—Ammonia and carbonate of ammonia are contraindicated in acute inflammatory conditions in consequence of their stimulant action; but they, as well as the chloride, are suited for subacute and chronic affections of the mucous membrane of the organs of respiration, especially those attended with dryness and scanty and tenacious expectoration; as in subacute and chronic catarrhs, in chronic laryngitis and bronchitis, especially when the secretion is thick and at the same time abundant; in chronic ulceration of the larynx, and whooping-cough. The stimulating and solvent action of these remedies often render them of the greatest service in cases of bronchitis in which there is much difficulty of breathing, owing to the presence of masses of secretion blocking up the bronchial tubes.

The vapour of ammonia and its carbonate, more or less diluted with water, has been found very beneficial in cases of coryza, aphonia, chronic hoarseness, and in relaxed conditions of the throat generally. By its irritant action it excites a flow of watery mucus and so relieves congestion, dilutes adherent secretion, and promotes expectoration. It has also been used with advantage in croup.

Chloride of ammonium has been found extremely valuable in cases of chronic pharyngitis with dryness, it promoting secretion.

Carbonate of ammonia is sometimes insufflated in the form of powder in nasal catarrh and in chronic coryza.

Soda Caustica, Sodii Chloridum, Sodæ Carbonas.

Caustic soda is much more solvent and escharotic, in consequence of its stronger affinity for the water of the tissues, than is the solution of strong ammonia; carbonate of soda is also more powerful as a solvent than the corresponding salt of ammonia, but bicarbonate of soda, containing less alkali, is a milder preparation.

Properties.—Chloride of sodium, or salt is stimulant, solvent, and indirectly nutritive by the aid it affords to the digestion of food by increasing the free hydrochloric acid of the gastric juice. The solution of soda and the carbonates are antacids and good solvents of albuminous liquids and substances, and hence they promote expectoration and allay irritation.

Administration.—In the form of spray.

Dose.—Of the liquor sodæ P. B. 10 to 40 minims in 1 ounce of water, of the chloride or carbonate 10 to 15 grains in the same quantity of water, should be employed for each inhalation, and should be repeated two, three, or four times a day, according to the requirements.

Indications.—The chloride is especially indicated in cases in which it is desirable to promote digestion and nutrition. In dry conditions of the mucous membrane of the pharynx, larynx, and bronchi, by promoting secretion it lessens engorgement and reduces swelling. The carbonates are to be preferred in those cases in which the mucous secretion is of considerable tenacity, as its solvent action is much greater than that of the chloride. Its employment is indicated in subacute and chronic catarrh, with either dryness or with tough and adherent secretion; in pharyngitis with dryness of the mucous membrane, especially the granular form, that condition which has been termed 'clergyman's sore throat,' but which is common to many who have to exercise the voice much; it will also be found softening and soothing, relieving the irritation and cough in some cases of chronic laryngitis, tracheitis, and bronchitis; it is affirmed by Oertel, that it is very serviceable in dry asthma and in emphysema of long standing; finally, it has been employed in croup and diphtheria, but there are more effective applications.

At the International Congress of Hygiene held at Geneva in 1882, Dr. J. G. Partajas stated, that he had found the effects of the application of bicarbonate of soda in the initial stage of tonsillitis to be both certain and rapid. The soda is best applied in substance and frequently repeated. The remedy occasions nausea and much secretion of mucus, whereby the congestion and threatened inflammation are greatly relieved. The same remedy continued for some weeks will, it is said, bring about a gradual resolution of enlarged and hypertrophied tonsils.

Liquor Potassæ, Potassæ Carbonas.

The principal preparations of potash employed in inhalation, if we except chlorate of potash, are the solution and the two carbonates.

Properties.—Liquor potassæ is antacid, sedative, especially to the mucous membrane of the stomach; alterative, increasing tissue metamorphosis, and deobstruent; in large doses it increases the alkalinity of the blood and renders the fibrin more soluble.

The carbonates are more powerful and more solvent than the

corresponding salts of soda; they are alkaline, antacid, alterative, but not to the same extent as the solution. They increase secretion and promote expectoration by their solvent action. Powerfully depressant, and in large doses diminish the force and frequency of the heart.

Administration.—In the form of spray.

Doses.—Of the liquor potassæ, P. B. 10 to 30 minims in 1 ounce of water; of the carbonate, 10 to 15 grains in the same quantity of liquid.

Indications.—These preparations may be used for nearly the same purposes and in the same diseases as the soda preparations.

Dr. Walshe has highly recommended liquor potassæ in plastic or fibrinous bronchitis, it helping the softening and expulsion of the casts. The carbonates have been found to be extremely useful in small and repeated doses in whooping-cough.

Liquor Calcis.

This is a saturated, aqueous solution of lime, each ounce containing about half a grain of the alkali; it spoils quickly on exposure to the air, in consequence of the formation of carbonate of lime.

Properties.—Astringent; lessening secretion; solvent to fibrinous deposits; possesses but little diffusive power and therefore enters with difficulty into the circulation.

Administration.—In the form of an atomized spray.

The inhalation of lime water may be combined with or followed by that of the vapour of hot water, or the lime water itself may be warmed before use.

Liquor calcis saccharatus P. B. may in some cases be used in preference to lime water. 1 ounce of this is equal to 12 ounces of lime water. Dose 30 to 60 minims, with water as a spray.

Indications.—It has been strongly recommended as a solvent of the false membranes of plastic bronchitis, croup, and diphtheria. Several experimenters, including M. Küchenmeister, bear testimony to the fact of the diphtheritic membrane being dissolved in lime water.

The direct inhalation of the vapours given off when *caustic lime* is slaked is reported to have proved very beneficial in many cases of croup and diphtheria; these vapours really consist of the steam of hot water, holding in suspension many small particles of lime. There would appear to be reason to believe, that the beneficial effects, which do undoubtedly occur in some cases, are much more due to the warm, moist vapour than to the lime; the action of the vapour, therefore, is very different from that of lime water. This method of treatment was first suggested by Dr. Geiger, of Ohio, and Dr. Cohen writes concerning it to the following effect: 'I would not like to be prohibited from employing it in the management of a serious case of membranous croup. In some dozens of cases in private and in consultation practice I have seen life apparently rescued through its agency. Of all the methods of treating croup advanced of late years I know of no other that has held its ground so well as this. The general plan pursued by myself is to keep up a continuous evolution of steam from boiling water, and to administer the lime in the manner indicated by Dr. Geiger, for ten minutes at a time or thereabouts, whenever the respiration evinces the presence or formation of membrane, repeating it at intervals of half an hour, an hour, two hours, or longer, according to circumstances, and recurring to the remedy whenever the respiration is impeded by the exudation. I find a large piece of stiff writing-paper, loosely folded into a funnel-shaped cone, one of the best means of directing the vapour towards the mouth of the patient.'

Tetramethylammonium Hydroxide

Is formed when freshly precipitated oxide of silver is added to a solution of iodide of tetramethylammonium $(N.(CH_3)_4I)$. The formula is $N(CH_3)_4OH$. On evaporation of the solution *in vacuo* a crystalline mass is formed, which rapidly absorbs water and carbon-dioxide from the air.

It acts as a powerful caustic, is strongly alkaline, and generally resembles the fixed caustic alkalies in its behaviour; on neutralization with acids, salts are obtained which, like the iodide, are not decomposed by caustic potash.

Indications.—Has been suggested as a solvent for the false membrane in diphtheria.

Tetraethylammonium Hydroxide.

Obtained by gradually adding freshly precipitated oxide of silver to a weak solution of iodide of tetraethylammonium. The filtrate is evaporated first on a water bath and then *in vacuo* until a semi-solid, deliquescent mass is obtained, which closely resembles caustic potash in its reactions.

Indications.—Same as the preceding.

Neurin.

Synonymous with choline, $C_5H_{15}NO_2$, obtained from the brain or nervous tissue and from the bile of the ox and pig. According to Würtz and Baeyer it is trimethyl-oxethylammonium hydrate. J. Mauthner (Liebig's 'Annalen,' clxxv. 178) states, that an aqueous solution of neurin dissolves blood fibrin to a clear liquid which does not blacken lead acetate, is not precipitated by alcohol, and yields a precipitate of fibrin on the gradual addition of an acid.

Indications.—Has been recommended as likely to be effective in dissolving the false membrane of croup and diphtheria. Oertel recommends neurin, as also tetramethylammonium hydroxide and tetraethylammonium hydroxide, in 3 to 5 per cent. solutions. These are said to have the property of not only loosening the membranes, but also of acting antiseptically. Oertel, however, has not tried them sufficiently to give an authoritative opinion.

Succus Caricæ Papayæ.

The milky juice of the papaw tree. It is said to possess the remarkable property of making meat washed with it very tender, the leaves and fruit even having the same effect on the flesh of poultry and pigs which are fed with them. A substance is obtained from the juice, resembling the flesh or fibre of animals, papayotin.

Papayotin or papain is a whitish powder prepared from the juice of the papaw tree; like pepsin, it has the property of digesting fibrin, the presence even of carbolic acid not checking its action. So powerful is it, that it will peptonize 200 times its own weight of pressed fresh blood fibrin. Papayotin may be obtained of Messrs. Hopkin and Williams at 4s. per drachm.

Indications.—Recommended as a solvent of the false membrane in diphtheria. Dose: 1 to 4 grains suspended in ½ ounce of water as a spray, or a stronger solution applied with a brush. Rossbach has proposed and has himself employed, apparently with some amount of success, succus caricæ papayæ L. and papayotin as disintegrators and solvents of adventitious membranes. Further observations are, however, required to establish their true value.

Acidum Lacticum.

An odourless, syrupy liquid, formed by the fermentation of milk.

Properties.—Solvent of fibrinous exudations and false membranes.

Administration.—In the form of a spray, one part of the acid in from 10 to 15 parts of water, repeated, if necessary, several times a day, or a stronger solution, if a brush be used.

Indications.—Strongly recommended as a solvent of the false membrane in diphtheria, but Oertel considers it to be inferior to lime water. He states that it loosens the membranes, but does not dissolve them. For the spray, to dissolve the false membrane in diphtheria, 30 minims of the strong acid, spec. grav. 1040, should be dissolved in ½ ounce of water and repeated as may be required. It is said, that in doses of 10 minims two or three times a day in ½ ounce of water it allays cough and quenches thirst in phthisis.

REFRIGERANTS.

Potassæ Chloras.

It was formerly supposed, that this remedy became disintegrated in the system, that the oxygen was set free, and that it thus promoted oxidation processes in the system. It is now known, however, that most of the salt is recoverable from the urine unchanged.

Properties.—It is refrigerant, diuretic, and increases the secretion of the saliva. Locally applied, it exerts a powerful but unexplained action upon mucous membranes.

Administration.—As an atomized liquid, 10 to 20 grains being dissolved in ½ ounce of water, or the powder insufflated.

Indications.—It is one of the best remedies for a spongy condition of the gums, ulceration of the same, for aphthæ, and thrush or stomatitis.

Dr. Leonard Sedgwick states, that it quickly relieves stuffing of the nose, rawness of the throat, and hoarseness of the voice due to cold. If it be taken early and repeatedly, it will even stop a cold altogether.

Dr. Lloyd Roberts recommends it to relieve the dryness of the throat frequently left by diphtheria and scarlet fever.

It has been recommended by Dr. Spendler in phthisis, large doses being employed in the form of a drink. He states that it checks diarrhœa and prolongs life.

Potassæ Nitras.

Properties.—Refrigerant, diuretic, and sedative, reducing the frequency of the pulse.

Administration.—Is employed in the form of spray, or the fumes and vapours of the burning nitrate are inhaled.

Dose.—For the spray, 20 to 30 grains, dissolved in 1 ounce of water. For the fumes 20 to 30 grains.

The nitrate is prepared for burning in two different ways; either in the form of papers which have been dipped in a saturated solution of nitre and then dried, or as pastilles, with the addition sometimes of chlorate of potash, stramonium leaves, &c. The composition of the fumes or smoke, varies according to the nature of the substances with which it is combined; the principal constituents are ammonia, carbonic acid with some carbonic oxide, cyanide of potassium, and cyanogen. If the papers or pastilles contain, as they sometimes do, chlorate of potash, then the fumes may also contain oxygen.

Since the nitre papers and pastilles are prepared in several ways and contain varying quantities of nitre, and since they are burned and inhaled in different ways, of course the effects must vary also, and one is not surprised, therefore, that the inhalation of the fumes of nitre, generally so beneficial, sometimes fails to afford the relief anticipated.

The paper or the pastilles used should always contain a known quantity of nitre, as much being burned as would correspond to some 20 or 30 grains of the salt. When the paper is properly prepared, the fumes should be white and not brown or black.

It is equally necessary, that the papers or pastilles should be burned in such a way as to ensure the inhalation of the fumes. If they be burned in a large room, or with an open door or window, it cannot be expected that they will prove effective. They should be inhaled by means of the apparatus which I have devised, and which will be found described and figured in Chapter III., or in some other equally efficient manner.

Indications.—Chiefly employed for the relief of the paroxysms of asthma, over which it very frequently exerts a speedy and beneficial effect. Oertel thinks, that the benefit derived is due to the ammonia contained in the vapour, it occasioning expectoration, which is always followed by relief.

SOLVENTS, ALTERATIVES, AND DEPURANTS.
Mineral Waters.

The action of the carbonated alkalies, soda, and potash, and of chloride of sodium having been already described, a few remarks may be appropriately made in this place on the general action and effects of mineral waters containing those and other more or less allied constituents.

Those waters in which carbonate of soda is the predominant ingredient, as the waters of Vichy and many others, exert of course a solvent and antacid action, and may be employed as sprays in the same conditions and maladies as the solution of carbonate of soda itself.

The same remark applies to those mineral waters which contain an excess of chloride of sodium, as the waters of Soden in Taunus, of Kissingen, &c.

Waters containing an excess of both carbonate of soda and chloride of sodium, as of Ems and Aix-la-Chapelle, stand in a different category and possess the properties of both those salts.

The good effects of many mineral waters are almost entirely due to the chloride of sodium which they contain, and at some foreign spas the air of rooms is impregnated with cloud-like sprays of the atomized water. At Reichenhall, in Bavaria, and some other foreign bath establishments, the outer air is charged with the salt in a different manner. According to the description given by Dr. J. Burdon Sanderson in the 'Practitioner' for October 1868, the patient lounges in the immediate neighbourhood of enormous hedges 40 to 50 feet high, composed of bundles of twigs arranged horizontally, so that their projecting ends form a kind of wall. The water is conveyed to these hedges by pipes, and is allowed to trickle over the bundles of twigs into reservoirs, whence it is conveyed into vats and undergoes further evaporation by the aid of heat. The air near these hedges is found to be very rich in salt, and the researches of Professor Vogel and Dr. von Liebig show that about 10 grains of salt are received into the lungs in the course of an hour, being only 2 grains less than the amount taken in when the water is nebulized in the inhalation rooms of the *Kurhaus*. Similar 'vapour promenades' exist in several other places, including Kreuzanch, Salzungen, and other health resorts.

Indications.—These inhalations are said to be of great value in chronic catarrhal affections of the respiratory mucous membranes, and Dr. Sanderson believes that the beneficial results are mainly due to their effect in increasing molecular disintegration throughout the system.

The sulphuretted waters owe their properties and efficacy mainly to the presence of the sulphides they contain. Three of the principal sulphuretted waters are those of Harrogate, Barèges, and Aix-les-Bains.

The inhalation of sulphuretted waters is indicated amongst other maladies, in consequence partly of their stimulant properties, in aphonia, tonsillitis, chronic catarrh of the nasal pharynx, granular pharyngitis, in bronchial catarrh, and in chronic laryngitis and bronchitis, in the absence of any considerable amount of irritative fever.

Although the mineral waters are best employed at the springs themselves, yet for inhalation purposes the bottled waters will answer nearly as well, and if these are not procurable, then approximate imitations of them may be readily prepared.

Sulphides of Potassium, Sodium, Ammonium, and Calcium.

The sulphuretted hydrogen in these compounds is very loosely united to the bases, so that it is readily liberated by the action of even weak acids; when these salts are introduced into the stomach, the secretions of which are nearly always acid, the gas is set free and makes its presence known by disagreeable eructations.

The sulphide which has been most employed is sulphide of calcium, but sulphide of ammonium is also sometimes used.

Properties.—They are alkaline, stimulant, sudorific, and expectorant, increasing the secretion of the mucous membrane of the bronchi; antiseptic; in strong solutions irritant to the skin ; but the most important property they possess is that of controlling suppuration, lessening it in some cases, and hastening its maturation in others, as in boils, carbuncles, chronic and scrofulous abscesses. They lessen and subdue at the same time the surrounding inflammation. Not only are these effects well established, but they are brought about in a very short time and with extremely small doses. The effects of sulphuretted hydrogen and the sulphides on the skin have been long known by the results produced by the employment of such sulphuretted mineral waters as those of Harrogate, Barèges, &c., both when outwardly applied, as baths, or taken internally. Usually the general health improves under the administration of the sulphides, but if they be too long continued they give rise to much depression and may even induce anæmia.

That remedies having such striking properties must also have a great value in some diseases of the lungs, especially those of a suppurative character, is undoubted, though hitherto they have been but little employed. It may be assumed that some of the beneficial effects of the action of these remedies is due to their antiseptic and antiparasitic properties.

Administration.—They may be inhaled in the form of spray, or administered internally in powders or pills; the sulphides of potassium, sodium, and ammonium are freely soluble in water, but the lime salt is almost insoluble, so that when a spray is administered a soluble sulphide may be selected. Dose of sulphide of calcium $\frac{1}{10}$ to $\frac{1}{2}$ grain in 1 oz. of water. It is less efficacious in larger doses.

Indications.—The administration of the sulphides is therefore indicated in unhealthy and suppurative conditions, especially of the throat, larynx, bronchi, and even of the lungs; in tonsillitis, pharyngitis, chronic laryngitis, and bronchitis; in phthisis of the larynx and lungs. Sulphide of calcium is said to aid in the detachment and expulsion of the false membrane in diphtheria.

NAUSEANTS AND EMETICS.

Cephaëlis Ipecacuanha.

This drug owes its efficacy mainly to the presence in the cortex of the root, of an alkaloid which has received the name of emetina or emetin; it is soluble in alcohol, but sparingly so in ether or water.

Properties.—Irritant, when topically applied to mucous membranes or abraded surfaces, and also to the eyes, nose, and bronchi when the powder is inhaled or even smelt in some cases. In small doses diaphoretic, relaxant, and expectorant, increasing the secretion of mucus; in larger doses sedative to the vascular system, and in still larger doses depressant, emetic, and antiperiodic.

Administration.—The powder of ipecacuanha is seldom used alone; with some susceptible persons its inhalation will produce all the symptoms of a severe cold, sneezing, and running from the eyes and nose, and the attacks may even simulate those of hay asthma or bronchitis. Usually employed in the form of a spray, for which the vinum is a convenient preparation. In order to ensure some of its beneficial effects, it is not necessary that nausea should be produced; indeed, in some cases this, and more particularly vomiting, should be avoided. With this view the ipecacuanha should be diluted in a small quantity of liquid and slowly inhaled; the patient should avoid as much as possible swallowing the spray; in fact, he should expectorate any unabsorbed liquid from time to time. Sometimes it will be sufficient to inhale once a day, at others night and morning, or even more frequently according to the requirements of the case. When the depressant or emetic effect is desired, larger quantities of the drug must be used.

Doses.—As a relaxant and expectorant, 10 to 30 minims of the wine in 2 drachms of water; as a depressant, 1 drachm of the wine; and as an emetic, 2 to 4 drachms. When the emetic effect is required, the remedy should be taken in the ordinary way in place of being inhaled, and it may be preferable in some cases to employ the powder, 15 to 30 grains, instead of the wine. For the relief of irritable coughs, compound powder of ipecacuanha may be used, in doses of 2 to 4 grains, as a spray, it being suspended by means of mucilage.

Indications.—In sore throat unaccompanied by fever, the mucous membrane being very red and swollen; in hoarseness from congestion of the vocal cords; in bronchitis with tenacious secretion, especially when combined with an alkali; in the bronchitis of children with difficulty of breathing from accumulation of mucus; in this case the emetic effect should be induced, for which purpose the quantity required for children is considerable, as they are very tolerant of the remedy. In winter cough and bronchial asthma, ipecacuanha is of the greatest value, as shown by Dr. Ringer and Dr. William Murrell; also in bronchial asthma with emphysema. In some cases of bronchial asthma the effects produced vary;

sometimes considerable relief is afforded; in others a sense of constriction and tightening of the breathing is produced, with aggravation of the dyspnœa. In whooping-cough, the ipecacuanha spray will either cut short or relieve the paroxysms, and even stop the vomiting induced by the violent paroxysmal cough. It has been strongly recommended in hay asthma to relieve the dyspnœa.

Antimonium.

The principal officinal preparations of antimony in use in medicine, and suitable for inhalation, are pulvis antimonialis, consisting of 1 part of oxide of antimony with 2 of phosphate of lime, and commonly known as Dr. James's fever powder; antimonium tartaratum, and vinum antimoniale.

Properties.—Externally powerfully irritant. The oxide and the vinum are chiefly diaphoretic, but the latter in large doses is also nauseant and emetic; the tartrated antimony is diaphoretic, expectorant, nauseant, emetic, a vascular and nervine depressant, and in very small doses relaxant, promoting the secretion of mucus from the bronchi, stomach, and intestines. Eliminant of urea, carbonic acid, and in a less degree of uric acid and blood pigment. Acts on the skin and epidermis somewhat in the same manner as arsenic.

Administration.—Chiefly as an atomized spray.

Doses.—As a diaphoretic and expectorant from 15 to 30 minims of the wine in 1 ounce of water, as a depressant 1 to 2 drachms, and as an emetic 2 to 4 drachms in the same quantity of liquid; of James's powder 2 to 4 grains suspended in mucilage, if used as a spray; of tartrated antimony $\frac{1}{16}$ to $\frac{1}{6}$ grain, and if as an emetic, from 1 to 2 grains.

Indications.—In the acute catarrh of children in very small doses; very effective in tonsillitis, in acute bronchitis, shortening and mitigating the attacks; also in chronic bronchitis, when the expectoration is tenacious and difficult to bring up.

Ringer states that it is invaluable in cases such as the following: in the wheezing and quasi-asthmatical breathing of children from 6 to 12 years old, which is apt to occur in winter and even in summer on exposure to cold. He directs, that one $\frac{1}{2}$ grain of tartar emetic should be dissolved in $\frac{1}{2}$ pint of water, and a teaspoonful given until nausea or vomiting is produced. In acute pneumonia;

under the influence of tartar emetic, the expectoration will soon lose its rusty character, the inflammation will be stopped, and the frequency of the pulse and number of respirations will be greatly reduced.

Mineral Astringents.

Alumen.

Solubility of sulphate of alumina and ammonia in 10 parts of water; 10 parts in 8 of boiling water. When heated, it parts with its water of crystallization, is reduced to one-half its weight, and acquires slightly escharotic properties, owing to its strong affinity for water.

Properties.—Astringent and styptic, acting both directly and remotely, coagulating the albumen with which it comes into contact in the same way as tannin; it also contracts the vessels, lessens the quantity of blood in them, diminishes secretion, and condenses the tissues.

Administration.—Used either in the form of powder or as an atomized spray. The solution should be strong, and should contain 10 to 15 grains per ounce. May be prescribed with tannic acid or medicaments containing it, but not with alkalies and their carbonates.

Indications.—The uses for this substance are much the same as for tannic acid, but its action is milder and hence less effective. In epistaxis the powdered alum may be strongly inhaled by the nostrils, while in stomatitis and aphthæ it should be applied directly to the abraded or ulcerated surfaces. It is often very useful in chronic coughs, and in hoarseness, especially where the throat is in any degree implicated; it sometimes checks the vomiting, which is apt to occur in phthisis, and it is an excellent remedy in whooping-cough when the acute stage is over and in the absence of any complications. Good as an emetic in croup in $\frac{1}{2}$-drachm doses.

Powdered alum has been chiefly inhaled or insufflated in coryza, in aphthæ, tonsillitis, in œdema of the glottis, and in croup and diphtheria, the quantity employed being from 10 to 15 grains, sometimes mixed with sugar.

Dr. Cohen speaks very highly of the prompt effects of sprays of alum in superficial inflammation of the tonsils, pharynx, and larynx, and in aphonia.

Liquor Ferri Perchloridi Fortior.

Of the numerous preparations of iron all are tonic and many of them astringent; they act especially on the red blood-corpuscles, which are increased in number, a portion of the iron being incorporated with the hæmoglobin. The greater part of the protosalts of iron, when introduced into the stomach and upper part of the small intestines, become converted into persalts, and these again pass into the state of a sulphide of iron. It is this that causes the black colour of the fæces when iron is administered for any length of time. The sulphide is insoluble and inert, and no doubt some portion of the beneficial effects of the administration of many iron salts is due to the acid liberated by their decomposition. It is a curious fact, that when injected into the blood, much of the iron is eliminated by the mucous and serous membranes, including the mucous membrane of the air passages.

There are several preparations of iron which may be usefully employed in inhalation, but the three following in particular : the stronger solution above named, liquor ferri perchloridi, which is only one-fourth the strength, and the tincture. The strong solution is usually somewhat acid, but by cautious evaporation it may be rendered almost neutral.

Properties.—Most of the preparations are strongly astringent and powerfully hæmostatic. They constringe the blood-vessels, contract the tissues, and coagulate the albumen on abraded or ulcerated surfaces. They are also stimulant and somewhat irritant, preventing the union of wounds by the first intention, and Ringer considers on this account 'carbolic acid will probably supersede perchloride of iron; for this acid, properly employed, does not hinder the immediate closure of the wound.'

Administration.—Best applied in the form of an atomized liquid, mixed sometimes with glycerine or almond mixture. These additions have the advantage of rendering the taste pleasanter and of causing the remedy to adhere for a longer time to the mucous membranes with which it comes into contact.

Doses.—The strong neutral solution may be applied to diphtheritic patches by means of a brush without dilution; of the weaker solution and of the tincture 10 to 30 minims as sprays in $\frac{1}{2}$ to 1 ounce of water, either with or without glycerine.

Indications.—The more astringent preparations of iron are

particularly valuable in the treatment of many affections of the throat and lungs, both from their local and constitutional effects, they augmenting the number of red corpuscles in the blood.

Thus they are especially useful in some of those affections which are characterized by a deficiency of red blood, as anæmia, including the anæmia of phthisis; also in many of those affections in which there is profuse secretion, as in some forms of bronchitis and bronchiectasis, in the colliquative sweating and diarrhœa of phthisis; in hæmoptysis and other hæmorrhages from the organs of respiration.

It is, however, in disorders and diseases of the throat that the effects of the application of perchloride of iron are most marked, because these parts are so readily reached by the spray; these effects being not only local but in part constitutional, so much of the iron becoming swallowed; in simple relaxed sore throat; in cases in which there is much secretion of unhealthy mucus or pus; in ulcerations of the throat, in hæmorrhages of the same, in chronic aphonia, and in laryngeal phthisis. Then, again, it is specially serviceable in some specific diseases of the throat, in croup, diphtheria, and scarlet fever.

Ferri Sulphas.

This preparation is but seldom employed in inhalation, but is used chiefly as a tonic in the form of pills. It is powerfully astringent, and may be used as a spray containing 2 to 3 grains in $\frac{1}{2}$ ounce of water. The solution of the perchloride of iron is, however, preferable, as it is even more astringent and far less irritating.

Indications.—Internally administered it is often very serviceable in the chronic diarrhœa of phthisis. As a spray its employment is indicated in much the same cases as these for which the perchloride is used.

Plumbi Acetas.

Properties.—Desiccant, astringent, and sedative; lessens greatly mucous discharges, and arrests hæmorrhage in some cases. Topically applied, unirritating, calming, and forming on mucous and abraded surfaces an albuminate; in the bowels it becomes converted into a sulphide of lead and renders the motions slate-coloured.

Administration.—To be inhaled in the form of an atomized spray, the dose being from 3 to 8 grains two or three times a day in ½ to 1 ounce of water. The effects of the remedy should be watched; it is sometimes given with excess of acetic acid, which renders it less injurious. Sulphuric and tannic acids are incompatible, since they form with this salt insoluble compounds. Acetate of lead and some others of the lead salts have the advantage of being less irritating than most of the other astringent metallic salts. In place of the acetate, the liquor plumbi acetatis P. B. may be employed, in the proportion of about 10 minims to an ounce of water.

Indications.—In affections of the throat, accompanied by much discharge; in bronchitis with profuse secretion; to diminish expectoration in phthisis; in hæmoptysis and hæmorrhages from small vessels. Few remedies are more effective in tubercular disease of the intestines with accompanying diarrhœa, than the acetate, which should be given, however, in the form of pill and combined with opium. It may be taken for long periods in from 3 to 5 grain doses, without any ill effects being produced.

The insufflation of the powder of acetate of lead is chiefly indicated in chronic nasal catarrh with profuse secretion.

Zinci Sulphas.

Properties.—Desiccant, astringent, antispasmodic, tonic, emetic, irritant, and escharotic, forming an insoluble compound with albumen.

Administration.—In the form of an atomized spray, containing from 2 to 4 grains per ounce.

Indications.—It lessens secretion from mucous membranes and acts as a stimulant to ulcerated surfaces; it also acts as an antispasmodic in croup and whooping-cough; it is given as an emetic in croup and even bronchitis; it diminishes expectoration in phthisis, excess of secretion in chronic bronchitis, and it frequently checks the colliquative sweating of phthisis.

The powders of sulphate and oxide of zinc, more or less weakened with starch or any other inert powder, are used in nearly the same affections as powdered alum or tannin.

Caustic Astringents.

Argenti Nitras.

Properties.—Tonic, antispasmodic, astringent; contracting the vessels and constringing the tissues; uniting with the albumen of abraded and ulcerated surfaces and forming an albuminate of silver, the nitric acid being set free. When applied in a concentrated form it is irritant, and may excite inflammation and superficial destruction of the tissues.

Administration.—In the treatment of throat and lung affections nitrate of silver is sometimes employed in the solid form, or in concentrated solutions, by means of curved brushes, or as an atomized spray. When used in the form of a spray, the nitrate should be dissolved in as little water as possible, as a considerable portion of it will inevitably become swallowed and so produce effects which are not desired. The swallowing of the salts, however, may be to some extent avoided by timely expectoration, and it is to be remembered that even when it reaches the stomach it is speedily converted into an albuminate, in which condition it ceases to be irritating. The quantity of nitrate for each inhalation may vary from 2 to 3 grains, according to the nature of the case, and it should be dissolved in not more than 2 to 4 drachms of water. Ringer points out, that nitrous ether is one of the best solvents or vehicles for nitrate of silver, because it dissolves out of the tissues to which the nitrate is to be applied any fatty matters which they may contain, and hence the action is more uniform; but it must be remembered that the effect is stronger. Great care is required in the application of the spray, lest it should stain the skin of the lips and mouth. This may be avoided by causing the spray to pass through a small funnel-shaped cone, or by the use of a spray apparatus, the delivery tube of which is sufficiently long to allow of its entering the mouth. The powder of nitrate of silver when insufflated requires to be mixed with a considerable portion of extraneous matter, the quantities being regulated by the nature of the case.

Oxide of silver is strongly astringent, and does not possess the escharotic properties of the nitrate.

Indications.—Nitrate of silver is chiefly employed as a local remedy, especially in affections of the throat, larynx, and bronchi.

Useful in relaxed throat. In superficial inflammation of the throat, simple or erysipelatous, the application of a strong solution will sometimes stop the inflammation at once. By its stimulant power it will alter the condition of unhealthy ulcers, producing healthier action and better granulation. Spraying the throat with nitrate of silver often gives great relief in whooping-cough, lessening its frequency and violence, although it may excite spasm at first. By its astringent action, when topically applied, it will often arrest small bleedings, but it is apt to excite irritation and even inflammation. Nitrate of silver is often applied to the false membrane of diphtheria, but it is liable to do more harm than good by its irritant action.

Cupri Sulphas.

Properties.—In small doses tonic; astringent, especially to mucous membranes; styptic and escharotic; combines with albumen. Externally applied, it is milder in its action than nitrate of silver.

Administration.—May be employed in the form of a spray containing $\frac{1}{2}$ to 1 grain in 1 ounce of distilled water.

Indications.—It acts as a stimulant to indolent ulcers, as an escharotic to some morbid growths, and by its styptic action arrests bleeding from small vessels. Internally administered, it is often beneficial in the diarrhœa of phthisis, and it is believed to exert a specific action on the larynx. It is sometimes used in croup as a spray.

VEGETABLE ASTRINGENTS.

Acidum Tannicum.

Gall nuts of good quality contain from 30 to 40 per cent. of tannic acid or tannin, with a comparatively small percentage of gallic acid. It contains the elements of glucose and gallic acid, and is hence regarded as a glucoside. It strikes a bluish black colour with persalts of iron. In passing through the system, it becomes converted into gallic acid and glucose; is eliminated by the kidneys and is found in the urine as gallic and pyrogallic acids. The urine sometimes, after having been exposed to the air for a short time, becomes of a dark colour, owing not only to the presence of the acids named, but to a third substance, the nature of which is not well understood.

70 parts of this acid are soluble in 8 of water and in 8 of rectified spirit, while 1 part is dissolved by 3 of glycerine. It is soluble in olive oil.

Properties.—Astringent and styptic; it contracts the vessels and tissues, checks secretion, and when applied to ulcerated surfaces it precipitates albumen, as also gelatine, thus forming a protective coating.

Administration.—It may be used either in powder or as a spray when dissolved in a mixture of water and glycerine. The solution should be more or less concentrated, according to the nature and urgency of the case. When the spray is freely employed much of the acid will of course be swallowed.

Doses.—Of the acid, 3 to 15 grains; of the glyceride of tannin, ½ to 1 drachm, diluted to ½ ounce. This glyceride contains 1 part of acid to 4 of glycerine.

Indications.—It is especially useful in affections of the nose, pharynx, throat, and even larynx in their subacute and chronic stages, and when accompanied by infiltration, thickening, granulation, or ulceration, or with much mucous or purulent discharge; in the ulcerations and excoriations which are so apt to occur in the nasal passages after measles and scarlet fever; in aphthæ; in subacute and chronic catarrhs; in œdema glottidis and in bronchitis, particularly in children.

Many of the above conditions and affections of the throat often give rise to very troublesome cough, which the application of the tannin frequently greatly relieves. It is also often very serviceable in the cough of bronchitis and phthisis, where this is aggravated by a congestive and irritable condition of the pharynx, uvula, and throat; in the tickling and explosive cough arising from an elongated uvula, and in some cases of simple whooping-cough.

Lastly, it is very effective when the application is sufficiently strong, in hæmorrhages from the nose and throat, but in those of the larynx, bronchi, and lungs it is less efficacious, and it is better to have recourse to other more powerful remedies.

Tannic acid has also been used with benefit in croup and diphtheria, but here again more effectual means may be adopted.

In place of pure tannin, recourse may be had to some of the

VEGETABLE ASTRINGENTS. 173

many tannic and gallic acid containing remedies, but the tannin itself is for throat affections usually the most effective.

Tannin in powder is used in much the same cases as alum, either alone or with alum. It is said to be capable of astringing and finally destroying nasal polypi, and should it fail to do this, it would yet be of value in checking the discharge, in correcting any accompanying fœtor, and probably also in diminishing the size of the polypi. Used also in bleeding from the nose.

Acidum Gallicum.

Gallic acid is naturally contained in the powder of galls in the proportion of about 5 per cent., but this proportion is greatly increased at the expense of the tannic acid by the treatment which the powder is subsequently made to undergo. Gallic acid may also be produced by the action of either acids or alkalies on tannin, grape sugar being formed. Nearly soluble in cold, very soluble in boiling water, rectified spirit, ether, and 1 in 20 parts of cold glycerine, but nearly 1 in 5 when the glycerine is heated. It gives a bluish black colour with persalts of iron.

Properties.—Astringent, but less so than tannin, and believed to be more effective as a remote astringent. It does not precipitate either albumen or gelatine. As the tannic acid becomes converted in its passage through the system into gallic acid and glucose, the quantity of the former acid which is effective is diminished.

Administration.—Chiefly in the form of an atomized spray, the glyceride being a convenient preparation for the purpose.

Dose.—Of the powder 3 to 15 grains, of the glyceride $\frac{1}{2}$ to 1 drachm diluted to $\frac{1}{2}$ ounce. This glyceride is the same strength as that of tannic acid.

Indications.—As a remote astringent in nearly the same cases as tannic acid; in diarrhœa; in the sweating of phthisis; in hæmorrhage.

Eucalyptus rostrata.

A red gummy exudation from the bark of Eucalyptus rostrata, about 90 per cent. being soluble in cold water.

Properties.—Astringent, adhering with tenacity to the mucous surfaces with which it is brought into contact, its superiority in some cases over catechu and kino being, it is believed, due to this peculiarity.

Administration.—May be inhaled as a spray.

Either the extractum liquidum may be used or the tinctura. The liquid extract consists of 1 part of gum in 2 of water, the insoluble matter having been removed by straining; the tincture of 1 part of gum in 4 of rectified spirits of wine; this contains all the constituents of the gum, and does not become turbid when mixed with water.

Doses.—Of the extract 10 to 20 minims, of the tincture 20 to 40 minims in ⅓ ounce to 1 ounce of water. Should be slowly sprayed and inhaled.

Indications.—Useful in relaxed throat; in epistaxis, a spray of the liquid extract may be injected or inhaled through the nostrils; in aphonia.

Hamamelis Virginica.

The leaves and bark of Witch Hazel, or Winter Bloom, are very astringent, and contain a peculiar acrid essential oil. The astringency is not, however, due to either tannic or gallic acid; it forms the basis of Pond's extract and hazeline.

Properties.—Powerfully astringent, but to what the astringency is due is not known; hæmostatic.

Administration.—In the form of an atomized spray.

Dose.—From 2 to 5 minims of the tincture of the bark in ½ to 1 ounce of water, or from 30 to 60 minims of the liquid extract as prepared by Messrs. Allen and Hanbury.

Indications.—In passive hæmorrhages, in epistaxis and hæmoptysis.

Claviceps Purpurea.

Ergot is a fungus which is apt to infest more particularly the paleæ of rye, and it owes its properties to the presence of ergotin, which is soluble in water. The preparations mostly employed are the infusion, the liquid extract, the tincture, and ergotin itself, which is now prepared by dialysis and is almost pure.

Properties.—Produces contraction of the unstriped muscular fibre of the arteries and veins, and especially of the uterus, and hence it is much employed to arrest hæmorrhage, for which purpose it is one of the most effective of remedies.

Administration.—As an atomized spray.

Doses.—Of the liquid extract, 10 to 40 minims; of the infusion, 1 to 2 ounces; of Bonjean's ergotin, 1 to 3 grains.

Indications.—Chiefly in cases of epistaxis and hæmoptysis. When the infusion is employed much of it doubtless reaches the stomach. When the bleeding is profuse the larger doses should be resorted to, and frequently repeated, every hour if necessary.

NON-OXIDIZING ANTISEPTICS.

Acidum Carbolicum.

Phenol, or carbolic acid, C_6H_6O, is obtained from coal tar and has the following properties:—When pure it melts at 107·9° F. to a colourless fluid, which is slightly heavier than water. It boils at 359·6° F., and distils without decomposition. The crystals readily absorb moisture from the air, a hydrate, C_6H_6O,H_2O, containing 16·07 per cent. of water, and melting at 63° F., being formed. One part of the absolute acid requires 10·7 parts by weight of water for complete solution. The solutions do not redden litmus, but the acid forms definite salts with strong bases. It is miscible in all proportions with alcohol, glacial acetic acid, and glycerine; also when anhydrous, with ether, benzene, carbon disulphide, and chloroform. Smell and taste are much less marked in the pure than in the commercial acid.

When carbolic acid melts at 95° F. it is said to contain cresol, which enters very largely into inferior kinds of liquid carbolic acid. The cresol, however, possesses much the same properties as carbolic acid, but is more irritating.

Nothing is said in books about the volatility of phenol at ordinary temperatures, yet from the powerful odour emitted one would be led to infer that it must be volatile to some small extent. With a view to test this point the following experiments were made: 0·5 grm. exposed uncovered to the air, at a temperature of about 48·0° F., lost 0·0035 grm. in two hours and 0·0071 grm. in four hours; 0·5 grm., placed under a bell glass and in air dried by sulphuric acid, lost 0·0025 grm. in two hours and 0·0054 grm. in four hours. These experiments show that the volatility of phenol at ordinary temperatures is exceedingly slight; it increases, however, as shown by some further experiments, which it is not necessary to detail in this place, in proportion as the temperature is raised.

Properties.—Antiseptic, antifermentive, antiparasitic; coagulates albumen; whitens and acts as a caustic to the skin.

Fermentive substances or ferments may be divided into the chemical and the organized; the former, which include pepsin and ptyalin, are only destroyed by this acid when in a concentrated state and after prolonged contact, while the action of the latter, as of yeast and many bacteria, is arrested by much weaker solutions. Very precise statements as to the strength of the solutions capable of destroying each particular ferment have been made, but to these considerable latitude must be allowed. Vibriones and bacteria are said to be destroyed by solutions containing only 0·1 per cent. or 1 in 1000 parts. Yeast requires a 0·2 per cent. solution, and must be in contact with it for 24 hours. Parkes states, that while carbolic acid rapidly arrests the growth of fungi it will not completely destroy them, and he cites experiments in proof of this statement. The activity of cow-pox lymph is said to be destroyed by a 2 per cent. solution of the acid, and putrid lymph needs 5 or more per cent., but ½ per cent. is sufficient to prevent the decomposition of non-infective material, such as prs. Fresh meat, blood, and urine require as much as 0·1 to 0·5 per cent. to prevent decomposition (Lemaire and Plugge). But albumen and meat, according to Hoppe-Seyler, require a 2 per cent. solution. Putrefaction of meat and other animal substances may be retarded by suspension in an atmosphere containing carbolic acid.

It being a non-oxidizing agent, it does not destroy sulphuretted hydrogen or other malodorous substances, but by its antiputrescent property simply prevents their formation.

Carbolic acid quickly passes into the urine, accumulation in the system being thereby precluded, unless the dose administered be large and frequently repeated. The urine may be sometimes of the natural colour at first, but becomes dark when kept; at others it may be dark, and of a brownish or blackish green hue when first voided.

The presence of carbolic acid in the urine may be detected not only by the colour but by tests; the odour of the acid is said to be rendered perceptible by the addition of sulphuric acid, while with chloride of iron the urine becomes of a blue colour.

Administration.—Carbolic acid may be inhaled either as vapour or as an atomized spray or 'nebula.'

The vapour may be so evolved by natural evaporation at an ordinary temperature as to be almost gaseous or aëriform, and it may be equally diffused throughout the air of a room. See the account of the Inhalation Chamber in Chapter IV.

The vapour may also be inhaled in several other ways :—

In an oral or oro-nasal inhaler, at the temperature of the air; this, as has been already explained, is of little or no utility whatever.

It may be effectively inhaled by means of the globe inhalers, combined especially with the use of the accompanying water bath.

It may be inhaled in a still more concentrated form by means of an apparatus devised by me for the vaporization of medicaments in a condensed form (fig. 19, page 73).

Or, the vapour may be inspired as it is given off by hot water, for which purpose the eclectic inhaler or the apparatus just referred to, or the chamber inhaler No. 2, may be employed.

The air of the whole room occupied by the patient may be thoroughly charged with the vapour, by means of either the chamber inhaler No. 1 or No. 2.

Lastly, the acid may be inhaled as fumes. Messrs. Corbyn, Stacey, and Co. prepare antiseptic cigarettes composed of eucalyptus leaves and oil with carbolic acid or other antiseptics.

For the inhalation of the spray, either a good hand air spray or Siegle's steam spray apparatus may be used. With the air spray, the atomized liquid will be cold and with the steam spray more or less warm, according to the extent of the admixture of outer air and the distance from the spray producer. This fact, therefore, should not be lost sight of in using these atomizers. Preference should always be given to an apparatus which the patient can himself employ.

Again, it must not be forgotten, that whatever apparatus be employed, and whether the carbolic acid be inhaled in the form of vapour or spray, there will always be a considerable loss of the acid or other medicament used, external to the mouth, and also that much of that which is inhaled is usually swallowed and passes into the stomach.

Doses.—With respect to the dose of carbolic acid, this of course must vary with the method of inhalation adopted and the nature of the malady.

If chamber inhalations be resorted to, the quantity of phenol must be very large in order to sufficiently charge the air; if the acid be inhaled with the vapour of hot water, the quantity must also be considerable, because so much is lost in the uninspired vapour and a farther quantity will remain in the inhaler. Again, if an oral or oro-nasal inhaler be used and the air be not artificially warmed, then it must be remembered that by far the greater part of the acid will be found in the inhaler on the completion of the inhalation; if, however, the air which passes through the inhaler be artificially warmed, then a much larger quantity of acid will become vaporized. When sprays are used there is no vaporization of the acid, but the solution is simply broken up into minute particles or atoms.

Doses of the hydrated acid for an oro-nasal inhaler, if used with the water bath, 15 to 30 minims; for the spray apparatus, 5 to 10 minims; for that for concentrated vapours, 10 minims; and for inhalation in the vapour of a pint of hot water maintained at a temperature of about 140° to 150° F., 20 to 40 minims.

It cannot be doubted, that up to the publication of my papers on the comparative inefficiency of inhalation by oro-nasal and some other forms of apparatus the quantities of carbolic acid and other similar disinfectants used were much too small, and that far larger quantities are required, and indeed are now not unfrequently prescribed.

Vapours and solutions of considerable strength may doubtless be applied usually with impunity to the throat, and there can be no question but that the mucous membrane of the air passages in chronic affections will bear the application of much stronger vapours and atomized sprays than have hitherto been generally employed.

When, therefore, the medical man prescribes inhalation, he should always have clear ideas in his mind as to the quantity of the acid likely to reach the affected parts, and this, as we have seen, will vary according to the manner in which the inhalation is practised, the apparatus used, and the temperature. As a rule he need not entertain any fear as to the vapours or the atomized liquids being too strong, at all events in chronic affections, though of course reasonable limits must be observed in the quantities employed.

Dr. Sansom has particularly recommended the sulpho-carbolate of soda, a soluble, stable, and odourless salt, for internal administration, on the ground that a larger quantity of carbolic acid may be introduced into the system and with less liability to produce gastric irritation. He states that it is decomposed in the blood and that it is chiefly eliminated by the lungs. Dose, 10 to 15 grains dissolved in water.

A valuable combination is camphorated carbolic acid, made by rubbing up 12 parts of the acid with 4 of camphor and 1 of water; it is sometimes applied to the false membranes of diphtheria. Carbolate of ammonia is another useful combination.

Liquor carbonis detergens is an alcoholic solution of coal tar, and of course contains carbolic acid as one of its chief constituents.

In cases in which it is desired, that the effects on the mucous membranes should be more persistent, the carbolic acid may be combined with glycerine, and where a strong astringent effect is desired, sprays may be employed of the mixed glycerides of carbolic and tannic acids. Dose of these glycerides, 10 to 20 minims of the first and 20 to 40 minims of the last in $\frac{1}{2}$ to 1 ounce of water.

Not unfrequently a variety of other remedies are inhaled at the same time with carbolic acid, as alcohol, chloroform, ether, iodine, opium, stramonium, and hyoscyamus. Most of these, when employed in oro-nasal inhalers, are, as already pointed out, absolutely useless on account of their non-volatility.

Indications.—The inhalation of carbolic acid is indicated in the following conditions :—

In cases in which the throat is relaxed, pale, and anæmic.

When it is coated with unhealthy and sometimes even offensive secretion.

Where there is, in addition, more or less superficial ulceration.

Again, where the fermentive or putrefactive changes are dependent on the presence of fungi or bacilli.

In diseases distinguished by the presence of certain special organisms, as diphtheria and phthisis.

In some of the above divisions, the remedy acts mainly by its stimulation, in others by its antiputrescent and antifermentive powers, while, lastly, in other cases its effect depends upon its antiseptic and germicide properties.

It has been beneficially employed in the following affections: in relaxed throat and in anæmia; in chronic tonsillitis with unhealthy secretion or suppuration; in mycosis of the air passages; in chronic bronchitis with abundant expectoration;

bronchiectasis with fœtid expectoration ; in broncho-pneumonia ; in gangrene of the lung with offensive expectoration; in phthisis without or with cavities; in whooping-cough, croup, and diphtheria. The sulpho-carbolate of soda has been highly recommended in the dyspepsia of phthisis accompanied by flatulency.

The renal secretion should always be watched during the inhalation of carbolic acid, and as soon as this becomes decidedly discoloured, the treatment should be modified or suspended until the darkening becomes lessened. A watch should likewise be kept on the general symptoms, including particularly the pulse and temperature.

The symptoms of the accumulation of the acid in the system are giddiness, deafness, noises in the ears, formication, great feeling of weakness, perspiration, lowering of temperature, and lessening of the frequency of the pulse, pain in the throat and air passages, nausea, and even vomiting. As small a quantity as 80 grains has proved fatal, but much larger amounts are usually needed.

Resorcin.

Derived from either carbolic acid or benzol. It is crystalline, melts at 110° F., is then easily volatilized, and is very soluble in water.

Properties.—Powerfully antiseptic, coagulates albumen, and acts as a caustic on the skin. A 1 per cent. solution prevents the putrefaction of the blood and urine.

Administration.—May be inhaled as an atomized spray.

Doses.—5 to 15 grains in $\frac{1}{2}$ to 1 ounce of water.

Indications.—In whooping-cough and diphtheria. 'It is an effective remedy in diphtheritic affections and produces no injurious consequences' (Messrs. Martindale and Westcott). M. Moncorvo, of Rio di Janeiro, has reported in favourable terms on the employment of resorcin in whooping-cough, and he considers that the benefit derived in this disease is due to its germicide properties.

Hydroquinone, which is isomeric with resorcin, is obtained chiefly from coal tar, is soluble in 20 parts of water, and is even a stronger antiseptic than resorcin.

Creasotum.

Creasote, as is well known, is the product of the distillation of wood tar ; it is a very variable admixture of several phenol-like bodies : guaiacol, $C_7H_8O_2$, boiling at 392° F. ; creasol, $C_8H_{10}O_2$, boiling at 422° F.; phlorol, $C_8H_{10}O$; and methylcreasol, $C_9H_{12}O_2$, being the chief constituents. The composition of creasote varies with its origin. Carbolic and cresylic acids occur in pine wood creasote. Guaiacol predominates in Rhenish creasote. Morson's English creasote consists chiefly of creasol, and boils at about 422° F. Creasote is sparingly soluble in water; it is insoluble in glycerine, whereas carbolic acid is very soluble therein; it possesses a smoky taste and smell, and is a powerful antiseptic; it preserves animal matters without causing disintegration, as phenol is apt to do, and is less powerfully caustic than the latter. Creasote is miscible in all proportions with alcohol, ether, benzine, chloroform, acetic acid, &c., like phenol, with which it is often adulterated. According to Messrs. Martindale and Westcott, there are two kinds of creasote, the hydrated and anhydrous.

½ gramme exposed to the air under the same circumstances as the phenol, at 48° F., lost 0·0090 gramme in 2, and 0·0113 in 4 hours; another 0·50 gramme placed under a bell glass with sulphuric acid lost 0·0084 gramme in 2 hours and 0·0106 in 4 hours. It thus appears, that creasote is somewhat more volatile than phenol, but still is of very slight volatility.

The chemistry, therefore, of creasote is very different in some particulars from that of carbolic acid.

Properties.—Stimulant, astringent, styptic, and antiseptic; it coagulates albumen and whitens the skin. In these respects it resembles very closely carbolic acid.

It passes through the body unchanged, and may be detected by its odour in the blood and intestines; it also makes its way readily into the urine, which sometimes becomes darkened, owing, it is stated, to the presence of carbolic acid.

Administration.—When it is desired that the creasote should reach the air passages and lungs, the most effectual methods are: by oro-nasal respirators when the vaporization is aided by an increase of temperature; by the spray producer, by the apparatus for the concentrated vapours, or in the steam of hot water.

For the inhalation of Vapor Creasoti, the Pharmacopœia directs that '12 minims of creasote be mixed with 8 ounces of boiling water in an apparatus so arranged that the air may be made to pass through the solution for inhalation.' As I have shown, this proceeding is of very little value.

Doses.—With respect to the quantities of creasote that should be used, these should be less than in the case of carbolic acid, although usually considerably more than is generally employed, particularly when the vapour is inhaled from oro-nasal respirators. For inhalation from a medicated respirator, if artificial heat be employed, from 5 to 10 minims; with the vapour of hot water by means of an apparatus by which the temperature can be *maintained* at a given point, from 10 to 20 minims; and for inhalation by means of a spray, a quantity of the solution should be inhaled containing 3 to 6 minims of creasote.

Indications.—The medicinal properties of creasote approximating so closely to those of carbolic acid, the indications for its employment are almost the same, though there seem good and practical reasons for giving the preference to creasote in the following affections: in most winter coughs, in whooping-cough, in chronic bronchitis with profuse expectoration, in bronchiectasis with abundant and offensive secretion, and in lung gangrene.

Cresoline.

This is obtained from cresol, which is itself obtained from creasote; it is closely allied in its properties to carbolic acid.

Benzinum.

Benzine, or benzol, is an oily, carbonaceous, inflammable substance, C_6H_6, occurring in the light oil obtained in the distillation of coal tar. It is a colourless, strongly refracting oil; specific gravity 0·85 at 60° F., and boils at 176° F. It possesses a strong and peculiar odour, is insoluble in water, but readily soluble in alcohol and ether, and it dissolves sulphur, phosphorus, bromine, iodine, and a variety of other substances.

This oil must not be confounded with the benzine or benzoline obtained by the fractional distillation of crude petroleum, which boils at 170° and which is a paraffin or a mixture of paraffins having the formula $C_nH_{2n\times2}$.

Properties. — Antifermentive, antiparasitic, antiputrescent, anæsthetic, and antispasmodic.

Administration.—Inhaled in the form of vapour by means of a medicated respirator; by the apparatus for concentrated vapours, or sprinkled on a piece of sponge or cotton-wool and inhaled by the nostrils. For the respirator and for the cotton-wool, the quantity must be considerable, as its volatility is not very great. (See Table II. page 23.)

Doses.—For the respirator 40 to 60 drops, and for the sponge or cotton-wool, 1 to 1½ drachm.

Indications.—In decomposed and putrid conditions of the throat and bronchi; in putrid bronchitis, bronchiectasis; in chronic phthisis. 'On account of its anæsthetic and antispasmodic effect it has been recommended in whooping-cough' (Oertel).

Pix Liquida.

There are two kinds of tar, one the bituminous liquid or pitch obtained by the destructive distillation of the wood of Pinus sylvestris and other species of pine, and the other coal tar.

The following analyses, gathered from Watts, represent their composition:—

Wood Tar.—The chief liquid constituents of wood tar are methylic acetate, acetone, hydrocarbons, toluene, xylene, and cumene, methol (a mixture of volatile oils boiling between 100° and 265° F.), eupione, creasote, and a number of indefinite oxidized compounds, including picamar and capnomor. Amongst the solid portions are resinous matters more or less resembling colophony; also paraffin, naphthalene, anthracene, chrysene, retene, pyroxanthin, pittacal, and cedriret.

Wood tar possesses powerful antiseptic properties, due mainly to the creasote which it contains.

Coal Tar.—The more volatile portion of coal tar, called *light oil* or *coal naphtha*, consists mainly of benzine and its homologues, i.e. toluene and xylene, together with a number of bases containing carbon, hydrogen, and nitrogen, such as pyridin, &c.

The less volatile oil, or *dead oil*, contains phenol and cresol; also aniline, picoline, chinoline, and other volatile bases, and a number of solid hydrocarbons, including naphthalene, anthracene, chrysene, &c. Tars from peat and lignite are intermediate in composition between wood tar and coal tar. Coal tar has great value as the source of aniline colours, and of phenol, picric acid, &c. Heavy coal oil or dead oil is remarkable for its antiseptic properties, and is commonly used for the preservation of timber, &c.

It appears, therefore, from the above analyses, that wood tar and coal tar differ considerably in their composition, and doubtless also to some extent in their medicinal properties. The first kind owes its activity mainly to the presence of creasote, and the second to carbolic acid or phenol, which as a germicide is less powerful than creasote.

Liquor carbonis detergens is an alcoholic solution containing all the constituents of the tar, including benzine, naphthol, toluene, and phenol.

Of the many constituents of liquid pitch some are soluble in water and others only in rectified spirit. Tar water is made by stirring a pint of wood tar with ½ gallon of water for 15 minutes and decanting the liquid. Eau de goudron of the French is prepared by digesting 1 part of wood tar with 30 of water for 8 or 10 days.

Pitch is the solid resinous matter obtained by the distillation of the liquid pitch; there are three kinds, Archangel and Swedish pitch, and that obtained from coal tar. Pitch is chiefly used in the form of pills, which are said to have the effect of increasing the weight of the body.

There is still another form of what may be termed liquid pitch; oil of cade, oleum cadinum, obtained by the dry distillation of the wood of Juniperus oxycedrus. This is much less disagreeable in smell and taste than ordinary liquid wood tar. There are some other species of tar which are imported from Germany, but which are chiefly employed in skin affections, as oleum betulae pyroligneum, or birch tar; oleum fagi pyroligneum, or beech tar; and oleum rusci pyroligneum, or butcher's broom tar.

Properties.—Diuretic, stimulant, astringent, and antiseptic, exerting a special influence upon the mucous membrane of the bronchi.

Administration.—May be inhaled in the form of fumes or as an aqueous or feebly spirituous atomized spray. If the fumes of hot tar be inhaled, then care must be taken to neutralize with carbonate of soda any acetic acid present in the tar before using it. The tar, after neutralization, may be employed either diluted or undiluted. It is always best to dilute it with water, as in this state it may be used for a longer period, and when undiluted it quickly dries up and is apt to burn. It should be heated by means of a spirit lamp.

Tar water is too weak to be employed in most cases as a spray.

Most of the apparatus recommended for generating the fumes of tar are cumbrous and unsuitable. In some cases it is advised to place the tar in an iron saucepan on the fire, a somewhat dangerous proceeding in the case of so inflammable a substance. For the direct inhalation of the vapours the following simple arrangement will suffice :—In a porcelain evaporating dish about 4 inches in diameter, and holding about 10 ounces of liquid, from 1 to 2 ounces of tar should be placed with 6 ounces of water. This should be put upon a thin sheet of copper, perforated in the centre to receive the dish, both being made to rest on a stand with a lamp beneath the dish; over this latter a funnel of glass, tin, or other suitable material, even of stout paper, is placed for the concentration and conveyance of the vapours. If hot water be added to the tar in place of cold, the fumes will be developed in a few minutes after lighting the lamp.

For charging the air of a room, a somewhat similar apparatus may be used on a larger scale; the porcelain dish must have a diameter of 8 or 9 inches, and the funnel of course is dispensed with. If it be desired to slacken the rate of the evaporation of the water, the dish may be placed on one of the open copper water baths so much used in every chemical laboratory.

Doses.—For the vapour 1 to 2 ounces of tar in water. The inhalations should be repeated 2 or 3 times a day.

Indications.—It is indicated in chronic affections of the mucous membranes of the organs of respiration, with profuse secretion, either with or without decomposition or offensive odours; in chronic bronchitis, and bronchiectasis. It has likewise proved very beneficial in laryngeal and in lung phthisis. Both Dr. Ringer and Dr. Murrell have found tar administered in the form of 2-grain pills every 3 or 4 hours to be extremely effective in winter cough, and in diminishing the proneness to take fresh colds. If these pills are so serviceable, there is no reason why the fumes should not be equally beneficial. If the remedy be employed too frequently, it may prove over stimulating and occasion headache, and lessen the secretion too suddenly.

Salicinum, Acidum Salicylicum, Sodæ Salicylas.

Salicin is a neutral crystalline glucoside, obtained from the bark of *Salex caprea* and other species of Willow, from the bark of the Poplar, and from the flower buds of Meadow Sweet. Treated with strong sulphuric acid a bright red colour is developed. 1 part requires 20 of water for its solution.

Salicylic acid is obtained in two ways : artificially, by passing carbonic acid through a strong solution of carbolic acid and caustic soda at a high temperature, the salicylate of soda, thus formed, being decomposed by the addition of a strong acid; or naturally, from the decomposition of salicylate of methyl, present in oil of Winter Green, *Gaultheria procumbens*; this is treated with caustic soda, distilled, and the distillate decomposed with hydrochloric acid. The natural acid, which is considered the best, is prepared from salicin or from the oil of winter green.

Very insoluble in cold water, but readily soluble in 9 parts of boiling water and in rectified spirit; but little soluble in olive oil or glycerine. Its solubility is greatly increased by the addition of ace·ate of potash or borate of soda. This acid gives rise to a violet colour with persalts of iron, and this is the test whereby it is readily distinguished.

It quickly makes its way into the urine, as salicylol, salicyl hydride or oil of spiræa ulmaria, when it gives with persalts of iron a purple red colour.

Salicylate of Soda.—An alkaline salt, soluble in equal parts of water and in 4 parts of rectified spirit.

Properties.—Salicin is very similar in its action to sulphate of quinine, but it is a less powerful tonic; is antipyretic, antiperiodic, and antiseptic; less effectual than carbolic acid as an antiseptic. In large doses, frequently repeated, may produce dulness of hearing and even temporary deafness.

Salicylic Acid.—Similar to those of salicin, but is said to be more powerful as an antipyretic and three times as effective as carbolic acid in the prevention of fermentation. It is less tonic, however, than salicin, and sometimes even acts as a depressant. Both as a tonic and antipyretic, inferior to quinia.

Salicylate of Soda.—Somewhat less powerful than the acid, especially as an antiseptic.

Administration.—Like so many of the other medicaments employed in lung inhalation, being non-volatile, salicin and its preparations cannot be used with an oro-nasal inhaler, but must be in the form of an atomized solution. In some cases, salicin and salicylate of soda are to be preferred to the acid, on account of their greater solubility.

Doses.—Salicin and salicylic acid, 5 to 10 grains, and salicylate of soda, 10 to 20 grains, in 1 oz. of water as sprays.

Indications.—Employed principally in affections of the throat and lungs, mainly on account of their antiseptic and antipyretic properties, and may be used in many of the same cases as carbolic acid and creasote. By their antiseptic property they will often correct fœtor of the breath and of the expectoration.

Berthold has employed the acid in stomatitis or thrush; he dissolved one part in a little alcohol, making up with water to 250 parts.

Dr. Hunt has found salicylate of soda very effective in acute tonsillitis, between which affection and rheumatism a connection may often be traced.

The powder of salicylic acid, combined sometimes with bismuth, has been advantageously inhaled through the nostrils in cases of fœtid coryza and hay fever.

Acidum Boracicum.

This acid is soluble in the proportion of 1 part in 30 of cold and 3 of boiling water and in 4 parts of glycerine.

Properties.—Antifermentive, antiseptic, and non-irritant, but inferior to carbolic and salicylic acids in power; it is said, however, to be equal to benzoic acid. It does not destroy the action of pepsin, ptyalin, or pancreatin, and thus it does not prevent the conversion of starch into glucose.

Administration.—Must be inhaled in the form of a spray, from 10 to 15 grains being employed for each application, dissolved in about 2 drachms of glycerine and 6 of water.

Indications.—Has been specially recommended as a solvent for the false membrane of diphtheria by Dr. Simpson, of Highgate.

Sodæ Biboras.

Found native in several parts of the world; made in Tuscany by neutralizing the boracic acid which occurs in the lagoons, with carbonate of soda. Soluble in 22 parts of cold and 2 parts of hot water and in equal parts of glycerine; insoluble in alcohol and becomes solid when mixed with mucilage; glycerine also aids its solubility in water. It is slightly alkaline.

Properties.—Antacid, diuretic, refrigerant, antiseptic, antifermentive, checking the action of diastase, emulsin, myrosin, and also yeast. Sedative when applied to the mucous membranes.

Administration.—Not being vaporizable at ordinary temperatures, it must be inhaled in the form of an atomized liquid, some 10 to 30 grains being dissolved in about 1 ounce of a mixture of glycerine and water for each inhalation, which should be frequently repeated. Borax is frequently used mixed with sugar or honey in aphthæ and superficial ulcerations of the mouth, throat, and pharynx.

Indications.—In superficial excoriations and ulcerations, as in aphthæ of the throat; in thrush; in catarrh attended with hoarseness.

Chinolinum.

May be obtained as a derivative from quinine or cinchonine, or it may be prepared synthetically. It forms salts with tartaric and salicylic acids. The tartrate is a powerful antiseptic and germicide. A 1 per cent. solution completely destroys the coagulability of the blood, and weaker solutions render propagating fluids sterile.

Properties.—Antiseptic, germicide, and powerfully antipyretic.

Administration.—Both chinolinum and its tartrate may be employed as sprays. The dose of the former is from 2 to 4 minims for each spray dissolved in a little spirit and made up with water to $\frac{1}{2}$ ounce, or a 5 per cent. solution may be applied directly to the throat by means of a camel's-hair brush. The dose of the tartrate is 5 to 10 grains in $\frac{1}{2}$ ounce of aqua chloroformi.

Indications.—Checks the onset of dangerous symptoms in diphtheria, and is said in many cases to cause the membrane to be thrown off in 24 hours.

The above particulars are taken from the 'Extra Pharmacopœia' of Messrs. Martindale and Westcott.

Oxidizing Antiseptics.

Potassæ Permanganas.

1 part of this acid is soluble in 16 parts of pure distilled water.

Properties.—It is a powerful oxidizer, and attacks very energetically all organic substances. It consequently arrests fermentation and putrefaction, and is also an effective deodorizer. Its action when applied to moist organic surfaces, as to mucous membranes, is so energetic that its oxidizing power is speedily destroyed and the application has to be frequently renewed. If applied in a too concentrated condition, it may exert irritant and even caustic effects. It is very destructive to bacteria.

Administration.—In the form of an atomized spray. From 1 to 2 grains dissolved in $\frac{1}{2}$ to 1 ounce of water may be used at each application, the inhalation being frequently repeated in the more severe cases.

Indications.—It is in certain affections of the throat that this remedy is the most effective; those in which the secretions are putrescent, purulent, and offensive, or in which false membranes have been thrown out, especially in diphtheria. It is said to be useful in whooping-cough and in putrid bronchitis, but there are other remedies which are generally to be preferred in these cases.

STIMULANT, BALSAMIC, AND ANTISEPTIC REMEDIES.

Thymus Vulgaris.

Thymol is a camphoraceous body contained in oil of thyme, but obtained principally from the fruit of *Ptychotis Ajowan*. It has been synthetically prepared from cuminol, a constituent of cumin oil, and is a solid crystalline substance with a pungent taste and aromatic odour. It melts at 111·2° F., and does not easily resolidify unless touched by a solid body or a crystal of thymol. It boils at from 428° to 446° F.; it is slightly heavier than water; in the fused state rather lighter. It is but little soluble in water, requiring about 800 parts for solution. Thymol is a powerful antiseptic, its preservative power being ten times as great as carbolic acid, according to Buchholz, and four times as great, according to Willmott. 1 in 2000 parts of water, according to Buchholz, prevents the development of, and 1 in 200 parts arrests the growth of, bacteria. Lewin states that 1 in 1000 parts of water acts as a very powerful antiseptic, that it arrests fermentation in a solution of sugar and yeast better than either carbolic or salicylic acid, and that it also arrests the putrefaction of animal matters. It acts as a caustic on the lips and mucous membrane, but does not irritate the skin like carbolic acid. It is soluble in ether, strong acetic acid, and in its own weight of rectified spirit, but is very sparingly soluble in glycerine. 0·5 gramme exposed to the air at 48° F. gained 0·0003 gramme at the end of 2 and the weight had not altered after 4 hours. In an experiment in dry air under a bell glass, there was also no alteration of weight even at the end of 4 hours. These figures show that thymol is not volatile at a temperature of 48° F.

Properties.—Stimulant, powerfully antiseptic, antifermentive, and antiputrefactive.

Administration.—Although thymol possesses scarcely any volatility at ordinary temperatures of the air, yet when melted and in the fluid state it rapidly volatilizes. It may be inhaled in the vapour of hot water, or as a spray from its solution in cold water containing a small portion of rectified spirit. The vapour may be inhaled either by means of the globe inhaler aided by the water bath, or perhaps still more conveniently by the apparatus for the inhalation of concentrated vapours. For diffusing the vapour of thymol in the air of the room occupied by an invalid, the chamber inhaler No. 2. may be employed. The heat used to melt and volatilize thymol should be no more than is absolutely necessary, or the vapours may be too rapidly diffused and so become irritating.

Doses.—For the globe inhaler 5 grains of thymol; for the apparatus for concentrated vapours 3 grains, and for the chamber inhaler 40 to 60 grains; for the vapour of hot water 5 to 10 grains and for the spray 2 grains of thymol.

Indications.—In dry nasal catarrh; in anæmia of the throat; in relaxed throat; in aphonia; in chronic bronchitis with offensive expectoration; in tubercular affections of the larynx, bronchi, and lungs unattended with inflammation or great irritation; and in diphtheria.

Oleum Terebinthinæ.

Oil of turpentine is obtained from the resinous substance which exudes from several species of pine or fir, notably Pinus sylvestris, P. palustris, P. tæda, and P. pinaster. It is imported chiefly from America and France. It has a specific gravity of 0·864, mixes readily with other oils, is soluble in alcohol and ether, and is an excellent solvent for many bodies, including resin, fats, sulphur, and phosphorus. Ordinary oil of turpentine contains resin, from which it can be separated by distillation; the proportion of resin increases if the oil be kept for a long time. It passes unchanged into the blood, and may be detected therein and in the cutaneous excretion; but by the time it has reached the bladder it has undergone some chemical changes, and the urine has acquired a fragrant odour, which has been compared to that of violets.

STIMULANT, BALSAMIC, AND ANTISEPTIC REMEDIES. 191

Properties.—Stimulant, astringent, diuretic; applied to the skin, counter-irritant, rubefacient, vesicant; in large doses purgative.

Administration.—May be employed in the form of vapour or in that of a spray. If as vapour, by the globe inhaler and water bath, or the apparatus for concentrated vapours (see Table, page 23). If employed as a spray, the oil must either be dissolved in a little rectified spirit, or it must be suspended in water by means of mucilage. Since the odour of turpentine is much objected to by many persons, recourse may be had to the milder and pleasanter oleum pini sylvestris or ol. juniperi. Oleum pini sylvestris, or fir-wood oil, is obtained by the distillation of the leaves of Pinus sylvestris. It has a specific gravity of 0·868 and an agreeable and refreshing odour, on which account it is much to be preferred to turpentine. May be employed in the same manner and in the same doses as turpentine.

Doses.—For the spray 10 to 20 minims in mucilage, for the globe inhaler and for the apparatus for concentrated vapours 30 to 40 minims, repeated as may be required.

Indications.—Mainly employed for its stimulant and astringent properties. It is specially indicated in chronic bronchial affections with relaxed mucous membranes and profuse secretion; in hæmorrhage of the lungs, whether it be in small quantity or profuse, and in that of phthisis. If profuse, the larger and even 1-drachm doses should be inhaled in the form of spray, and repeated in an hour or so, according to the case.

The inhalation of turpentine requires to be carried out with great care, in consequence of its irritating after-effects, which may set up acute inflammation of the bronchi and even lungs.

Terebena.

Terebene is obtained by the action of sulphuric acid on oil of turpentine and subsequent distillation. Unlike turpentine it possesses a very agreeable odour and is colourless, this last particular serving to distinguish it from less pure preparations of terebene.

Properties.—Stimulant, antiseptic, deodorant, and disinfectant.

Administration.—May be inhaled by means of the globe inhaler, or in the vapour of hot water, or as an atomized spray.

Doses.—For inhalation in hot water, 30 to 40 minims; and

for the spray, 5 to 15 minims suspended in 1 ounce of water containing mucilage.

Indications.—The same as those for oil of turpentine. If administered internally in 5-minim doses, it is said to destroy the infective properties of the sputa of phthisis when swallowed, and to lessen the risk of intestinal complication.

Eucalyptus globulus.

The leaves of the above, as well as of the other species of eucalyptus, are studded with glands which are filled with a volatile oil having a sp. gr. of about 0.900. It possesses a peculiarly grateful and penetrating odour, which is evolved from the leaves while on the trees, and still more freely when they have been gathered and carefully dried out of the sun. The oil is obtained by distillation, and contains a terpene, boiling at 150° to 151° F., and another terpene called eucalyptine, boiling at 172° to 175° F., together with cymene ($C_{10}H_{14}$), and a camphor-like body, $C_{10}H_{16}O$. The first-named terpene is present in small amount only (Faust and Homeyer). According to Cloez, eucalyptol is a mixture of 70 per cent. of eucalyptine and 30 per cent. of cymene.

Properties.—Stimulant, antispasmodic, febrifuge, and antiseptic; ozonizes the oxygen of the air. It is said to be three times as powerful as carbolic acid in preventing the development of bacteria. 'It lowers reflex excitability by acting on the cord and its prolongations, it reduces the temperature of the body somewhat in health, and has a very decided antipyretic influence on the septic fever produced artificially in dogs by the injection of putrilage into their veins' (Garrod). It lessens secretion from mucous membranes, improves its quality, and aids the expectoration.

Of the oil of eucalyptus received into the system, part escapes by the breath, while a small portion appears unchanged in the urine, and it seems probable that a further portion enters the bladder in an altered condition.

Administration.—Although comparatively volatile at ordinary temperatures, yet it is not sufficiently so for inhalation unless the volatility be aided by heat. The globe inhaler and water bath, the apparatus for concentrated vapours, or the chamber inhaler No. 2 should therefore be used (see Table 2, page 23). But the oil may also be inhaled in the form of spray, it being dissolved in a little spirit and then diluted with water. For the globe inhaler 30 to 40 drops, dissolved in spirit and water, will usually be sufficient for each inhalation; for the apparatus for concentrated vapours 20 to 30 drops; for the chamber inhaler 1 to 2 drachms; for the spray, 10 to 20 minims of the oil, and of the tincture of the leaves from 30 to 40 minims. The tincture is prepared by macerating 1 part of the fresh leaves in 2 of rectified spirit. A very pleasant way of imparting the odour of eucalyptol to the air of the room, is to place from 1 to 2 ounces of the coarse powder of the leaves in a shallow dish, changing the powder every other day.

Indications.—In aphonia; in relaxed conditions of the mucous membrane of the throat; in chronic catarrh; in chronic bronchitis with abundant secretion; in chronic phthisis.

Cubeba Officinalis.

The unripe and dried capsules of cubeb pepper contain a volatile oil, to which the odour and taste of cubebs are due, a crystalline principle, cubebine, similar to piperine, and resin derived from the oxidation of the oil. The preparations best suited for inhalation are the volatile oil or the tincture, which contains all the active constituents of the pepper.

Properties.—Diuretic; stimulant to mucous membranes, and expectorant.

Administration.—As an atomized spray, vapour, or as fumes.

Doses.—Of the oil, from 5 to 15 minims dissolved in spirit and made up to ½ or 1 ounce with water, or the oil may be suspended by means of mucilage; of the tincture, ½ to 1 drachm in ½ to 1 ounce of menstruum; as sprays; as vapour, 20 to 30 minims in a globe inhaler with water bath. Savar's cubeb cigarettes consist of cubebs, stramonium, and cannabis.

Indications.—Very useful in subacute and chronic catarrhs, promoting expectoration and so lessening cough. In catarrhs associated with emphysema; in chronic bronchitis; said to be very serviceable in croup, the false membrane quickly disappearing under its use.

Camphora officinarum.

A concrete, volatile oil slightly soluble in water, forming camphor water, but freely in spirit, spirit of camphor; in acetic acid; in volatile and fixed oils, but not in alkalies.

Properties.—Diaphoretic, stimulant, afterwards sedative, antispasmodic, antiseptic; lowers the pulse and temperature in septicæmia.

Administration.—In powder, in vapour, or as an atomized spray.

Doses.—Of the spirit, 10 to 30 minims as sprays in almond emulsion, milk, or mucilage; of the compound tincture, or paregoric elixir, which contains opium and benzoic acid as well as camphor, from 15 to 30 minims.

For the inhalation of the vapour of camphor, an oro-nasal inhaler may be employed charged with the spirit, or the camphor may be inhaled in the vapour of hot water. For vapour inhalation the compound tincture should not be used, as the opium in this is non-volatile, and it is best suited for use as a spray. Of the powder of camphor, 1 to 4 grains insufflated.

A very simple way of inhaling camphor was suggested by M. Raspail: the coarsely powdered camphor should be placed in a large quill or piece of cane, loosely closed at each end with a little cotton-wool; one end of the tube should then be introduced into the mouth or nose and the air strongly inspired.

Indications.—The inhalation by the nostrils of the powder of camphor, or of a strong camphorated spray, will sometimes stop colds in the head, if employed at the outset; or if it does not arrest them, it will frequently give much relief. In aphonia, in hoarseness, in spasmodic affections of the organs of respiration, in whooping-cough, and in the coughs of bronchitis and phthisis. 'Camphor inhalations are sometimes useful in that troublesome chronic complaint characterized by seizures of incessant sneezing and profuse watery running at the eyes and nose, the patient remaining well in the intervals. The attacks may occur daily, beginning early in the morning, and may last for a few minutes only or may persist for several hours; and they may occur at any hour of the day, recurring several times daily. Sometimes several days intervene between the attacks, which may last 24 hours or even longer. They are generally accompanied by severe frontal headache, and in some instances an itching of a point inside the nose, denotes the imminence of the attack. This affection lasts for years' (Ringer). It evidently closely resembles hay fever.

The powder of camphor is sometimes inhaled as a snuff, mixed with tannin, in nasal catarrh or coryza. Monobromate of camphor has been found useful as an antispasmodic in whooping-cough and in asthma, in doses of 3 to 6 grains. If employed as a spray, it may be dissolved in a little spirit made up with water to ½ or 1 ounce.

Acidum Benzoicum, Sodæ Benzoas, Ammoniæ Benzoas.

Benzoic acid is obtained by sublimation from the balsamic resin, which exudes from incisions in the bark of *Styrax benzoini*. 1 part of the acid is soluble in 220 of cold and 12 of boiling water, and in 4 parts of rectified spirit; it is also soluble in the caustic alkalies, and borax aids its solubility, 1 part of borax and 1 of the acid being, according to Squire, soluble in 100 parts of water. The benzoates of soda and ammonia, unlike benzoic acid, are freely soluble in water, and are on this account in some cases to be preferred.

Properties.—Stimulant and expectorant, lessening secretion from the mucous membranes, particularly of the bronchi and bladder, and retarding somewhat fermentation and putrefaction. The benzoates of soda and ammonia are less stimulant and act also as diuretics.

In passing through the system, benzoic becomes converted into hippuric acid, in which form it is excreted by the kidneys and appears in the urine.

Administration.—Employed in the form of the free acid, or more frequently as benzoates of soda or ammonia, or the balsamic resin benzoin itself may be used. These preparations are usually inhaled as sprays, or in the vapour of hot water. In the case of the spray, the solubility of the resin requires to be aided by warm water or by a little alcohol.

Doses.—From 3 to 5 grains of the acid in 1 ounce of the menstruum for each spray, and 5 to 10 grains of benzoate of ammonia in 1 ounce of water; but the amount of benzoate of soda used is very much larger, from 100 to 300 grains per day, the quantity of the solvent being of course very greatly increased; Oertel states, that from 1000 to 1200 grammes of a 5 per cent. solution of benzoate of soda can be inhaled without any ill effects, but that as a rule, when there is a large amount of decomposing secretion to be dealt with, from 200 to 500 grammes of a 5 to 10 per cent. solution may be used, though 'for complete cleansing and disinfection of the affected parts as much even as 1000 grammes (=15400 grains) or more may be necessary.' Notwithstanding the employment of so large a quantity, the results obtained are sometimes negative, but still he regards the remedy as about equal in antiseptic power to salicylic acid and thymol, though inferior to carbolic acid.

It is questionable how far it is correct to speak of the inhalation of such large quantities of a medicated solution, of which it is obvious that only a very small portion can make its way into the lungs, and that by far the greater part must be swallowed. If the attempt be really made to inhale so large a quantity of fluid, then the operation must necessarily extend over several hours.

For the inhalation of benzoate of ammonia in vapour, from 20 to 40 grains should be added to 10 ounces of hot water in an eclectic inhaler, a uniform temperature being maintained.

When benzoin itself is inhaled as a spray, 20 grains must be dissolved in a mixture, consisting of a very small quantity of liquor potassæ, a little alcohol and water. If used as a pastille, the formula of Roumier as given by Oertel may be adopted: vegetable charcoal 0·5, benzoin 0·25, iodine 0·1, balsam of Tolu 0·05, nitrate of soda 0·1 gramme.

Indications.—Since benzoic acid and its compounds are stimulant, expectorant, lessen secretion, and prevent putrefaction, they are remedies which may be advantageously employed in cases in which there is profuse secretion and expectoration, either with or without offensive odour. In whooping-cough, chronic bronchitis, bronchiectasis, and in chronic phthisis; in diphtheria, in which case the spray should be of considerable strength and should be freely used.

BALSAMIC REMEDIES.

Balsamum Peruvianum.

Balsam of Peru transudes from incisions in the trunk of *Myroxylon Pereiræ.* According to Frémy, black Peru balsam is composed of variable quantities of a volatile oil, cinnamein or cinnamate of benzyl, meta-cinnamein, a crystallizable substance isomeric therewith, cinnamic acid, and resin; it also contains styracin or cinnamate of cinnyl, freely soluble in rectified spirit.

Properties.—Stimulant and expectorant, acting especially on the mucous membrane, lessening the secretion therefrom.

Administration.—May be inhaled as an atomized spray, which may consist of a spirituous solution or an emulsion with mucilage. Only as much spirit should be used, as is necessary to dissolve the balsam.

Doses.—For the spirituous solution or for the emulsion, 10 to 20 minims of the balsam in 1 to 2 ounces of water.

Indications.—Chiefly in chronic affections of the organs of respiration accompanied by excessive secretion; in chronic catarrhs, in chronic bronchitis, and in asthma.

Balsamum Tolutanum.

Balsam of Tolu exudes as a soft solid of an aromatic and balsamic odour from incisions in the bark of *Myroxylon toluiferum*. Balsam of Tolu is a mixture of volatile oil, free acid, and resin. The oil is a mixture of tolene and cinnamein. according to Deville, but according to E. Kopp and Scharling, the oil consists wholly of tolene. The free acid is cinnamic acid only, according to Frémy and E. Kopp, or a mixture of cinnamic and benzoic acids according to Deville and Scharling. E. Kopp considers, that the benzoic acid is a product of decomposition and does not pre-exist in the balsam.

Administration.—The balsam, the syrup, and the tincture may all be used as sprays.

Doses.—Of the syrup as a spray, from 1 to 2 drachms in 1 ounce of water, and of the tincture as a spray $\frac{1}{2}$ to 1 drachm in the same quantity of water.

Properties and Indications.—Nearly the same as those of Peruvian Balsam.

Ammoniacum.

A gum resinous exudation from *Dorema ammoniacum*; sparely soluble in water, with which it forms a white emulsion. It contains neither benzoic nor cinnamic acid, but according to Buchholz consists of—resin 72, soluble gum 22·4, bassorin 1·6, and 4 per cent. of insoluble matter.

Properties.—Stimulant, antispasmodic, and powerfully expectorant. Possesses but little action on the nervous system, in which respect it differs from assafœtida.

Administration.—$\frac{1}{2}$ to 1 ounce of the mistura ammoniaci as an atomized spray. Apt to nauseate in large doses.

Indications.—In chronic catarrh, and bronchitis with profuse secretion, and when unattended with fever, especially when these affections occur in advancing age. In asthma with wheezing.

Assafœtida.

A gum resinous exudation from incisions in the root of *Narthex assafœtida.* It contains no benzoic or cinnamic acid, but consists of resin 65, soluble gum 19·4, bassorin 11·2, volatile oil 3·6, sulphide of allyl, the most important constituent, and malate of calcium 0·3 per cent. It forms an emulsion with water.

Properties.—A powerful nervine stimulant and antispasmodic.

Administration.—As a spray, the emulsion, the tincture, or the resin may be used.

Doses.—Of the emulsion from ½ to 1 ounce and of the tincture from ½ to 1 drachm in ½ to 1 ounce of water. Of the gum resin 10 to 30 grains, suspended in mucilage.

Indications.—In spasmodic and nervous affections of the organs of respiration, in hysterical aphonia, in whooping-cough and spasmodic asthma with wheezing or chronic bronchitis. Dr. Garrod is of opinion, that the value of the drug is chiefly due to the volatile or sulphur oil which it contains, and which is not present in either ammoniacum or galbanum.

Galbanum.

A gum resin obtained from an umbelliferous plant, *Ferula galbaniflua;* it contains neither benzoic nor cinnamic acid, but consists according to Meissner of a volatile oil, said to be isomeric with oil of turpentine 3·4, of resin 65·8, gum 27·6, mucilage 1·8 per cent., and insoluble matter.

Properties.—' Supposed to act as assafœtida, but to be much less powerfully antispasmodic. Galbanum is probably more allied to ammoniacum in its action, and may be given as a stimulating expectorant.' Garrod.

Administration.—Inhaled as a spray, from 10 to 20 grains being rubbed up with 1 ounce of water for each inhalation.

Indications.—In chronic bronchitis and other affections of the mucous membrane of the air passages, in which a stimulant expectorant is indicated.

Olea Essentialia.

A great variety of essential oils are occasionally employed in lung inhalation; they nearly all act upon the same principle as the fir-wood oils, namely, that of the stimulation of the mucous membrane. The principal of these oils, of which it is unnecessary to give any detailed account, are the following :—

Oil of aniseed, a mild stimulant; of cajaput, a more decided stimulant; of sweet flag, a powerful stimulant; of carraway, a gentle stimulant; of cloves, a strong stimulant; of cassia and cinnamon, both mild stimulants; of juniper, a decided stimulant, it being also diuretic; and of marjory, rosemary, sage, all stimulants.

NERVINE STIMULANTS.

Calcis Hypophosphis, Sodæ Hypophosphis.

These salts consist of phosphorus in combination with lime or soda. Subjected to a red heat they emit phosphuretted hydrogen and become inflammable. They possess to a considerable extent the properties of phosphorus itself, only that they are less stimulant and irritating. The lime salt is soluble in 8 parts of water, the soda salt in 2 parts of water and 2 of glycerine.

Properties.—Nervine stimulants, tonic and alterative, increasing appetite and promoting digestion and nutrition.

Administration.—Usually given by the stomach, but may be inhaled as an atomized spray. Hypophosphite of lime may either be given in the solid form or dissolved in a little weak hydrochloric acid; when given undissolved, it becomes acted upon and brought into a state of solution by the acids of the stomach; or it may be given in the form of syrup, or the soda salt may be administered in glycerine, but the lime combination is that generally employed.

Doses.—2 grains, gradually increased to 6 or 7 grains a day, or 3 grains 3 times a day. In large doses they may produce 'weakness, sleepiness, headache, giddiness, noises in the ears, loss of appetite, colic, diarrhœa, and even bleeding from the nose and lungs' (Ringer). Best given uncombined.

Indications.—In nervous and general debility, in phthisis, in which they often lessen cough, diminish expectoration, check sweating, stop the diarrhœa, and, it is said, will even sometimes effect a cure.

Calcis Phosphas.

This salt is regarded as necessary to growth and nutrition. It is soluble in weak acids, including acetic acid, and in the acids of the stomach. Its solubility in the blood is explained by the free carbonic acid and chloride of sodium therein contained.

Properties.—Promotes cell growth and thus increases nutrition; it forms an essential constituent of new tissues.

Administration.—Should be given shortly after food, either in powder or dissolved in an acid.

Doses.—From 1 to 2 grains 2 or 3 times a day. Owing to its little diffusive power and to the fact that it is soluble only in acids, part of the phosphate administered remains undissolved and becomes incorporated with the fæces. There is no utility, therefore, in administering large doses.

Indications.—In defective nutrition and cell growth. Valuable in chronic bronchitis, in chronic phthisis where there is no considerable amount of fever, and in the diarrhœa of phthisis.

Cocaine.

The alkaloid of *Erythroxylon Coca*; acts chiefly on the nervous system, diminishing its sensibility, and exerts an anæsthetic effect on mucous membranes; this last property renders it of much value in some affections of the organs of respiration attended with irritation, pain, or spasm, and also in cases where operative procedure is necessary.

Dose.—Of the hydrochlorate $\frac{1}{6}$ to $\frac{1}{2}$ grain in the form of a spray. To produce anæsthesia of the nasal, pharyngeal, and laryngeal mucous membranes, a 5 to 10 per cent. solution of cocaine, to be applied with a brush.

ALTERATIVES.

Arsenicum.

Arsenic is employed in medicine in the form of arsenious acid, which is volatilized by heat, or in that of some of its salts, including arseniate of soda. This acid is soluble in 100 parts of cold, but much more so in boiling water; it is, however, on cooling, deposited in the form of the usual octahedral crystals; it is also soluble in the carbonates of soda and potassa.

Properties.—Externally escharotic; internally a nervine stimulant; acts particularly on the epithelial layers of the sk'n, and mucous membranes; in small doses alterative and tonic; it is said to check oxidation and so lessen change of tissue. In large doses it is irritant. It often greatly improves appetite, digestion, and nutrition.

Administration.—May be inhaled in the form of vapour or fumes, or as an atomized spray. The fumes of arsenious acid or arseniate of soda may be conveniently inhaled, by means of the apparatus for concentrated vapours. A piece of paper specially prepared, containing a given quantity of the acid or its arseniate, may be placed in one of the capsules, and the little lamp ignited; or cigarettes may be smoked. For the spray, either liquor arsenicalis, Fowler's solution, which consists of 1 part of arsenic in 120 parts of the solution, or the liquor sodæ arseniatis, which is of the same strength, may be used. Dr. Garrod states, that from his trials he has come to the conclusion that the action of arseniate of soda is milder than arsenious acid, less liable to produce irritation of the mucous membrane, and equally effectual.

Doses.—For the fumes, a piece of paper should be burned containing $\frac{1}{4}$ to $\frac{1}{2}$ grain of crystallized arseniate of soda. It is safest always to commence with the smaller doses, some patients being much more susceptible than others to the action of the remedy. For the spray, 2 to 5 minims of liquor arsenicalis, or of the solution of arseniate of soda from 5 to 10 minims for each inhalation in from 2 to 4 drachms of distilled water.

Indications.—In chronic coryza, in paroxysmal coryza, with or without fits of sneezing; in some forms of hay asthma; in asthma, especially accompanied by emphysema, with much wheezing on taking cold; in asthma with derangement of digestion; in general tuberculosis; in chronic phthisis; in the diarrhœa and bowel ulceration of advanced phthisis. 'In emphysema, arsenic often gives great relief to a class of emphysematous persons who on catching cold are troubled with a slight wheezing at the chest, difficulty of breathing, especially on exertion or at night time, and are obliged to be partially propped up in bed.' Again, 'arsenic generally relieves the wheezing with oppressed breathing which affects some children for months and even years' (Ringer).

Some of the therapeutical effects of arsenic depend doubtless upon its peculiar action on the epithelial structures of the skin and mucous membranes, both of which it stimulates, and others upon its nervine and tonic properties, whereby assimilation and nutrition are increased.

The beneficial results of arsenic quickly manifest themselves in some cases, while in others it is necessary that the remedy should be continued for several days.

Iodum.

Iodine is a non-metallic element. It is prepared from the ashes of seaweed, or wrack. It volatilizes at ordinary temperatures in the open air, especially in the presence of moisture; it melts at 224·6° F., and boils at from 347° to 356° F.; the fumes are very irritating; it is very sparingly soluble in water, 1 in 7000 parts only, but is much more soluble in alcohol and ether, and in aqueous solutions of the soluble iodides and chloride of sodium. It resembles closely chlorine and bromine in its chemical relations.

Half a gramme exposed to the air at 48° F. lost 0·0124 gramme in 2 hours and 0·0178 gramme in 4 hours; 0·5 gramme in dry air under a bell glass, lost 0·0103 gramme in 2 hours and 0·0145 gramme in 4 hours. (See page 23.)

Properties.—Locally applied, strongly irritant and vesicant; stimulant to the lymphatics, absorbent, alterative, powerfully antiputrescent, and regarded by some as a specific for the poison of lues.

Administration.—Inhaled in vapour or in solution as an atomized spray. There are various contrivances by which the fumes of iodine may be inhaled; one of the most convenient for the purpose is that for concentrated vapours. The capsule containing the iodine must be placed upon the water bath, which should be gently heated; or the tincture may be used, diluted with water; or iodide of potassium in water, acidulated with dilute sulphuric acid. The addition of a little tincture of conium will serve to render the vapours less irritating. Care must be taken not to inhale the vapours in too concentrated a form, as in this state they are very irritating. The vapours of iodine may be inhaled at the ordinary temperature of the air if either the ethereal tincture or the crystals be employed; the first may be inhaled from an ordinary oro-nasal inhaler, and the second by placing the crystals in a small piece of hollow cane, as first suggested by Dr. A. P. Merrill, of New York, closing the ends with a little cotton-wool; the tube is of course placed in the mouth and the air drawn

ALTERATIVES. 203

through it. Iodine is often prescribed with other remedies, such as carbolic acid, between which and iodine, however, chemical action takes place, whereby its properties are modified. The fumes do not become equally or widely diffused in the air, but speedily condense, so that when volatilized in a room and without any special apparatus the patient should not be far away. Again, when the fumes come into contact with the mucous membrane of the mouth and fauces, they become quickly absorbed and transformed into an iodide, so that if, directly after an inhalation, the saliva be tested with a solution of starch, it will give no evidence of the presence of iodine until an acid has been added, whereby the iodine is set free.

Cigarettes and tapers containing iodine are also occasionally used, and these, if well prepared, would be likely to prove more effective than the inhalation of the fumes given off in a chamber or by an oro-nasal inhaler. It must be remembered that, owing to the fumes being very irritating, the dose prescribed must be small.

Doses.—For the vapour or fumes, from $\frac{1}{4}$ to $\frac{1}{2}$ grain; when from an uncovered plate, 1 to 2 grains; but, in place of employing the iodine itself, it will be more convenient to take corresponding doses of the tincture of iodine, which may be also used for the spray, dissolved in $\frac{1}{2}$ to 1 ounce of water. The tincture contains 1 part of iodine in 40 of rectified spirit.

The Pharmacopœia contains a formula for the inhalation of the vapour of iodine. 1 drachm of the tincture is directed to be added to 1 ounce of water, mixed in a suitable apparatus, and, having applied a gentle heat, the vapour that rises is to be inhaled. The nature of the apparatus is not stated, and the term 'suitable' is very indefinite. As ordinarily employed, the greater part of the tincture is of course lost.

When the spray proves too irritating, or it is desired to bring the system fully under the effects of the remedy, iodide of potassium, 3 to 6 grains, may be substituted for the iodine, or the iodides of ammonium or sodium, which are less depressing. In many cases, iodide of ethyl or iodoform are to be preferred to iodine, as they are both non-irritant.

Indications.—From what has been stated as to the properties of iodine, it is evident that it and the soluble iodides exert a remarkable effect upon mucous surfaces, on the glandular elements which these contain, as well as on the glandular system generally, and it is, therefore, in ailments affecting the mucous membrane of the respiratory track that this remedy often proves of great value.

Iodine vapours, when they come into contact with the mucous membrane, become speedily converted, as already noticed, into an iodide and cause the discharge of a watery secretion. This relieves congestion and hypertrophy, dilutes any viscid mucus which may be present, and thus promotes expectoration and relieves cough.

Iodine inhalation has been found very serviceable in aphonia resulting from excessive use of the organs of speech; in chronic and syphilitic ozæna; in chronic catarrh, and in the habitual catarrh of old people; in discharges of a thin watery liquid from the nose, with fits of sneezing, such as sometimes occur in hay fever, iodine inhaled through the nostrils will often stop the discharge at once; in some affections of the mucous membrane of the throat with unhealthy and putrefactive secretions; in ulcerative conditions of the same; in œdema glottidis; in chronic bronchitis with excessive or putrescent secretion; in bronchial asthma; in ulcerations of the larynx in laryngeal phthisis, and in phthisis in nearly all stages; (Ringer states that the tincture is often of signal benefit in chronic forms of phthisis when expectoration is abundant and the cough troublesome; an inhalation night and morning will generally lessen the expectoration and allay the cough); in scrofulous conditions of the throat; in syphilitic ulcerations and affections of the throat, larynx, bronchi, and lungs; in croup and diphtheria. Again, it is very serviceable in the hoarseness, cough, and wheezing which is apt to follow measles on exposure to cold. Laennec had such confidence in the therapeutic effects of iodine, that he was in the habit of surrounding the beds of his consumptive patients with seaweed, while M. Piorry attached vials of iodine to the frames of the bedsteads, so that the vapours might be inhaled during sleep. The quantity given off was sufficient to cause the starch in the curtains to become more or less blue.

Hydrargyrum.

Mercury at any temperature above 40° F. becomes slightly vaporized and solidifies at 39° F. below freezing point.

Properties.—Increases most secretions, is sialogogue, cholo-gogue, alterative, absorbent, deobstruent, antisyphilitic.

ALTERATIVES. 205

Administration.—May be inhaled as a vapour, or in the smoke of a cigarette, or as a spray. For the vapour, the metal itself may be taken, or calomel, hydrargyri subchloridum, or sulphide of mercury, hydrargyri sulphuretum; for the spray, grey powder, hydrargyrum cum creta, liquor hydrargyri perchloridi, lotio hydrargyri flava (yellow wash), or l. hydrargyri nigra (black wash). The vapours of the metal, of calomel, and of the sulphide or cinnabar will be most conveniently inhaled by means of the apparatus for concentrated vapours. If inhaled together with the vapour of hot water, the fumes are much less irritating. When the perchloride is subjected to considerable heat it is stated that hydrochloric acid is evolved, and when sulphuret of mercury is similarly treated sulphurous acid is formed, and the presence of these sometimes occasions irritation. Several contrivances have been devised for the combined inhalation of the fumes of mercury and the vapour of hot water, as those of Bumstead and Lee. Occasionally, the grey powder or the red oxide of mercury are locally applied, or are insufflated, the effect produced being greater; they are usually mixed with sugar or some similar substance. The indications for the use of the above powders are much the same as for the other preparations of mercury; they are specially indicated in syphilitic ulcerations of the tongue and throat; the effect produced is probably more lasting than when sprays or fumes are inhaled.

Preparations of mercury, such as the nitrate, or perchloride and sulphuret, may sometimes be inhaled with advantage from the fumes of cigarettes, as was originally recommended by Trousseau, or of pastilles. After a few inhalations a mild stomatitis ensues, when the remedy may be discontinued, at all events for a time.

Doses.—For the vapour of calomel 10 grains; and of the sulphide of mercury, 20 to 30 grains; for the spray, of the solution of perchloride of mercury from 30 to 60 minims in $\frac{1}{2}$ ounce of distilled water; of grey powder $\frac{1}{2}$ to 2 grains suspended in about $\frac{1}{2}$ ounce of thin mucilage; of the yellow wash 2 to 4 drachms, and of the black wash $\frac{1}{2}$ to 1 ounce. In employing these preparations, care should be taken not to inspire too deeply when it is intended that the remedy should reach the throat only.

Indications.—In quinsy with great enlargement of the tonsils, and much difficulty in swallowing; even when suppuration is threatened, according to Ringer, its maturation and evacuation appear to be effected more quickly by the hourly administration of ⅓ of a grain of grey powder. In mumps, in ill-conditioned affections and ulcerations of the throat, larynx, bronchi, and lungs; in the syphilitic ozæna of children; in syphilitic affections and ulcerations of the throat and air passages.

ANÆSTHETICS.

Æther Sulphuricus.

Ether is prepared from alcohol by the action of sulphuric acid, which abstracts an atom of water from the spirit. Pure or absolute ether has a specific gravity of 0·720, but ordinary or medicinal ether of 0·735. This should not contain less than 92 per cent. of pure ether or oxide of ethyl, the remainder being made up of water and a little alcohol. It is soluble in 10 parts of water and freely in spirit, while it dissolves a great variety of substances. Spiritus ætheris consists of 1 part of ether to 2 of rectified spirit.

Properties.—A powerful diffusible stimulant; expectorant; antispasmodic; anæsthetic. Unlike chloroform it does not reduce blood pressure. Applied externally to the skin, it acts as a refrigerant, on account of its extremely rapid evaporation. It stimulates the secretions of the salivary glands and pancreas, whereby it aids the conversion and digestion of farinaceous and fatty substances, and for this reason it is often prescribed with cod-liver oil. Unlike chloroform, it is a cardiac stimulant.

Administration.—Usually employed in the form of vapour or as an atomized spray.

Doses.—Of ether 15 to 30 minims, and of the spirit of ether 30 to 60 minims. The ether may be inhaled from a sponge or a globe oro-nasal inhaler; if from sponge, the vapour should be conducted to the nose and mouth by a cone of paper or other suitable material. Ether is sometimes combined with other but little volatile substances, and even with some which are non-volatile, under the idea that they aid evaporation. As appears from the experiments which have been already quoted, the ether does promote to a small extent the volatility of substances which are themselves but little volatile at the ordinary temperature of the air, but it has no effect on non-volatile medicaments.

Indications.—Often very beneficial in hysterical aphonia; in the dyspnœa of chronic bronchitis, of phthisis, and of asthma, especially when this is accompanied with flatulent dyspepsia; in whooping-cough and in diphtheria.

Ethyl Iodidum.

Ethyl iodide, C_2H_5I, is formed by heating together alcohol and hydriodic acid, also by the simultaneous action of iodine and phosphorus on alcohol.

It is a colourless, strongly refracting liquid, possessing a peculiar ethereal and pleasant smell, boiling at 148° F. and having at 60° F. a specific gravity of 1·940. But little soluble in water, miscible with alcohol and ether; it burns with difficulty, giving off iodine vapours, and is also decomposed by light, the iodine being set free.

Administration.—In the form of very thin glass capsules coated with silk; each capsule contains 5 minims, the ordinary dose. The capsule is broken, and its contents, absorbed by a piece of cotton-wool, are inhaled, or an oro-nasal inhaler may be used, or a spray producer.

Properties.—Anæsthetic, antispasmodic, and stimulant, rousing the respiratory centres. It increases bronchial secretion.

Indications.—Very effective in spasmodic affections of the throat, larynx, and lungs; in spasmodic and bronchial asthma.

Iodoformum.

Iodoform or tri-iodomethane, $C.HI_3$, is obtained by the action of iodine upon alcohol in the presence of the caustic or carbonated alkalies. A yellow crystalline substance, the large lemon-yellow, six-sided crystals melting at 119° F., subliming when strongly heated, and becoming partially decomposed with the evolution of iodine vapours; possesses a sweetish taste and a saffron-like odour.

Insoluble in water; soluble, according to Messrs. Martindale and Westcott, in 12 parts of chloroform, 10 of ether, of sp. gr. 0·735, and 80 of rectified spirit; in 14 of oil of eucalyptus.

Properties.—Somewhat anodyne and anæsthetic; powerfully antiseptic. Although it consists principally of iodine it is non-irritant.

Administration.—As an atomized spray, suspended in mucilage or milk, or dissolved in ether or oil, or the powder mixed with starch may be insufflated. Dr. Macnaghten Jones has had some cigarettes made by Messrs. Corbyn and Co. which contain $\frac{1}{8}$ to $\frac{1}{4}$ grain of iodoform; much too small a quantity. Lozenges are also made containing variable amounts of iodoform with glyco-gelatine.

Doses.—1 to 3 grains, finely powdered twice a day; in some cases more frequently.

Indications.—In chronic, fœtid, or syphilitic coryza; in syphilitic affections of the tongue and throat; in bronchitis and in phthisis, in which latter complaint Dr. R. Shingleton Smith, of Bristol, at the recent International Medical Congress at Copenhagen, stated that it increases the weight of the body, produces a fall of temperature, diminution of cough and expectoration, cessation of night sweats, improved appetite, and even diminishes in some cases the number of bacilli contained in the sputa, or causes their total disappearance. As appears from the Proceedings of the Hygienic Congress at the Hague, Dr. J. Sormani, of Pavia, has also obtained favourable results from the treatment of phthisis by iodoform. On the other hand, Dr. G. Hunter Mackenzie states, he has found from his observations, that even when the remedy is pushed to such an excess as to produce mental excitement, in no single case could it be said to have any effect on the bacilli.

Chloroformum.

Chloroform, terchloride of formyl, CH,Cl_3, is obtained from a weak spirit by distillation with chloride of lime and caustic lime, or from chloral by the action of a caustic alkali. It is a heavy liquid, having a specific gravity when pure of 1497 to 1500, the boiling point being 140° F. It usually contains about half a per cent. of alcohol. It is only slightly soluble in water, but mixes with alcohol and ether in all proportions. It is decomposed by fixed alkalies.

Properties.—Sedative, antispasmodic, narcotic, anæsthetic, reduces blood pressure.

Administration.—May be inhaled in the form of vapour or in that of an atomized liquid. The preparations chiefly employed are the aqua chloroformi, the spiritus chloroformi, 1 in 20 of spirit, or chloroform itself. For the inhalation of the vapour, 15 minims of chloroform may be placed on a small piece of sponge and inhaled, or an oro-nasal inhaler may be used.

Doses.—For the spray, not less than 1 ounce of the chloroform water; of the spirit of chloroform, which is a better preparation, 20 to 60 minims in ½ to 1 ounce of water may be employed. If the chloroform be used in the form of spray, from 2 to 6 minims may be dissolved in a little spirit, and made up to ½ ounce with water, or suspended in mucilage. Chloroform and other volatile substances are frequently prescribed with slightly volatile and even non-volatile medicaments, under the idea that their volatility is thereby greatly promoted. As will be seen by reference to the experiments on p. 28, this effect, even in the case of substances which do possess some degree of volatility at a medium temperature of the air, is so small as to be of but little practical utility.

Indications.—It is in spasmodic and convulsive affections especially, and in those in which there is increased sensibility and excitability, that this remedy is particularly indicated, its effects being in some cases heightened by the addition of a small quantity of hydrochlorate of morphia. In convulsive coughs, as whooping-cough, combined with small doses of morphia, chloroform is often very efficient. Ringer points out that, 'cough very often indeed arises from a morbid condition of throat, and even when due solely to lung disease, the application of the mixture just recommended to the throat and parts about the glottis is often beneficial, in accordance with a general fact that remedies applied to the orifices communicating with certain organs, as the nipple, rectum, and throat, will by nervous communication act on the organs themselves.' In laryngismus stridulus; in chronic phthisis, especially in fibroid phthisis; in asthma, whether simple or complicated; in angina pectoris.

Chloral Hydras.

Hydrate of chloral is produced by the action of dry chlorine on absolute alcohol, aldehyde being first formed and afterwards chloral, which on exposure to moisture becomes converted into the hydrate. The hydrate forms a crystalline, non-deliquescent salt, which is readily soluble in water, alcohol, ether, and in 4 parts of chloroform; it is decomposed by the action of alkalies, with the formation of chloroform.

Properties.—Antispasmodic, hypnotic, anæsthetic, and a vascular depressant. It therefore relieves spasms and convulsions, and produces sleep; but, since it weakens the heart, it should not be given in cases in which the action of that organ is enfeebled. The sleep induced by chloral is calm and refreshing, and it does not occasion headache, impair appetite, or give rise to constipation.

Administration.—May be inhaled in the form of a spray, either alone or combined with bromide of potassium, or with hydrochlorate of morphia.

Dose.—5 to 10 grains for each inhalation, to be repeated more or less frequently according to necessity. As a hypnotic, 15 to 30 grains. When it is required that the drug should act chiefly on the mucous membrane of the throat, larynx, or bronchi, the spray should contain about 1 drachm of glycerine; and as some of the liquid will usually be swallowed, it is not well that the hydrate of chloral should be dissolved in more than ½ ounce of water.

Indications.—In whooping-cough, in asthma uncomplicated with heart disease, in emphysema, and in chronic phthisis.

'The shortness of breath affecting the emphysematous on catching cold often yields to chloral' (Ringer).

It not only produces sleep in phthisis, but it allays cough and checks sweating, and these results are accomplished without appetite and nutrition being affected.

Aldehydum Dilutum.

Aldehyde is formed by the oxidation of alcohol; it possesses a strong ethereal odour. The dilute acid of The Throat Hospital Pharmacopœia contains about 15 per cent. of aldehyde, the remainder consisting of spirit.

Properties.—Sedative, resembles chloral in its physiological action, diminishes the frequency of the pulse but strengthens the action of the heart, and does not occasion headache.

Administration.—May be inhaled in the vapour of hot water.

Dose.—1 to 2 drachms of the dilute acid, T. H. P. in 1 pint of water, the temperature of which should be maintained at 140° to 150° F.

Indications.—In recent catarrhs and ozæna.

Amyl Nitris.

An ethereal liquid produced by the action of nitric or nitrous acid on amylic alcohol. Specific gravity 0·877 and boiling point 205° F. Insoluble in water, but freely soluble in rectified spirit.

Properties.—An arterial relaxant, lessening greatly arterial tension in the arterioles. It quickens the pulse and causes the heart and carotids to beat strongly, giving rise to a feeling of fulness in the head; the face flushes, in consequence of the relaxation of the capillaries allowing of the entrance of more red blood; it lowers temperature, which it is believed to do by lessening oxidation; and these effects are further explained by the union of the nitrite with the hæmoglobin of the blood corpuscles.

Administration.—Best in the form of vapour, but it may also be inhaled in a spray, the nitrite being dissolved in a few drops of spirit. As there are great differences in the toleration of this remedy it is safest to commence at first with a minim dose, increasing it if required up to 5 minims.

Doses.—The little hermetically sealed glass tubules, encased in cotton-wool and silk, containing from 1 to 5 minims of the nitrite, will be found most convenient. In inhaling the remedy in the form of vapour there is of course a great loss, which does not take place when the spray is inhaled; consequently the dose in the latter case should be smaller, from $\frac{1}{2}$ to 1 minim to commence with. Dr. Brunton is of opinion that the vapour is more effective than the spray. If too large a dose be administered it will occasion pallor of the face, giddiness, and sickness.

Indications.—Very valuable in angina pectoris, especially uncomplicated with marked disease of the heart; also in the paroxysm of spasmodic asthma; in non-inflammatory croup and whooping-cough. Its effects on the arterial system being so marked, it should be employed only with the greatest caution in old people whose vessels have lost a portion of their resiliency through degeneration or ossification.

Nitroglycerinum

Is obtained by dropping glycerine into a mixture of strong sulphuric and nitric acids, kept cool by iced water. Its specific gravity is 1·600, and it is slightly volatile.

Properties.—Nearly the same as those of nitrite of amyl, but its effects are more lasting.

Administration.—The most convenient way is by means of Mr. Martindale's tablets, which consist of nitroglycerine dissolved in oil of theobroma and combined with chocolate. The tablets are made of various strengths; the most suitable for general use are those containing $\frac{1}{100}$ grain of the remedy. In lieu of the tablets a 1 per cent. solution in almond oil may be prepared, which is safer and more stable than an alcoholic solution. Of this the dose to commence with is 1 minim, gradually increased and repeated as may be necessary.

Indications.—In spasmodic asthma and in angina pectoris.

ANODYNES AND NARCOTICS.

Papaver Somniferum.

The principal constituents of crude Opium are morphia, codein, narcotina, papaverin, thebaia or paramorphia, narcein, meconin, meconic acid, apomorphia, with gummy, fatty, and extractive matters. Of these constituents none are, it is believed, volatile at the ordinary temperature of the air.

Of the above principles meconin is the more purely hypnotic, while thebaia exhibits convulsive properties. Morphia is both soporific and convulsant, the former property predominating in the human subject. The salts of methyl-morphia possess only the hypnotic properties of the morphia. Codeia possesses a feebly soporific property only and is useless as an anodyne. Narceia is narcotic, but inferior in power to morphia and of very little value. Papaverin is soporific and narcotic. Meconin is a mild hypnotic and produces but little effect when administered by the mouth. Dr. Garrod has given it in $\frac{1}{2}$-drachm doses without the production of any narcotic symptoms, and says that it probably acts as a tonic and antiperiodic. Apomorphia is a powerful emetic.

Properties.—In small doses opium is excitant to vascular and nervous systems, increasing the fulness and frequency of the pulse; in larger doses it relieves pain and causes sleep; it diminishes the secretions, except that of the skin, which it promotes. It reduces the number of respirations and so lessens oxidation.

Administration.—When crude opium is submitted to distillation in water, even after the addition of an acid, no volatile alkaloid distils over, and hence it may be inferred that in its natural state it does not contain any volatile active principle. But when it is subjected to a process of combustion the fumes are found to possess decidedly anodyne and narcotic properties, possibly either from the greatly increased temperature or from new compounds being formed. These particulars are of practical importance, since they show that it is useless to employ opium or any of its preparations in oral or oro-nasal inhalers, as is so frequently done in ignorance. To obtain the beneficial effects of this remedy, therefore, it must be employed in one of two ways, either in the form of fumes obtained by the combustion of an opium cigarette, or in that of an atomized liquid or spray. It is possible, that by means of the cigarette a larger quantity of some of the more active ingredients makes its way into the lungs than by means of the spray, and no doubt when the latter is employed much of the opium is swallowed and enters the system through the stomach. It is well in some cases to combine the opium with gum or glycerine, to render it more adherent to the mucous surfaces. When, therefore, the spray is used the amount of liquid in which the opium is dissolved must not be too considerable, and when the topical effects only are required care should be taken to avoid swallowing the liquid as much as possible.

Doses.—One of the best preparations for use as a spray is liquor morphiæ hydrochloratis in doses of from 10 to 20 minims, in $\frac{1}{2}$ ounce of water; the effect of this will be heightened in some cases by the addition of 15 to 30 minims of spiritus chloroformi. When opium cigarettes are smoked each of these should contain $\frac{1}{2}$ to 1 grain of the powdered drug.

Indications.—The chief indications for the use of this remedy are irritation and pain; it will therefore be found of much service in irritable and painful affections of the throat and larynx, and in the troublesome cough of bronchitis and phthisis; it should not, however, as a rule be employed when the mucous membrane of the throat and air passages is in a dry condition.

Hyoscyamus Niger.

The chief constituent contained in the leaves of Hyoscyamus, or Henbane, is hyoscyamia or hyoscyamine; this is met with in two forms, crystalline and amorphous, the former being the purer. Like many alkaloids it is destroyed by caustic potash.

Properties.—Hyoscyamus resembles in its properties and effects belladonna and stramonium, but is milder, and, like opium, is employed to diminish pain and allay irritation when this either disagrees or its administration is unadvisable. It is narcotic and sedative, is often employed to produce sleep when opium fails; it diminishes secretion from the mucous surfaces like opium, and so occasions dryness of the throat, but to a less extent.

Administration.—Being non-volatile, it is best inhaled in the form of a spray. The preparations adapted for inhalation are the tincture and succus of The Pharmacopœia, but as the taste of these is disagreeable, the alkaloid may be substituted for them.

Doses.—The quantity for each inhalation will be from 20 to 40 minims of the tincture, of the succus the same, and of the pure crystallized alkaloid $\frac{1}{120}$ to $\frac{1}{40}$ of a grain, which may be dissolved in a little dilute sulphuric acid. Dose of the amorphous alkaloid, $\frac{1}{20}$ to $\frac{1}{6}$ of a grain.

Indications.—It has been well spoken of in painful, irritative, and spasmodic affections of the throat, larynx, and bronchi, but is inferior to opium; in whooping-cough and in chronic laryngitis and chronic bronchitis, especially with much cough. Since it checks secretion, it should not be inhaled when the mucous membrane of the throat and bronchi is already dry.

Conium Maculatum.

Conium, or Hemlock, owes its properties to the presence of a volatile oil, to which its peculiar odour is due, to a volatile alkaloid, conia, the odour of which resembles that of a mouse, and which is freely soluble in ether and alcohol and slightly so in water; its most important constituent is methyl-conia and a small quantity of another substance, conhydrine, which is volatile, subliming in the form of needles, and strongly alkaline. Conia, as ordinarily met with, is said always to contain a considerable proportion of methyl-conia.

Properties.—These are mainly due to the presence of the two alkaloids conia and methyl-conia, the action of which is different. The chief effect of conia is sedative and exerted on the terminations of the motor nerves, giving rise to more or less weakness or paralysis of the voluntary muscles, according to the dose; but this action is greatly modified by the presence of the methyl-conia, which, according to Fraser and Crum Brown, not only acts on the ends of the motor nerves, but exerts an influence on the cord itself, first exciting and then abolishing its reflex function. Conium therefore has a powerful effect in allaying muscular spasm.

Administration.—Since conium contains active volatile principles it may be inhaled in the form of vapour, or as a spray. It may be inhaled in the vapour of hot water to which a little liquor potassæ has been added to separate the alkaloid, as indicated in The British Pharmacopœia ; but, as I have elsewhere shown, the quantity of the extract directed to be used is far too little to produce the smallest therapeutical effect.

Doses.—Some 5 to 10 grains of the extract may be used when the remedy is inhaled in the vapour of hot water. If the conium be inhaled by means of an oro-nasal inhaler 5 grains may be taken, but it is questionable whether, unless artificial heat be employed, sufficient of the alkaloids are volatilized to produce therapeutical effects. If the alkaloid were obtainable in the pure state and at a moderate price, this would be the easiest and most certain way of employing the remedy ; the dose of the alkaloid is $\frac{1}{4}$ to 1 grain. Conia may be obtained of Messrs. Hopkin and Williams at 2s. 6d. per drachm. For the spray, the succus is the best preparation to use, in doses of 30 to 60 minims for each spray, or the tincture may be employed in 20 to 40 minim doses.

Indications.—In most spasmodic affections of the organs of respiration, as in spasmodic croup and in the coughs of catarrh, chronic laryngitis, bronchitis, and phthisis. It is also very serviceable for the relief of pain in laryngeal phthisis and in cancer of the organs of respiration.

Hydrobromate of conia has also been found serviceable in spasmodic affections of the organs of respiration, in doses $\frac{1}{3}$ to $1\frac{1}{2}$ grain.

Aconitum Napellus.

The leaves and roots of Aconite, or Monk's Hood, contain aconitic acid, aconella, which resembles narcotine in its composition and properties, and the alkaloid aconitia, upon which the properties of the plant almost entirely depend. Aconitia is alkaline, uncrystallizable, soluble in ether and alcohol, and freely in dilute acids. The principal preparation used in inhalation is the tincture, but a solution of the alkaloid itself is sometimes employed.

Properties.—Locally applied, it produces tingling and numbness, and relieves pain. The same effects are produced by the internal administration of aconite. It dilates the arterioles, according to Dr. Fothergill, and so lessens the supply of blood to inflamed parts. It lessens considerably the frequency and force of the pulse, the number of respirations, and reduces abnormal temperature, it is believed, by lessening oxidation and tissue change. It diminishes the secretions of the mucous membrane and skin, and hence it dries the throat and lessens perspiration.

Administration.—Inhaled chiefly as a spray, but the effects vary with the dose and the frequency of administration. As a febrifuge the dose at first should be small and often repeated until the desired results are produced, namely, free perspiration and diminished temperature.

Doses.—$\frac{1}{2}$ to 1 minim of the tincture diluted in $\frac{1}{2}$ ounce of water may be inhaled every 15 minutes for the first 2 hours, and then every hour or two hours as may be necessary. This plan of giving frequently repeated small doses will be found much more successful, as well as safer, than the administration of from 5 to 15 minims at intervals of 4 or 6 hours. The dose of the alkaloid to commence with should not exceed from $\frac{1}{200}$ gradually increased to $\frac{1}{60}$ of a grain; it should be prescribed in a solution of definite strength. With children, aconite sometimes fails to produce perspiration. Price of aconitia about 5s. per drachm.

Indications.—The power of aconite to lessen fever and to subdue inflammation is surprisingly great; it should be administered, however, as soon as possible after the symptoms have set in. It is of proved efficacy in simple, non-specific inflammatory or febrile affections, in acute pharyngitis and tonsillitis; if taken in time these affections very generally yield in 24 to 48 hours, the redness and swelling quickly subsiding, and mucus or pus appearing on the inflamed surfaces. In severe and feverish colds with chilliness and hot and dry skin; in spasmodic or catarrhal croup. It is very valuable, as pointed out by Ringer, in a class of cases characterized by a great susceptibility to catch colds of a severe character on the slightest exposure; these may affect at first the nose only, then the throat, and may extend at length to the bronchi and lungs. In acute pneumonia and pleurisy; in asthma, especially in that dependent upon or aggravated by cold, particularly in children, and in bronchial asthma with fever. In non-febrile affections it is useless, while in specific febrile diseases it exerts but little influence on the course and duration of the malady, although it may control somewhat the severity of any attendant inflammation. It is said to be very useful in the epistaxis of children, and to lessen expectoration in phthisis.

Atropa Belladonna.

The recent researches of Ladenburg have thrown great light on the alkaloids of Atropa Belladonna, Datura Stramonium, and Hyoscyamus niger. He has shown, that while belladonna owes its properties mainly to the presence of atropia, which is contained in all parts of the plant, but especially the root, it also contains hyoscyamia, or 'light atropine.' Atropia is antagonistic to morphia.

The preparations most suitable for inhalation are the tincture prepared from the leaves, the solution, and sulphate of atropia.

Properties.—Checks secretion from the glandular structures of the mucous membrane, except that of the intestines, which it increases, and from the sudoriferous, salivary, and mammary glands; hence it dries the throat and mouth, checks perspiration, and diminishes or arrests the secretion of saliva and milk. It at first increases the frequency, fulness, and force of the pulse, and Dr. J. Harley regards it as a powerful heart tonic; but if the remedy be pushed too far weakness follows, and a depressent or paralyzing effect is produced on the spinal cord. Dr. Brown-Séquard points out, that both belladonna and ergot act especially on the unstriped muscular fibre, belladonna on that of the eye, on the vessels of the heart, and the muscular fibre of the intestines and bladder, whilst ergot excites to contraction the vessels of the cord and uterus. Powerfully narcotic. Topically applied it relieves pain and checks sweating.

Administration.—Should be inhaled as an atomized spray, prepared principally from the tincture, but also from liquor atropiæ when properly diluted.

Doses.—Of the tincture, 5 to 20 minims; of the liquor, to commence with, 1 minim in 2 to 4 drachms of distilled water; of the sulphate, $= \frac{1}{100}$ to $\frac{1}{40}$ grain, increasing the dose as may be required up to $\frac{1}{20}$ grain. The best plan is to make a solution of the sulphate.

Belladonna is much better borne by children than by adults.

Indications.—In acute tonsillitis and other inflammatory affections of the throat, combined with aconite, if there be fever. Often a very effective remedy in uncomplicated whooping-cough, especially after the subsidence of the acute stage. It will frequently relieve the breathing and cough of bronchitic asthma; given in phthisis to allay cough, to check sweating, and to promote sleep.

In prescribing belladonna, we must not forget that it is antagonistic to opium.

Datura Stramonium.

Stramonium, or Thorn Apple, owes its efficacy partly to the presence of an alkaloid, daturine or daturia, which is probably identical with atropia, but it also contains, according to Ladenburg, hyoscyamia, or 'light daturine.' These alkaloids are present in the dried leaves, ripe seeds, and seed vessels.

Properties.—Sedative, antispasmodic, anodyne.

Administration.—Daturine is said to volatilize only at 140·6° F., and then to undergo partial decomposition, so that stramonium cannot be inhaled with effect by means of oro-nasal inhalers or by the vapour of hot water. It is, however, volatilized when the leaves or the extract of the seeds are burned, and hence stramonium is very commonly smoked in the form of cigarettes, or in pipes like tobacco. The leaves are sometimes used alone or mixed with opium and nitrate of potash. About 20 grains of the leaves or about 10 of the powdered root may be smoked at a time, care being taken to inhale deeply so that the fumes may really reach the lungs. The quality of the drug varies a good deal. An allied species of datura, D. Tatula, is sometimes either mixed with D. Stramonium or is employed alone, and Ringer states that it will sometimes succeed when the other fails; the reason of this is that it is stronger. Both are apt to lose their power when their use has been long continued.

Dr. Cohen makes the following remark under the head of Stramonium : 'As ordinarily smoked it is doubtful whether the remedy gains access into the lungs; but if the smoke as it issues from the mouth after the puff of the smoker is drawn back by an act of inspiration, it will then be inhaled into the lungs, and the effect will be more certain and more prompt; otherwise it does not enter the larynx, and the effect is in great measure due to absorption by the mucous membrane of the mouth and pharynx.'

The extract may be used either in the form of papers saturated with the same and burned, or the extract may be added to a cigarette, or, lastly, it, as well as the tincture or sulphate, may be inhaled as a spray.

Doses.—The quantity of the extract to be employed at each inhalation is ¼ grain, gradually increasing the dose if necessary; of the tincture 10 to 20 minims; and of the solution of the sulphate, a quantity, to commence with, containing $\frac{1}{100}$ to $\frac{1}{40}$ grain of the alkaloid, increased if necessary to $\frac{1}{10}$ grain.

Since stramonium, like opium and belladonna, is very drying to the mucous membrane of the throat and bronchi, it should not be used where this is already greatly deficient in natural moisture. Should dryness come on during the use of the remedy, this may be taken as an indication that its employment should be suspended for a time. Stramonium, usually with nitrate of potash, enters into the composition of a variety of so-called cures or specifics for asthma, as, for instance, Himrod's, Bliss's, and the Green Mountain cures.

Himrod's cure contains stramonium, aniseed, probably tobacco, and nitrate of potassa. Mr. Clayton, F.C.S., who at my request examined it, did not detect the presence of lobelia, theine, or arsenic. Whether, as some have asserted, it contains belladonna or not is uncertain, since the reactions for daturine and atropine are the same. A very good imitation of Himrod's cure may be made by mixing together equal parts of the powders of nitrate of potash, aniseed, and stramonium; to this mixture about half as much lobelia may be added as stramonium, but in most cases this is better omitted.

Joy's cigarettes also contain stramonium; a sample was found on analysis to be free from both arsenic and theine.

Indications.—Stramonium is useful in some convulsive coughs, but the malady in which it is specially serviceable is asthma, particularly the nervous and spasmodic forms, in which the lungs are free from any obvious or detectable disease. When smoking is resorted to, it should be commenced, if possible, before the onset of an attack, as if not employed till the attack be well advanced it usually fails to give the desired relief. Stramonium is of little service in asthma arising from disease of the heart.

Grindelia Robusta.

This remedy is much used in America for the relief of spasmodic affections of the organs of respiration, as bronchitis, whooping-cough, and asthma. The two chief preparations are the extract and the fluid extract. The dose of the latter is from 10 to 20 or even 30 minims, repeated every half-hour or hour until relief is obtained. It may be used as a spray, the resin being suspended in milk or mucilage. It is said to afford great relief in asthma, the attacks of which it will even sometimes ward off. This remedy was first introduced into this country by Mr. Martindale, of New Cavendish Street, London.

Cannabis Indica.

The chief active constituent of Indian hemp is cannabin, a resinous principle soluble in alcohol and ether, and obtained from the dried flower-tops of the female plant, which, however, also contain two volatile oils, to the presence of which the irritating and intoxicating properties are said to be due, and from which tannate of cannabin is entirely free.

Properties.—A nervine stimulant and excitant, producing much mental exhilaration, followed by sleep and even stupor; antispasmodic, anodyne, and soporific. In its anodyne and soporific properties it resembles opium, but it does not affect the secretions like it, and does not give rise to constipation or loss of appetite. It is not often, however, that it is employed to produce sleep.

Administration.—The two preparations used for inhalation are the extract and the tincture; the first is the most suitable for the preparation of cigarettes, in which form the remedy is most frequently employed. If inhaled as a spray the tincture or the tannate may be used, the resin being kept in suspension by means of a little alkali or gum.

Doses.— Of the extract, $\frac{1}{4}$ to 1 grain for each cigarette, and of the tincture for the spray, 5 to 20 minims, the latter corresponding to about 1 grain of the extract; of the tannate as an antispasmodic 1 to 3 grains in mucilage, and as a soporific 4 to 8 grains in pills. The quality of the flower-tops varies greatly, and this variation explains in many cases the comparative inefficiency of the remedy. Care should therefore be taken to obtain it from a reliable source. It is probable that the alkaloid, cannabin, or its tannate would yield the most certain results.

Indications.— In some spasmodic and convulsive diseases, especially in asthma.

Lobelia Inflata.

The pharmaceutical preparations of Indian tobacco are the simple and ethereal tinctures, both being prepared from the dried plant gathered when in flower. It contains a volatile oil of an acid nature, lobelic acid, and an alkaloid which is the active constituent, lobeline, and which cannot be volatilized without decomposition; it is liquid, lighter than water, and forms crystalline salts with the mineral acids; it is very soluble in both alcohol and ether.

Properties.—Expectorant, diaphoretic, depressent, antispasmodic, and emetic. It will thus be seen that it resembles very closely in its properties Nicotiana Tabacum.

Administration.—The herb may be smoked and the fumes inhaled, but the preparation usually prescribed is the ethereal tincture. In consequence of some persons being more susceptible to its action than others, it requires to be administered at first with great caution, particularly where the action of the heart is weak; very small doses frequently repeated should be employed. The ethereal tincture is usually directed to be inhaled in combination with other remedies by means of an oro-nasal inhaler; but this method of inhalation is probably useless since the alkaloid is non-volatile. A more certain method is in the form of the atomized spray.

Doses.—The ordinary doses of the simple or the ethereal tincture range between 10 and 30 minims. It is well to begin with the smallest dose, repeated every 15 minutes till the symptoms are relieved, but if this quantity is well borne and no great faintness or sickness is produced, the treatment may commence with a larger dose, from $\frac{1}{2}$ to 1 drachm, at longer intervals, but sometimes a single dose will be sufficient.

Indications.—Prescribed principally in bronchitic asthma, unassociated with disease of the heart. It should only be given at the onset of the paroxysms of dyspnœa, as its action is merely temporary and palliative. If well tolerated and the heart be not weak the doses for asthmatics should be considerable. It sometimes proves useful in bronchitis accompanied with paroxysmal dyspnœa; often very serviceable in laryngismus stridulus, in croup, and in the spasmodic stage of whooping-cough, although in this latter complaint it occasionally fails. It is said to be very useful in simple bronchitis, though not adapted for all cases, but only for those in which the depressent action of the remedy would be calculated to afford relief. Should the remedy cause faintness, nausea, or even vomiting, these symptoms soon disappear and are never really dangerous. 'Lobelia inflata always allays the dyspnœa which accompanies capillary bronchitis in emphysema' (Ringer).

Nicotiana Tabacum.

Tobacco owes its principal properties to the presence of a liquid oily alkaloid, nicotine, and a volatile oil named nicotianin. The alkaloid is heavier than water, volatilizes at 480° F., and is of course contained in the fumes of tobacco. In strong Virginian tobacco as much as 6 or 7 per cent. may be present.

Properties.—Nauseant, expectorant, emetic, powerfully depressent; applied externally, irritant.

Administration.—Inhaled solely in the form of smoke or fumes; seldom used except by habitual smokers. The leaves are sometimes mixed with those of stramonium.

Indications.—Not unfrequently will afford relief in a paroxysm of asthma, but non-smokers must employ the remedy with great caution.

Cimicifuga Racemosa.

Cimicifuga racemosa, formerly Actæa racemosa or Black Snake root, a remedy much employed in America, owes its activity to the presence of a resinous substance, cimicifugin; soluble in spirit.

Properties.—A nervine tonic, antispasmodic, and sedative; lessens the force and frequency of the pulse; produces contraction of the uterus.

Administration.—As a spray.

Dose.—Of the tincture 20 to 40 minims in $\frac{1}{2}$ to 1 ounce of water; of cimicifugin 3 to 4 grains.

Indications.—Said to be useful in sore throat, 'simple and malignant, and in that troublesome chronic and obstinate disease in which the mucous membrane of the pharynx is quite dry and spotted over with inspissated mucus.' (Ringer) Given, it is affirmed, with much benefit in acute catarrhs and in influenza with headache and the feverish symptoms which ordinarily accompany acute and subacute colds.

Piscidia Erythrina.

A tincture and fluid extract of the bark of this plant, Jamaica Dogwood, are prepared.

Properties.—Sedative, antispasmodic, anodyne, and soporific. When given as a narcotic it does not lessen secretion, occasion headache, or cause constipation.

Administration.—1 drachm of the fluid extract is equal, according to Martindale, to 1 drachm of the bark; the dose ranges between 20 minims and 2 drachms.

Indications.—Said to be particularly serviceable in the cough of bronchitis and phthisis.

NERVINE SEDATIVES.

Acidum Hydrocyanicum Dilutum.

The dilute hydrocyanic acid P. B., or prussic acid, contains 2 per cent. of the strong acid, while Scheele's prussic acid contains as much as 4 per cent. of the anhydrous acid.

Properties.—Anodyne, sedative, antispasmodic, acting upon the nerve centres chiefly, but the acid also combines with the hæmoglobin, and this may interfere with the oxidation of the tissues.

Administration.—Used in the form of vapour or as an atomized spray. For the vapour, The Pharmacopœia directs that from 10 to 15 minims of the dilute acid should be added to 60 minims of cold water. Mix these 'in a suitable apparatus and let the vapour that rises be inhaled,' but the nature of the apparatus is not described. The spray may be used in the form of either mistura amygdalæ amaræ P. B ; aqua laurocerasi P. B., prepared from the leaves of the cherry laurel ; or the acidum hydrocyanicum dilutum may be employed.

Doses.—Of the bitter almond mixture from $\frac{1}{2}$ to 1 ounce may be used for each inhalation, of the cherry laurel water 5 to 20 minims in $\frac{1}{2}$ to 1 ounce of water, and of the dilute acid from 2 to 6 minims in the same quantities of water.

Indications.—It has been stated that hydrocyanic acid acts chiefly through the nervous system : its administration, therefore, is specially indicated in certain nervous and spasmodic affections of the organs of respiration; in nervous, irritable, and spasmodic coughs; in whooping cough and in asthma ; also in affections of a painful character, as in ulceration of the larynx accompanied with much pain.

Bromum.

Bromine occurs in sea-water, chiefly in the form of bromide of magnesium. It is obtained by passing chlorine gas through the liquor left after the common salt has been removed by crystallization; chlorine uniting with the magnesium, bromine rises to the surface and is separated by means of ether. Extremely volatile at ordinary temperatures, the fumes evolved being acrid and irritating. It boils at 117° F.

Properties.—Employed chiefly in the form of the bromides of ammonium and potassium, the properties of which differ somewhat from each other. These salts act as nervine sedatives, and hence are calmative, antispasmodic, and soporific, while the potash salt is also absorbent and deobstruent; on the other hand the ammonia compound, which is entirely volatile, has no depressent action. Dr. Weir Mitchell considers bromide of lithium to be more effective than either of the preceding, in consequence of its containing a larger proportion of bromine, especially where the hypnotic effect is desired.

'Bromide of potassium, like the chloride, paralyzes not only the central nervous system, but likewise the nerves, muscles, and heart, the central nervous system being affected sooner than the nerves, and the nerves sooner than the muscles, and therefore we conclude that these effects of bromide of potassium, which it possesses in common with all potash salts, are due solely to the potash, the bromide playing no part in their production' (Drs. Murrell and Ringer).

Dilute hydrobromic acid is sometimes employed to allay nervous irritability in place of bromide of potassium. It is very sour, and the dose required is large, 1½ to 3 drachms.

Administration.—In the form of spray, the vapours being too irritating.

Doses.—The dose for each inhalation of the ammonia salt should vary from 3 to 10 grains, and of the potassium salt from 5 to 15 grains dissolved in ½ to 1 ounce of water. The doses must be proportionately diminished in the case of children. The bromides when inhaled as sprays may sometimes be advantageously combined with glycerine.

Indications.—All the bromides lessen irritation and spasm of the nerves and muscles of the respiratory organs, and hence are calculated to afford relief in affections accompanied by irritation, pain, and spasm; in nervous, hysterical, and irritative coughs, in laryngitis stridulosa, in croup, in whooping-cough. It is only in the uncomplicated and simply spasmodic forms of whooping-cough that the bromides will prove useful; if there be any special cause or causes of irritation, as acute catarrh and inflammation accompanied by fever, they will probably fail; in spasmodic asthma, in which case the potash salt is perhaps to be preferred; often very useful in allaying the distressing reflex vomiting induced by the cough in advanced stages of phthisis. Bromide of potassium has been employed to produce anæsthesia of the pharynx and larynx, with a view to reduce the undue reflex sensibility of these parts, so as to allow more easily of the employment of the laryngoscope. It is, however, stated that while it diminishes reflex action of the upper part of the pharynx it does not render it anæsthetic, it being just as sensitive to pain as ever. The dose required for the purpose is large, and Dr. Mackenzie questions whether it is capable of so quieting the pharynx as to allow of the use of the mirror without difficulty, and he states that he finds ice more effective. By Cocaine the difficulty is now overcome.

Vascular Sedative.

Digitalis Purpurea.

Digitalis, or foxglove, owes its properties mainly to the presence in the leaves of the non-nitrogenous substance digitaline. The preparations chiefly used are the infusion, the tincture, or digitaline itself.

Properties.—Diuretic, a vascular sedative and heart tonic, diminishing the frequency of the pulse, but increasing its force. It does not escape readily from the system, in which, therefore, when regularly administered it is apt to accumulate. In large doses antipyretic. Much used on the Continent in non-specific and specific febrile affections.

Administration.—As an atomized spray, the active principle being non-volatile.

Doses.—Of the infusion 2 to 4 drachms; of the tincture 5 to 20 minims; of digitaline $\frac{1}{60}$ to $\frac{1}{30}$ of a grain.

Indications.—Given as a cardiac sedative, in almost all cases where there is excited action, whether it be of sympathetic origin or due to organic disease of the heart or great vessels; in pneumonia; valuable in epistaxis and hæmoptysis, especially when it reduces the frequency of the pulse, but considerable doses of the infusion will usually be necessary; in bronchitis associated with increased action or disease of the heart.

THE INHALATION OF POWDERS.

The inhalation of powders is very often resorted to in the treatment of affections of the respiratory track, and there are some occasions and affections in which they may be employed with decidedly beneficial effect, and when indeed they are to be preferred to vapours and sprays.

When the medicament is so strong that it cannot be applied to the parts without exciting too much irritation, it is customary to mix it with some indifferent or inert substance in variable proportions, which may be soluble or insoluble, as sugar of milk, white sugar, magnesia, or lime.

The treatment by the inhalation, or, as it may be termed, the insufflation of powders, is best suited for certain affections of the nose and its cavities; next for those of the throat and pharynx; and lastly, to a still smaller extent, for maladies of the larynx, trachea, and bronchi, and this mainly because the greater part of the powders used becomes arrested in the throat and before it reaches the glottis.

The particular affections in which these insufflations are indicated are coryza in several of its forms, including that arising from hay fever and also that which is accompanied by fœtid discharge; in aphthæ, in limited inflammations of the palate and throat, in aphonia, hoarseness, and in chronic affections of the larynx.

When the pulverized medicaments are intended to penetrate into the air passages, then the inspirations must be deep and prolonged.

The principal medicaments employed in the form of powder are alum, tannin, borax, bismuth, nitrate of silver, mercury in the form of bichloride or the red oxide, carbonate of ammonia, camphor, acetate of lead, sulphate of iron, sulphate of zinc, sulphate of quinine, salicylic acid, and morphia.

CHAPTER VII.

THE INHALATION TREATMENT OF DISEASES OF THE ORGANS
OF RESPIRATION.

Nasal Catarrh.

NASAL catarrh, or coryza, may be defined to be an inflammatory affection of the Schneiderian membrane, usually due to cold, and which may be either acute, subacute, or chronic. The inflammation may be limited to the mucous membrane of the nose; it may extend to the sinus of Highmore, to the frontal sinuses, or it may pass downwards to the throat and even to the larynx, trachea, and bronchi.

The Germans use the word 'catarrh' in a much more comprehensive sense than we do; in fact nearly every affection of the mucous membrane of the nose, throat, air passages, stomach, and bowels attended with a serous or mucous discharge is with them denominated 'catarrh,' no matter whether this discharge is occasioned by exposure to cold, whereby inflammation has been induced, or by inflammation and irritation arising from other causes.

In England, formerly the use of the word 'catarrh' was for the most part confined to the affection of the mucous membrane of the nose and throat due to cold only; now, however, the causes giving rise to nasal flux being better understood, the word is employed in a wider sense, although we have not yet learned to speak of stomach and intestinal catarrhs.

The causes of catarrh may be divided into exciting and predisposing, or constitutional.

One of the most frequent exciting causes is without doubt sudden exposure to cold, although the manner in which this operates is not exactly known. The local effect of cold on the mucous membrane is to produce contraction of the vessels, followed as soon as reaction sets in by an increased flow of blood to the part, and in some cases by increased secretion. Among other exciting causes are strong sunlight, which with many persons gives rise to sneezing, and a temporary flow of mucus; and the continued inhalation through the nostrils of irritating substances, as different kinds of dust and powder, including the pollen of grasses.

Among the predisposing causes are youth and old age, particularly, constitutional susceptibility, a strumous and rheumatic diathesis, and general debility.

Infants also are very liable to nasal catarrh or coryza, and when thus affected, in consequence of the narrowness of the meatuses, they experience great difficulty in breathing through the nose, so great indeed that they are often unable to suck and have to be fed by hand.

Catarrh sometimes prevails epidemically, but there is no conclusive evidence to show that it is catching, as is commonly supposed.

Not unfrequently a running from the nose follows the suppression of some habitual discharge, but these cases can scarcely be regarded as instances of veritable catarrh.

Again, the nasal complications which simulate catarrh more or less, and occur in scarlet fever, measles, &c., are not to be regarded as examples of catarrh, since they are due to specific poisons, and nothing of this kind has been discovered as yet in ordinary catarrh. It is extremely doubtful also, whether influenza is to be regarded as a simple catarrhal affection.

The symptoms premonitory of an acute nasal catarrh are lassitude, chilliness, fever, heaviness of the head, with perhaps some degree of frontal pain or headache; soon sneezing sets in, usually in paroxysms, the mucous membrane becomes dry and swollen, the nasal passages being narrowed, with, as a consequence, more or less difficulty of breathing through the nose and some alteration of the voice. Although the membrane is at first dry, this state is soon succeeded by the discharge of a thin saline and irritating liquid, which is often thrown off in great abundance.

These symptoms are characteristic of the acute stage, which after an interval of two or three days passes into the subacute and chronic stages; the fever disappears; the tumefaction is lessened; the discharge loses its watery and scalding character, gradually becomes thicker, is more mucoid and in some cases even muco-purulent and streaked or spotted with blood. As the catarrh draws to an end, the discharge becomes thinner and more transparent, but is usually less abundant and no longer possesses the same irritating qualities.

The chronic stage, in simple and typical cases, soon passes away, but occasionally it is indefinitely prolonged.

Sometimes the disease is limited to the anterior portion of the nasal passages; at others it may extend to the posterior part, or the mischief may involve the antrum or the frontal sinuses, as may be indicated by painful sensations in those situations.

The mucous membrane, in consequence of the discharge, becomes more or less denuded of epithelium, excoriated or superficially ulcerated.

Usually the affection terminates without leaving any permanent structural effects, though, if the chronic stage be greatly prolonged or the patient suffer from repeated attacks, it may result in permanent thickening or hypertrophy of the Schneiderian mucous membrane, or in atrophy, especially of its glandular structures, in ulceration, or in what would be usually called ozæna, that is to say, in a discharge possessing peculiarly offensive properties.

It is in the young, the aged, the debilitated, and the constitutionally susceptible, that colds are apt to recur. Colds are also liable to extend to the pharynx, along the Eustachian tube, producing in some cases deafness, or even to the larynx and bronchi.

When the chronic stage of the catarrh is prolonged, the character of the discharge will vary considerably; it may be thin, watery, and abundant or even excessive (rhinorrhœa), or scanty and thick.

When the chronic stage is greatly prolonged, a very careful examination should be made, with a view to discover whether the discharge may not be kept up by some other cause, as by ulceration, syphilitic or otherwise, or by the presence of polypi, adenoid or other morbid growths.

If the discharge be decidedly purulent, it may possibly depend upon leucorrhœal or gonorrhœal infection, but in such cases the irritation and discharge will usually extend through the lachrymal ducts to the conjunctiva.

Treatment.—The principles of treatment are just the same, whether the catarrh be confined to the nares, include the frontal sinuses, extend to the throat, or embrace all three parts. Catarrh of the throat is simply an extension, in most cases, of the disease from the nares; indeed, in catarrh in the head there is very generally a tendency to descend successively to the lower and deeper portions of the air passages, as the larynx, trachea, and bronchi, and even lungs. This descent is very apt to occur in delicate persons and in those who are disposed to lung mischief; the apices of the lungs becoming chronically affected, the catarrh not unfrequently leads to structural alterations, and thus in some cases lays the foundation for tubercular deposit.

The treatment must of course be modified according to the stage. In the premonitory stage it should be abortive, that is, measures should be adopted with a view to stop the catarrh at the onset and before it becomes fully developed. In the acute stage it should be on the principle of derivation, that is to say, the nasal congestion should be relieved by diverting it into other channels, as particularly the skin and bowels, while in the subacute and especially the chronic stages, local measures are particularly indicated.

The abortive treatment, to be effective, must be commenced at the very earliest moment possible, when the symptoms are premonitory rather than positive.

Various remedies have been employed with a view to arrest the development of a catarrh. Of these, some act on the nervous system soothingly by allaying irritation, others by stimulation, others antiseptically, and others again by derivation.

The chief soothing remedies are chloroform, inhaled through the nostrils, tincture or some other preparation of opium, applied as a spray or taken internally, or a powder of opium and bismuth insufflated or used as a snuff, or camphor, which may be used in powder, in vapour, or in spray.

The chief stimulant remedies are a solution of ammonia or carbolic acid, separately or combined, as they are stated to be in the popular remedy known as 'alkaram.' The fumes should of course be strongly inhaled.

The combination of ammonia and carbolic acid is an excellent one, for not only are the effects of both remedies obtained, but the volatility of the carbolic acid becomes greatly increased by the formation of carbolate of ammonia. Another stimulant sometimes used is acetic acid.

The derivative remedies which are specially suited for the

treatment of catarrh in the acute stage, are those which act either by reducing the fever, as aconite, or which act upon the kidneys, as chlorate of potash, on the skin, as spirit of nitrous ether, or on the bowels. Pulvis ipecac. Co., or Dover's powder, acts both as a sedative and sudorific.

One of the very best of all remedies for colds accompanied by feverish symptoms is aconite. This should be taken or inhaled, in the manner already pointed out in the chapter on the Therapeutics of Respiration.

Another general remedy given, it is affirmed with great benefit in catarrh and influenza in the acute stage, and while there is fever, soreness, and aching pains in the limbs and body, is actæa racemosa, or black snake-root.

At the same time that aconite or snake-root is administered, the vapour of hot water may be inhaled with advantage, simple or medicated, as with extract of conium; or warm, emollient, medicated sprays may be inhaled of infusions or decoctions of poppy-heads or marsh mallow. These may be rendered more effective in some cases by the addition of a small quantity of either extract of poppy or lettuce; or, lastly, a spray of almond emulsion may be used. These inhalations may be repeated three or four times a day.

When the subacute stage has been reached, a remedy much extolled in the early stage of cold in the head is chlorate of potash; if taken early enough it will, according to Dr. Leonard Sedgwick, stop many a cold, and it will quickly relieve the feeling of stuffing of the nose, soreness of the throat, and hoarseness of the voice.

The efficacy of emollient sprays will be increased by the addition of a few grains of chloride of sodium; or the fumes of chloride of ammonium may be inhaled in the manner and with the apparatus which has been already described and figured, but this last remedy is perhaps still better suited to the chronic stage.

The remedies best adapted to the third or chronic stage of catarrh, described by some writers as a separate affection, comprise some which act soothingly, some as mild stimulants, others as solvents for the thickened mucus, and again others which are astringent and which are designed to check the discharge when too abundant. The solvent remedies include particularly chloride of sodium or salt, and the carbonates of soda and potash; chloride of ammonium is mildly stimulant, while alum, gallic and tannic acids are all suitable astringents.

As the chronic stage is reached, the secretion, previously thin and watery, becomes gradually thick and tenacious. It is in such cases that the use of the solvent spray is specially indicated; but later on, when the chronic stage has become prolonged and the mucous membrane shows signs of relaxation, the astringent glycerides must be employed in the form of sprays, to complete the cure and to give tone to the debilitated structures.

Sometimes solid masses of hardened mucus are formed in one or even both nostrils, plugging them up more or less and so rendering respiration difficult and most uncomfortable; then the warm emollient alkaline injections should be strong and frequently repeated, and should afterwards be followed by efficient astringents.

In very obstinate cases it may be necessary to employ more powerful astringents, as sprays of acetate of lead, sulphate of zinc, or tincture of iodine.

If the discharge be fœtid, a weak solution of carbolic acid containing a little glycerine will usually remove this condition. This inhalation will be the more effective if preceded by the employment of the nasal douche.

Dry Nasal Catarrh.

The causes which lead to this condition appear to be but imperfectly understood. It is defined to be a dry state of the nasal mucous membrane with the formation of thin, dry, and adherent flakes of mucus, which often emit an offensive odour, and leading in many cases ultimately to atrophy of the turbinated bones and considerable enlargement of the nasal passages.

The flakes of mucus may present various tints of grey, yellow, or brown, and when they emit an offensive and characteristic odour the case is regarded as one of true 'ozæna,' to which indeed this term is now in general limited. Sometimes, in place of flat scales, masses of a considerable size and often of an elongated form, moist on the outside but dry within and variously coloured, are expelled from time to time. These masses are obviously of slow formation, and usually proceed from the upper part of the nasal fossa.

The adherence of the dry flakes to the mucous membrane gives rise to irritation and leads to attempts to remove them, which sometimes occasion slight hæmorrhage. Usually the subjacent membrane is entire, but occasionally superficial ulcerations may be detected; but when deep ulcerations are found to be present the case is in all probability no longer one of ozæna, but is dependent on struma or syphilis.

Dry nasal catarrh is essentially a chronic and persistent condition, occurring chiefly in children and but seldom in adults after middle life.

At first and for some considerable time the mucous membrane is somewhat hypertrophied, but after the complaint has persisted for months, or even years, it becomes thin, atrophied, and more like a serous than a mucous membrane, the turbinated bones themselves usually ultimately participating in the atrophy. The effect of the wasting is that the nasal passages become large and open, a condition which, according to some authorities, is noticed at an early stage of the affection and which is considered to be one cause of the drying of the mucus.

Although the scabs generally possess a peculiarly fœtid odour, there are some exceptions to this rule, and several explanations of the cause of the odour have been advanced, which, however, are not satisfactory: one is that it is due to the decomposition of the mucus, another that it is owing to fermentive changes therein. A very simple explanation, and one which seems probable, is the following: The surface of the otherwise dry flakes next the mucous membrane is always more or less moist, the air being at the same time in a great measure excluded; thus on this surface the precise conditions exist which are favourable to decomposition, and this of a peculiar kind. If an organic liquid, such as beef-tea, be partially excluded from the air, and if it be placed in such a situation that it retains only a moderate degree of warmth for some hours, it will often be found at the end of the time to have acquired a most disagreeable odour, while if a portion of the same liquid be freely exposed to the air and be allowed to cool naturally, it will remain sound and good for a considerable period. This fact is well known.

Now as to the causes of dry nasal catarrh. In some cases no specific or special cause is discoverable, but a considerable number of such cases are associated either with a scrofulous diathesis or with a syphilitic taint.

Treatment.—Dry catarrh is a peculiarly obstinate affection, and when it is accompanied by fœtor it is usually regarded as incurable, although it will still admit of great amelioration when suitably treated.

When the disease is traceable to any constitutional diathesis or malady, the treatment of this will occupy the first place in importance, but the local treatment is capable of affording very great comfort and relief.

The local treatment has for its principal objects the softening of the mucous scales, the prevention of their drying, and the correction of the fœtor.

The first object is aimed at by the inhalation of the vapour of hot water, by warm emollient and anodyne sprays, and particularly by the employment of solvent and alkaline sprays, composed of carbonate of potash, chloride of sodium, or borate of soda. These three remedies may be all used in combination together with a small quantity of glycerine.

The most suitable antiseptics for the fœtor are those of a non-irritating and non-drying character, and include sprays of quinine, salicylic acid, and permanganate of potash. Carbolic acid or resorcin would probably be too caustic, and iodine too drying.

With a view of remedying the dry condition of the nasal mucous membrane, Gottstein introduces into the nasal passages plugs or tampons of cotton-wool; these may be dry or variously medicated: by their presence they excite the mucous membrane with which they come into contact to secretion, and so bring about a moist condition. These appliances are, however, most disagreeable and trying, and a further objection which may be urged against them is, that in all probability they would not nearly cover all the surfaces affected, since the mischief sometimes extends into the frontal sinuses. It is not necessary that the tampons should be worn for more than an hour or two at a time, and no doubt they do in some cases afford considerable relief.

Dr. Morell Mackenzie states, that remedies which stimulate the mucous membrane certainly do good, and recommends the insufflation of a powder consisting of 1 part of red gum to 2 of starch, while Bosworth has expressed himself very favourably of a powder, composed of 1 part of sanguinaria to 3 of starch; but this application is much more irritating, and Mackenzie advises that it should be considerably diluted before being used.

Formerly almost any disease of the nasal passages attended with fœtid discharge would have been regarded as a case of ozæna, no matter from what cause the fœtor proceeded, whether from polypi or other morbid growths, from deep ulcerations or diseased bone; but the confusion attending such a use of the word has led to its restriction to cases of dry catarrh accompanied by fœtor.

Hay Fever.

This troublesome and distressing complaint, and which would appear to be on the increase, has received several names, none of which are altogether satisfactory, as hay asthma, hay fever, summer catarrh, and pollen catarrh. The asthmatical paroxysms are merely the consequences of the irritation set up in the Schneiderian and adjacent mucous membranes; the fever is often inconsiderable and not attended with any marked elevation of temperature. Again, the disease is not confined to the summer months, while the catarrh may be excited by other causes than by the pollen of certain plants.

The causes are divisible into the *predisposing* and *exciting*; the former include particularly race and constitutional predisposition.

With respect to race, it is a remarkable and hitherto unexplained fact that English and Americans are the chief sufferers.

The peculiar constitutional predisposition may be either hereditary or acquired. The precise nature of this predisposition is not known, but it is usually characterized by the possession of a nervous and excitable temperament. It is in this way that the prevalence of the disease among educated persons of the upper classes has been explained.

It seems to me to be probable that the condition, inherited or acquired, of the mucous membrane of the respiratory track in some persons may be an important predisposing cause or factor. It is well known that the mucous membranes of some persons are particularly prone to go wrong and to become variously affected, as with catarrh or diarrhœa; further, that the mucous membrane which has suffered frequently from catarrh, has its sensibility greatly increased, and so becomes more than ordinarily susceptible to irritation; and again, it can scarcely be doubted that a mucous membrane which has been repeatedly the seat of catarrh must suffer to some extent in its minute organization, and in this way, in some instances at least, a disposition or proneness to hay fever may be established. However this may be, when the mucous membrane of the eyes, lachrymal ducts, nose, pharynx, larynx, or bronchi, and especially the nasal fossæ, has been repeatedly subjected to hay fever, it is scarcely to be believed that it can altogether escape unharmed and return to its normal condition. Dr. Morell Mackenzie, it is true, states that he has never found 'anything more than general congestion,' and further that 'hay fever leaves no permanent structural lesion behind, and it cannot therefore be said to have any pathology.' Dr. Daly, of Pittsburg, on the other hand, believes that there is an intimate relation between the so-called hay fever and chronic nasal catarrh, and this view he supports by some well-marked cases which fell under his own observation. The patients had suffered from the complaint for 15 to 21 years; in two of the cases there was thickening of the turbinated bones, and in one a polypus, and as soon as the parts were restored to their normal condition the disease ceased. These views have since been supported by Roe and Hack in the 'Archives of Laryngotomy' for 1882, and they are important as bearing on the subject of treatment.

Another fact worthy of notice in the etiology of the disease is, that those who live in towns and afterwards go to the country are more liable to the disorder than habitual dwellers in the country.

We may now refer to the chief *exciting* causes.

These are dust of some kind or other, and strong light, especially sunlight; these, singly or combined, frequently give rise to fits of violent sneezing and coryza, especially in those whose Schneiderian membrane is characterized by extreme sensitiveness. Some forms of dust are of course much more irritating than others, and even the smell of the powder of ipecacuanha is capable in some rare instances of giving rise to symptoms which resemble those of hay fever. It has, however, been fully established by the researches and experiments of Blackley that the general exciting cause of hay fever is the dust or powder of pollen, almost always that of hay or of some of the cereals. Probably the chief reason why these kinds of pollen are most frequently the exciting causes of the catarrh is, that at certain times of the year they are far more abundant in the air than is the pollen of all the other kinds of flowers and trees put together.

There are, however, some other kinds of pollen which are particularly irritating to the Schneiderian mucous membrane; one is that of the rose, which in America has so frequently excited the disorder that it has there received the name of 'rose fever.' Another kind of pollen common in America, that of a species of wormwood or artemisia, is also known to produce similar effects, and Dr. Paget, of Great Crosby, states, he finds in his own case, when the complaint has once set in, that inhaling the perfume of almost any flower in full bloom will bring on an attack. It would be interesting to know, whether he finds the same effect to be produced by flowers which are destitute of pollen, as are many double flowers, including most cultivated roses.

The question may now be asked, What imparts to pollen its peculiarly irritating qualities? Is it the form of the envelope of the granule, the irritation due to the root-like prolongations which spring from it when in contact with moisture, or is it the contents of the granules, which become discharged from the tubules when these rest on any moist surface? It has been shown that pollen as a whole when applied to the skin is capable of setting up irritation therein. It can hardly be the form of the granules which occasions the irritation, since that of the Graminaceæ is of the simplest kind, and there are a great variety of other granules, as those of the natural order Compositæ, which are covered with spines and are really irritating.

Each pollen grain consists of a comparatively strong external membrane, in which are certain slits or holes, which are intended for the passage of the tubes which convey the granular fertilizing liquid. When the pollen falls on the stigmata, the grains, particularly those of the Graminaceæ, which are elongated and tri-lobed, swell up, become more or less rounded, the slits or apertures open out, thus allowing of the issue of the three tubes; these gradually elongate, penetrate downwards through the stigmata to the ovaries, where their contents are finally discharged, and the act of fertilization is accomplished. The grains of some plants give issue to a great many more tubes than the grasses.

The *symptoms* are either those of catarrh, or of catarrh and asthma combined, and either with or without fever. Although the most frequent cause of hay fever is pollen, there is reason to believe that the application of other powders, organic and inorganic, is capable, when these are inhaled continuously for a sufficiently long period, of inducing a precisely similar affection.

It is not necessary to enter into any lengthened description of the symptoms of this well-known but most troublesome affection; it assumes two forms, the catarrhal and the asthmatic, which are often combined, but which may exist more or less independently. The attacks are usually ushered in with considerable irritation and itching of the eyes and nose, with frequent fits of sneezing and discharge; the mucous membrane of the nares becomes so much swollen in the more severe cases as to almost entirely obstruct respiration by the nose. When combined with asthma, the seizure usually takes place in the day-time, and this, together with the sudden onset of the disease, the period of the year at which it occurs, and the attendant itching, serves in most cases to distinguish the affection sufficiently. Another diagnostic symptom is the puffiness of the eyelids.

The unsatisfactory nature of the term 'hay fever,' by which it is almost always distinguished, is shown by the general absence of constitutional disturbance and of those symptoms which in their aggregate denote fever.

Treatment.—To be successful, this must to a large extent be local, and inhalation should play an important part in it. Dr. Morell Mackenzie writes, 'I trust very little to local measures in the treatment of hay fever.'

The first thing to be done is to get away if possible from the influence of the pollen of the grasses and cereals, either by going to the seaside or by taking a sea trip, although these remedies, for some obvious reasons, are not always successful. One of these is that when the wind blows from the land it may convey the pollen far out to sea; another possible occurrence is, that the patient may carry the pollen about with him in his clothes.

The next step, when removal from the locality is not practicable, is to preclude as far as possible, on the principle that prevention is better than cure, the entrance of the pollen into the nares. This object is difficult of accomplishment, but various suggestions have been made to carry it into effect, and some of these have proved more or less effectual. One of these is, that the patient should plug his nostrils with cotton-wool or wadding by means of one of Gottstein's screw tampons. Of this plan one is tempted to observe that the remedy must be almost worse than the disease.

Plugging the lachrymal ducts with small glass rods has been recommended by Hannay, as also the employment of a clamp to close the nostrils. This last is more feasible and a far less disagreeable plan than those previously noticed, although one is led to remark of all these proceedings that even if they succeeded in completely preventing the entrance of the pollen into the nares, the channel by the mouth would still be open.

The glycerine nasal plug devised by Dr. M. D. O'Connell is less open to the objections which may be urged against the other nasal plugs proposed, and it acts on a totally different principle. It consists of a small piece of cotton-wool saturated with glycerine, which is to be introduced into one or both nostrils for an hour at a time. It causes a copious watery discharge, which greatly relieves the congestion. Dr. O'Connell found it very useful in his own case.

Another plan which has been recommended is to wear a three-ply fine gauze veil, and this, it is stated, has been found very useful. It should be open at both ends, attached to the hat by one extremity, and the other, after being passed over the face, is to be folded round the neck and shoulders.

It seems to me, that in place of aiming at the plugging or closing of the nostrils, the effort should be made rather to filter the air, and so to deprive it of pollen before it enters the nares. This object is, indeed, very imperfectly accomplished by the veil above referred to; but it will be still better effected by wearing the recently devised nasal respirator of Dr. George Moore, which has already been described, provided this is carefully and properly packed with cotton-wool; or the respirator of Dr. W. Williams.

A very obvious precaution, and one which is attended with a mitigation of suffering, especially when the eyes are affected, is to wear some well-fitting tinted spectacles.

A great variety of remedies, chiefly intended for inhalation, have been recommended for hay fever, and some of them have been lauded as specifics. They may for the most part be divided into five classes: powders, volatile stimulants, inhaled either as vapours or liquids; antiseptics, anodynes, and astringents.

The powders consist either of substances which are soluble in the nasal mucus, or insoluble, or both combined. One would say of these substances, so far as they are insoluble and consist of palpable particles, although some of them may be temporarily beneficial, that they are more likely to do harm than good by their mechanically irritating effects; their soluble constituents are of course less permanently irritating, and may in some cases give a brief relief, due in general to their strongly stimulant and anodyne constituents.

Amongst the remedies of this class Ferrier's snuff, Himrod's cure, and Dr. Granville's powder may be mentioned.

Dr. Ferrier's snuff consists of hydrochlorate of morphia 2 grains, subnitrate of bismuth 6 drachms, and powdered acacia 2 drachms; from $\frac{1}{4}$ to $\frac{1}{3}$ of this may be used as snuff in the course of 24 hours.

The general composition of Himrod's cure will be found described under the head of Stramonium.

The powder recently devised by Dr. J. Mortimer Granville, to which he has given the name Pulvis boracis co., consists of Boracis pulv. gr. xx.; Capsici pulv. gr. xv.; Ammoniæ carbonatis gr. x.

Dr. Granville states in the 'British Medical Journal' of June 21, 1884, 'that the powder acts by exhaustion of special irritability by intentional excitation,' and also 'that a third application generally, occasionally a fourth, at intervals of three or four hours will entirely cure the affection. I use the word "cure" advisedly.'

Of the volatile stimulant remedies that have been inhaled none has given greater, though only temporary, relief than the vapour of strong ammonia or its carbonate, and one or other of these forms the chief constituent of many of the preparations sold as cures for asthma.

The chief antiseptic remedies which have been employed are quinine, salicylic and carbolic acids. Quinine has been specially praised by Helmholz, who used it as an injection, but others have expressed disappointment at the results obtained. Mackenzie states, that in a few cases benefit was derived, in most cases no effect was produced, while some patients were actually made worse.

In the same way salicylic acid, 1 in 1000 parts of water, has been injected into the nares, and with the effect, it is affirmed, of cutting short the attack. Mixed with starch, salicylic acid in powder may be insufflated.

Carbolic acid probably acts less by its antiseptic properties, than by its effects in deadening for a time the sensibility of the nasal mucous membrane.

Of all the remedies employed there are probably none which are more generally beneficial, although the relief afforded is but temporary, than anodynes, especially opium and belladonna, in the form of atomized sprays. Mr. Phillips particularly recommends belladonna. Mr. Lennox Browne believes that this acts by its drying effect on the mucous membrane of the nares rather than by its narcotic properties. The combination of astringents with narcotics gives in many cases great relief, as does also brushing over the mucous lining of the nose with a strong solution of nitrate of silver or the inhalation of a weaker solution as a spray.

The sudden and violent fits of sneezing and the running from the eyes and nose constitute two of the most troublesome symptoms and accompaniments of hay fever; for these certain remedies have been specially recommended, and all of them doubtless often prove serviceable. One of these is the inhalation of a spray of sulphurous acid, another the inhalation of the vapour of iodine or of a spray of iodide of potassium, 5 grains to the ounce. Dr. Ringer states that arsenic never does any good in the paroxysmal sneezing of hay fever; cocaine would probably greatly relieve such attacks.

A local treatment of a totally different kind has recently been employed, namely, the galvano-cautery; the object of this application is to effect a permanent alteration in the condition of the hypersensitive mucous membrane.

Dr. Daly is of opinion that some local lesion, such as hypertrophy, is always to be found in some portion of the mucous membrane of the nasal track in hay fever, and in this opinion Dr. John O. Roe, of Rochester, New York, agrees, and Dr. Daly has employed the galvano-cautery in some cases with good effect. Mr. Lennox Browne also speaks favourably of the galvano-cautery. In connection with this treatment the difficulty may be pointed out of reaching by the means indicated the whole of the affected surfaces.

Again, Lefferts, of New York, has been in the habit of treating a hypertrophied condition of the nasal mucous membrane with either fuming nitric, glacial acetic, or chromic acid.

I will now briefly summarise the treatment to be pursued in most cases of hay fever.

First, one or more of the preventive measures already referred to should be adopted: removal to the seaside or a voyage, the wearing of tinted spectacles and also Dr. Moore's nasal respirator, or a veil or some other effective covering for the whole face, if the eyes are affected, and if not, for the mouth and nose only.

R

If these means are not sufficient then local treatment must be resorted to.

Every morning and evening the mucous track should be thoroughly cleansed with the nasal douche, for which tepid water may be used, or better still a strong decoction of poppy-heads rendered slightly alkaline with carbonate of soda.

If there be considerable tumefaction of the mucous membrane and the passages be obstructed with thickened mucus, a spray of carbonate of soda should be vigorously inhaled, about 10 grains of soda to 1 oz. of water.

If there be much irritation or pain, then either opium, morphia, or belladonna in spray or powder may be used.

When the tumefaction and irritation are greatly subdued, then astringent remedies may be used, or a solution of nitrate of silver, 3 to 5 grains to the ounce. This may be inhaled as an atomized spray; but if the solution be applied by means of a brush, then it should be stronger, 20 grains to the ounce.

Other atomized sprays which will often be found serviceable, acting partly by their drying effect on the mucous membrane and partly by diminishing its sensibility, are sprays of solutions of tannin or of acetate of lead. For the affection of the eyes a collyrium of acetate of lead or sulphate of zinc, 2 grains to the ounce, may be employed.

Should there be paroxysmal attacks of dyspnœa, these must be treated in the ordinary way, and chiefly by antispasmodics and anodynes. A little chloroform or ether, or better still the two mixed, or the fumes of nitre paper, should be inhaled from time to time. Dr. Hyde Salter found nauseating doses of ipecacuanha very serviceable in cutting short the dyspnœa of hay asthma, and preferred it to tartar emetic.

Many cases of hay fever occur, as already stated, in persons of nervous temperament, with occasionally accompanying debility. For this condition the administration of suitable tonics is indicated, as quinine, arsenic, or valerianate of zinc. Dr. Morell Mackenzie recommends the prolonged use of pills containing 1 grain of valerianate of zinc and 2 grains of the compound assafœtida pill; he advises his patients to begin taking them as the season for hay asthma approaches, and he states that the prophylactic and other beneficial effects are considerable.

Should all the measures adopted fail to afford the relief desired, then the galvano-cautery may be resorted to, if necessary; it is but little painful.

Hypertrophy of the Nasal Mucous Membrane.

As a consequence of repeated catarrhs, especially in the young and those of scrofulous diathesis, thickening or hypertrophy of the nasal mucous membrane is apt to occur; this may be confined to the anterior portion of the lower and middle turbinated bones, but not unfrequently the posterior parts are involved. When the thickening occupies the latter situation, mulberry-like, adenoid growths are prone to be formed; these may either present a reddish colour, when they are disposed to bleed, or they may be pale. The thickening is usually bilateral.

This condition is revealed especially by the narrowing of the nasal passages and the formation of adventitious growths, including polypi, which, if large, may give rise to reflex phenomena, as cough, spasm, or even asthma-like attacks.

The symptoms which first attract attention are those of an unusually severe and prolonged nasal catarrh, accompanied with considerable difficulty of breathing or even inability to breathe through the nose, and with alteration of the voice, which assumes a nasal character. In consequence of the difficulty of breathing through the nose the mouth is usually kept open, a symptom common to this affection and to the presence of polypi or other morbid growths in the nasal passages.

The diagnosis of this affection is on careful examination not difficult: the fact of the thickening affecting both sides alike, and being usually confined to the lower and middle turbinated bones, will serve to distinguish it from cases of polypi, although there would appear to be some connection between the two affections, since they not unfrequently coexist.

Treatment.—The treatment at first should consist in the use of soothing, anodyne, and mildly alkaline or solvent sprays, which tend to promote rather than to check secretion, and thereby to lessen congestion and to diminish the tumefaction and hypertrophy of the membrane. With this view, the nasal plugs steeped in glycerine of Dr. O'Connell would be likely to prove of service by encouraging exosmosis.

Should this object be accomplished to some extent, then unirritating astringents should be employed in the form of spray, as those of the glycerides of alum and tannin or sprays of acetate of lead. It will be necessary that these sprays should be injected into the nares both anteriorly and posteriorly, and their effect may be aided in some cases by the insufflation of astringent powders composed of tannin with a little powdered opium. In the absence of any acute symptoms, and when the mucous membrane seems to be relaxed, a spray of solution of perchloride of iron will be likely to prove beneficial. For the quantities of the medicaments generally employed the reader is referred, in order to avoid repetition, to the chapter treating in detail of the quantities of the several medicaments used in the inhalation treatment of the organs of respiration.

When the difficulty of breathing through the nose is very considerable, great relief will often be afforded by the use of elastic bougies; these preserve a passage for the air and promote absorption by the pressure they exert on the mucous membrane.

Should these several measures not be attended with success, then operative procedures will become necessary, and the thickened tissues or redundant growths will have to be removed by the electric cautery or the écraseur. But should an operation be particularly objected to, then the parts may be persistently treated with the well-known caustic paste, consisting of equal parts of caustic lime and soda, or with glacial acetic acid or nitrate of silver, insufflated by means of Bryant's apparatus.

Bleeding from the Nose.

Epistaxis is very apt to occur during childhood, at puberty, and in old age. The causes tending to its occurrence at these three periods are very different: in childhood the tissues are very soft and delicate; at puberty the whole system is in a state of excitement, and the epistaxis in most cases is congestive and salutary, while in advanced life nasal hæmorrhage is frequently due to degenerative changes.

Bleeding from the nose is prone to occur in two very opposite conditions of the system; in plethora, general and local, and in anæmia: in the first it is often congestive, and in the second it is due possibly to thinness or poorness of the blood. It is this latter condition which explains in part the frequency of its occurrence in purpura and scurvy.

Epistaxis is sometimes vicarious, and takes the place of other habitual discharges, particularly that of the menses.

Again, the tendency is sometimes hereditary, a circumstance to be explained by the transmission of certain peculiarities, either in the properties of the tissues or in the character of the blood.

Frequent local causes are irritation or injury of the nasal mucous membrane, arising from the inhalation of corrosive vapours, strong powders, or dust, and the concussion of blows.

Violent exertion and strong excitement will not unfrequently determine an attack of bleeding from the nose.

Other fertile causes are disease of the nasal passages, polypi, and other morbid growths, also ulcerations.

Then, again, maladies of an obstructive character, which impede the return of the blood to the right side of the heart, as in some diseases of the lungs, predispose to this form of hæmorrhage.

Lastly, it is a frequent concomitant of most eruptive fevers.

Thus epistaxis arises from a great variety of different circumstances and conditions. The reason that the mucous membrane of the nose should be so frequently the seat of hæmorrhage is explained principally by the peculiar character of its blood supply and the intercommunication between the veins of the nose and the sinuses of the dura mater.

Very generally the bleeding takes place from one nostril only, although in some exceptional cases it may appear to proceed from both, an occurrence which is to be explained by the blood passing from one nostril to the other behind the septum.

Treatment.—Bleeding from the nose, when it occurs in young people and unconnected with polypus, ulceration, or any other disease of the nasal passages, is usually salutary and is designed for the relief of some local or general congestion, and as soon as this is accomplished, the bleeding usually ceases spontaneously.

Before proceeding to treatment, however, the cause of the epistaxis should, as far as practicable at the moment, be determined, so as to serve as a guide to the employment of the most suitable remedies.

If interference be necessary, either on account of the persistency of the bleeding or the quantity of blood lost, usually simple means are sufficient to stop the hæmorrhage, as ice-cold water or ice to the face and back of the neck. If this does not arrest it, then strong sprays of one or other of the more effective astringents and styptics should be resorted to, as alum, gallic acid, and glycerine, or, best of all, perchloride of iron. These sprays should be gently thrown up and cautiously inhaled.

In some cases alum or gallic acid in the dry state may be insufflated.

But a variety of other remedies are available in epistaxis, the choice of which will depend in a measure on the cause and character of the hæmorrhage and on the presence or absence of accompanying fever.

One of the most valuable of these is aconite in small and frequently repeated doses, in cases in which the bleeding is attended with fever, until an effect is produced on the symptoms.

Digitalis is another remedy of proved value when it is desired to reduce the frequency of the pulse, or to give relief by the diuretic action of the drug.

Ergot, again, is very valuable, but is more suitable to non-febrile cases; it will often stop bleeding from the nose as well as hæmorrhage from other parts.

Ipecacuanha has been lauded for its power of arresting epistaxis of a congestive character, for which purpose the nauseating effects of the remedy must be produced.

Hamamelis virginica, or witch hazel, has proved itself a very effective remedy in the hands of Dr. Preston and others in cases of passive or non-febrile epistaxis.

For sprays the following quantities of the several medicaments may be used : of the solution of perchloride of iron P. B. from 20 to 40 minims in 1 ounce of water; of the glyceride of gallic acid 2 to 3 drachms made up to 1 ounce with water; of alum 20 grains in the same quantity of menstruum; of aconite 2 minims in 4 drachms of water every few minutes until the desired effect is produced; of digitalis 20 minims in 4 drachms of water every 2 or 3 hours; of tincture of ergot 30 minims every 15 minutes; of tincture of witch hazel 2 minims in 4 drachms of water, to be repeated in 20 minutes if required; of the wine of ipecacuanha 1 drachm, repeated if necessary, but this remedy will be best administered in the ordinary way by the mouth, and the same may be said of some of the other medicaments above named. These solutions should be injected slowly and with as little loss as possible. In the absence of an atomizer a small glass syringe may be employed.

Supposing the most suitable of the above remedies to fail and the case to be urgent, then recourse should be had to some of the usual surgical methods of treatment, as by plugging anteriorly one or both of the nostrils with tow or lint, first steeping it in a strong solution of perchloride of iron or tannin; or finally, should this fail, the posterior nares may be plugged in the usual manner.

It is not necessary to give any description of the general or constitutional treatment which is sometimes required. It will be evident, on a consideration of the very many and different causes which give rise to epistaxis, that with a view to the permanent cure of the affection the constitutional will be even more important than the local treatment.

Catarrh of the Naso-Pharynx.

The acute stage of naso-pharyngeal catarrh, or, as it is sometimes called, post-nasal catarrh, is usually very short, and it does not give rise to much inconvenience, but passes quickly into the chronic stage.

It is an inflammatory disease of the mucous membrane of the naso-pharyngeal space, involving particularly the follicles and attended with the secretion of a thick yellowish and tenacious mucus, which gives rise to a great deal of irritation.

The causes of this affection are in some respects obscure, but amongst them may be mentioned atmospheric vicissitudes, as extremes of cold and heat, damp and fog, irritating dust, previous attacks of ordinary catarrh, a catarrhal disposition, chronic dyspepsia, and in some cases the scrofulous diathesis.

The chief symptoms consist in some indistinctness of voice with more or less pain at the back of the throat and a troublesome sensation, as though some foreign substance were adhering to the mucous membrane; this is due to the presence of a tenacious and irritating mucus, which leads the patient to make frequent efforts for its removal.

On examining the throat it is seen to be covered with the mucus referred to; the membrane itself is more or less congested or granular, with, in advanced cases, slight excoriations, ecchymosed spots, or varicose veins.

The Eustachian tubes are apt to become involved in the mischief and to be blocked up with mucus, deafness being sometimes the result.

It is obvious from the foregoing description that this affection is allied to that which will be noticed hereafter under the name of Follicular Disease of the Throat.

In America this complaint is of very frequent occurrence; it is very widely diffused and is met with under a variety of dissimilar atmospheric conditions; but Dr. Morell Mackenzie considers dust, which is everywhere abundant on the American continent, to be a principal exciting cause.

The disease is of a peculiarly chronic and obstinate character.

Treatment.—One of the first things to be done is to cleanse the throat thoroughly, and to soften, dilute, and remove the thick and adherent mucus. This object is best effected by the employment of alkaline solvent remedies, as carbonate or borate of soda, or chloride of sodium, either separately or, preferably, in combination. With these carbolic acid may be combined, but this makes the applications very much more irritating. These remedies may be used either in the form of gargles or sprays, these latter being less liable to give rise to irritation.

After the throat has been well cleansed and the mucus removed, then unirritating astringents and antiseptics may be resorted to with advantage, as sprays of the glycerides of alum and tannin, of acetate of lead, sulphate of zinc, or bichloride of mercury.

Beverley Robinson, who has had great opportunities in America of treating this disease, has found benefit from the internal administration of sulphur, cubebs, or ammonia. While one or more of these remedies are given by the mouth the throat may still be sprayed with some sulphuretted mineral water or with cubebs or ammoniacum. The sprays should not only be applied to the postpharyngeal space, but should be made to enter the nares as well. If the force of the spray is not sufficiently great to remove the mucus, then gargles or a large syringe, or the nasal douche, may be employed.

When the health is manifestly deranged, general and constitutional treatment will be required. The diet must be carefully regulated and smoking avoided.

Dry Catarrh of the Naso-Pharynx.

This can scarcely be regarded as a distinct affection; it appears to be identical with dry catarrh of the nose, of which it is in fact in most cases an extension, though it is said that it sometimes originates independently. The masses of mucus which come away from time to time are different in form and consistency from those usually met with in dry nasal catarrh, owing to the difference of situation; instead of being flattened and dry they are thicker, more bulky, and moist and soft externally, but dry and condensed internally. Like those of dry nasal catarrh the masses often possess a fœtid odour.

Treatment.—This must be on the same principles as for dry nasal catarrh, the employment of solvent and disinfectant sprays to correct the fœtor being specially indicated.

Since this affection is often associated with a corresponding condition of the nasal mucous membrane, it will sometimes be necessary that the treatment should include the latter membrane as well.

It does not fall within the scope of this work to treat of the adenoid growths, polypi, tumours, and other surgical conditions of the nose and pharynx, and I will simply remark that inhalation treatment even in these cases will often prove of much service.

Catarrh of the Pharynx.

This is one of the most frequent causes of sore throat; it is sometimes merely an extension of an ordinary nasal catarrh, at others it has a separate origin.

The causes are very much the same as those of most catarrhs, one of the chief being exposure to cold and damp. It is not necessary to describe in detail the symptoms of an ordinary sore throat; they vary greatly in intensity and also in the extent and exact position of the parts affected. Sometimes the upper portion of the throat is chiefly involved, at others the lower; the uvula and epiglottis especially are liable to be implicated.

The mucous membrane is more or less red, sometimes even scarlet, swollen and œdematous, these conditions varying according to the severity of the attack. At first the membrane is dry and shining, afterwards coated with flakes of mucus, which give rise to a good deal of irritation and cough. If the uvula be involved it may become considerably enlarged and elongated, so as to be in itself a source of much irritation; and if the epiglottis be affected it may be red, swollen, and even œdematous. In some cases the mischief may even extend into the larynx.

When the attack is severe there will be considerable stiffness and pain, especially during the act of deglutition.

Sometimes catarrh of the pharynx prevails epidemically.

Treatment.—The treatment must be on the same principles as that for nasal catarrh. The acute stage quickly passes away, and it is the chronic stage which will chiefly demand local treatment.

At the very outset of the catarrh, an attempt may be made to stop the attack, and there are few, if any, remedies more likely to accomplish this object than opium or Dover's powder in stimulating doses. These may be used in the form of spray or powder, in which both their local and constitutional effects will be obtained.

In the acute stage great relief will be obtained by the derivative measures adverted to under the head of Nasal Catarrh, as also by the inhalation of the vapours of hot water or of those of a decoction or infusion of poppy-heads or camomile flowers; or in place of these, warm sprays may be used containing carbonate of soda and chloride of sodium with extract of poppies, lettuce, or conium.

In the subacute and chronic stages, alkaline solvent sprays are still more indicated, while in the chronic stage, and when the application of a gentle stimulant is indicated, the fumes of chloride of ammonium may be employed.

If there be much relaxation in consequence of repeated attacks, then astringent sprays must be used.

If the uvula or the epiglottis be involved, especially the latter, the condition of the patient will be a good deal aggravated, and these parts will require special treatment, as by strong alkaline sprays, to promote by exosmosis the escape of mucus, and in the more chronic stage the application of a solution of nitrate of silver.

When there is reason to believe that the chronic condition of the throat is kept up by a deranged state of the stomach and organs of digestion, these will require attention.

Inflammation of the Tonsils.

This is an acute inflammatory affection of the tonsils, popularly known as quinsy; it occurs in two forms: one is superficial and may be termed follicular, involving chiefly the mucous membrane and its follicles; the other extends deeper and affects the parenchymatous and glandular structures. It is a disease of middle life and is particularly apt to recur.

Among the predisposing causes may be mentioned the strumous and rheumatic diatheses, debility and general impairment of the health; among the exciting causes, the principal is exposure to cold and damp. One attack predisposes to others, and occasionally the disease has been known to prevail as an epidemic.

In the acute stage of the follicular form of the disease, the mucous membrane is at first red, dry, and shining, but soon it becomes moist, the lacunæ throwing off an increased amount of mucoid or muco-purulent secretion of a whitish or yellowish colour. This stage may be followed by ulceration of the follicles.

In the parenchymatous form of the disease at the outset the tonsil, for usually one only is attacked, is not only red and shining, but it becomes greatly enlarged, extending a considerable distance across the isthmus faucium, producing great pain in swallowing and sometimes difficulty in breathing, while the voice becomes more or less nasal. Soon an abundant sticky mucous secretion is thrown off, which adds greatly to the distress of the patient, the saliva at the same time often flowing freely from the mouth. Usually the affection runs a rapid course, and frequently terminates in the formation of an abscess.

The superficial affection is comparatively mild, and is unattended with much fever or with distressing local effects. When, however, it occurs in the cachectic, the symptoms may be of a typhoid character; the mucous membrane may present a dark and dusky hue and be more or less covered with ash-coloured ulcerations.

The parenchymatous disease is attended with a good deal of constitutional disturbance and much local distress; deglutition is most difficult, often impossible for solids, and the pain occasioned by the effort is very great.

So abundant is the thickened mucoid secretion, and so troublesome, that constant efforts are made to clear the throat and to get it away. The parenchymatous affection is ushered in with a good deal of disturbance, lassitude, chilliness, headache, quick pulse, and other symptoms. Should the case go on to suppuration, the fever will increase until that has occurred, an event which is often indicated by the occurrence of rigor. The fever usually quickly abates when the abscess has been formed, and on its bursting, the pain and difficulty of swallowing become suddenly relieved.

For some time after the opening of the abscess and escape of the pus the mucus or mixed mucus and pus continue to be abundantly formed and discharged, accompanied often with a highly offensive odour.

If now the tonsil be examined the redness will be seen to have greatly diminished, and the mucous membrane will be sodden or œdematous, while the aperture, from which the matter escaped, and which may be more or less ulcerated, will often be plainly visible.

Occasionally only, in depressed conditions of the system, sloughing takes place after the bursting of the abscess, and sometimes, especially after repeated attacks, the tonsils become hypertrophied; but in most cases hypertrophy of those organs is a separate bilateral affection, frequently associated with a scrofulous diathesis.

Treatment.—The measures to be taken in the superficial form of this disease are of a simple character. Two or three days' confinement to the house, a gentle aperient, the use of astringent sprays, or even lozenges, and simple diet will usually be sufficient.

In the parenchymatous form, the treatment must be more energetic, and the first endeavours must be directed to subduing the primary inflammation and swelling. For this purpose cold applications should be used, as iced water or the sucking of ice itself. Bicarbonate of soda freely applied to the tonsils in powder, will occasion a profuse discharge of mucus and thus afford great relief.

For the accompanying fever, aconite should be administered in drop doses of the tincture, at first every fifteen minutes, until the desired effects are produced, that is to say until the fever is abated, when the remedy may be continued for a time at intervals of two or three hours; the aconite will often stop the attack if given at the onset, as will also actæa racemosa, the action of which is somewhat similar to that of aconite.

But another remedy which exercises much power over the course of tonsillitis is guaiacum, either taken internally or applied locally every two hours; in the latter case a considerable part of the resin is swallowed, and therefore both local and constitutional effects are obtained. Like aconite, it will frequently arrest the disease at the commencement. It may be applied to the throat in the form of powder or as a spray of the ammoniated tincture, or lozenges may also be used from time to time. Brushing over the surface of the tonsil with a strong solution of nitrate of silver will, in some cases, also help to arrest the inflammation.

If, notwithstanding the means taken, the swelling and inflammation continue, warm applications must now be resorted to, such as the inhalation of the vapour of hot water, or warm emollient anodyne sprays containing either tincture of aconite or hyoscyamus; and if deglutition be both painful and difficult, then warm bread poultices, over the surface of which a teaspoonful of laudanum has been poured, may be applied to the throat, or the belladonna liniment may be employed.

If with the increase of the inflammation and swelling, much tenacious and slimy mucus be thrown out, then to the warm sprays, carbonate of soda or borax may be added.

Even when suppuration seems imminent, it may sometimes be averted, as pointed out by Ringer, who writes as follows:—'In a certain stage of tonsillitis the influence of mercury is most marked, owing probably to its absorption in the circulation. In quinsy or scarlatina, when the enlarged tonsils almost meet and block the passage, and when the difficulty in swallowing is nearly insuperable, with even danger of suffocation, at such a crisis $\frac{1}{3}$ grain of grey powder taken every hour greatly reduces the swelling in a few hours, and obviates the distress and danger; and even if an abscess has formed, its maturation and evacuation appear to be effected more quickly.'

In spite of all that can be done the inflammation too often goes on to suppuration, and the abscess either bursts spontaneously or will require to be opened. When it has burst, the treatment must be modified accordingly; sprays of astringent remedies must now be employed, as of tannin and glycerine, or of alum, and somewhat later a solution of perchloride of iron with glycerine. The astringents may be continued until the discharge has become greatly diminished and the parts have recovered, as far as may be, their usual condition. The use of gargles is in the inflammatory stage attended with difficulty and gives rise to pain, but they may be employed in the more chronic stage.

If the discharge be offensive, then antiseptic inhalations must be resorted to, as of carbolic acid; while if sloughing has taken place, arsenic will often be found to exert a very beneficial effect.

Generally, in consequence of the attack, debility ensues; this in many cases may be effectually remedied by the administration of bark and carbonate of ammonia, or of bark and dilute hydrochloric acid.

Follicular Disease of the Throat.

This is a special and well-marked affection, occurring in early and middle life, of the mucous follicles of the pharynx and adjacent parts, presenting itself in two forms essentially distinct, the hypertrophic and the exudative.

The principal causes of the malady are certain diatheses, as the strumous, rheumatic, and gouty, but the most frequent cause of all is over-exertion of the vocal organs, especially in persons who are not physically strong, and hence the liability of clergymen, public speakers, readers, singers, and others who have occasion to greatly exert their voices. Other causes are repeated attacks of catarrh from exposure to cold and to various kinds of mechanical irritants.

Many of the symptoms of the disease are common to some other affections of the throat, and they include in varying degrees stiffness, tenderness, irritation, cough, a dry condition of the mucous membrane, with, when the disease extends downwards to the œsophagus or larynx, pain in deglutition and hoarseness. In very advanced cases hearing may be impaired and taste perverted. When the dryness is extreme, the condition of the throat has been described under the name of *pharyngitis sicca*.

The character of the cough will vary much according to the condition of the mucous membrane and the extent of the disease. Should it extend upwards over the arches of the palate, involving the uvula, or downwards to the epiglottis, or even to the larynx, the cough and most of the other symptoms will be considerably aggravated; but the cough is, however, usually very troublesome and irritating in any case.

Although the mucous membrane in the hypertrophic form of the disease is more or less dry, yet it is frequently spotted over with scanty, thick, and tenacious mucus, which, acting as a foreign body, excites incessant efforts for its removal.

Again, as a consequence of the disease, the uvula is apt to become much elongated, thus acting as another exciting cause of the cough.

The symptoms are usually much more severe in the exudative than in the hypertrophic form of the disease. Indeed a considerable amount of hypertrophy sometimes exists without giving rise to any great irritation.

In the hypertrophic form, the disease is at first confined to the posterior wall of the pharynx, but as it proceeds it may pass upwards or downwards and involve the parts already enumerated. The mucous membrane is markedly granular; the granules are at first small, round, prominent, and isolated, but as the disease advances they increase in size and ultimately coalesce. The blood-vessels are enlarged and form a network between the granular-looking follicles. The exudative form of the affection, Dr. Morell Mackenzie states, 'generally commences in the tonsils or in their immediate neighbourhood, and advances to the posterior wall of the pharynx, the back of the tongue, the epiglottis, and the interior of the larynx.'

When the mucous follicles are acutely inflamed their secretion loses its transparent character and becomes white like milk; if the inflammation is more chronic it assumes a caseous appearance, while calcareous formations have in some cases been found to be present in the cavities of the enlarged follicles. Lastly, in some cases a viscid mucus is discharged, which adheres to the surface in patches. The follicles rarely become ulcerated.

In the exudative form of the disease there is rather a tendency to atrophy than to hypertrophy; while, according to Störk, in the hypertrophic form the changes are more in the epithelium of the enlarged follicles than in their cavities.

Treatment.—When the disease is fully established the treatment of so chronic a malady must necessarily be tedious and will require much perseverance, both on the part of the medical attendant and the patient. The measures to be taken will of course depend on the cause of the affection and the condition of the parts involved, and will include general and constitutional and more particularly local treatment.

In those cases in which the affection has been brought on or aggravated by excessive use of the voice complete and prolonged rest must first be enjoined.

If in the earlier stage there be much congestion, with a sensation of heat, pricking, soreness, and painful deglutition, the local treatment must be soothing and anodyne; ice, iced water, strong infusions or decoctions, allowed to grow cold, of marsh mallow or poppy-heads, with the addition in some cases of the extracts of lettuce or conium, should be applied in the form of atomized sprays.

If on the surface of the mucous membrane there be patches of tenacious secretion, exciting cough, then a combination of soothing and solvent remedies should be used, including the anodynes just mentioned, and chloride of sodium, carbonate or borate of soda.

Again, if in very chronic cases there be evidences of relaxation of the mucous membrane, then astringent sprays should be used, as of alum, or tannin and glycerine.

But, with a view to the eradication and cure of the disease, other and stronger measures must be adopted. One of these is the application to the granular follicles of some escharotic, such as a mixture of caustic soda and lime, the granulations being thereby destroyed in detail; that is to say, each successive application should be limited to a few granulations only at a time. Dr. M. Mackenzie speaks highly of this mode of treatment.

The curative treatment of the exudative form of the disease is less severe. After a thorough cleansing of the mucous membrane by means of some detergent and alkaline wash or spray, a strong solution of nitrate of silver, 40 grains to the ounce, may be applied to the follicles by means of a brush, or, better still, each may be touched successively with a pointed stick of the solid nitrate.

In the most obstinate hypertrophic cases heroic remedies have been recommended and tried. Lewin has suggested incisions or scarifications to relieve the hyperæmia; while to destroy the granulations the galvano-cautery has been used, and by this means, Oertel states, the best results have been obtained.

In many cases of follicular disease of the throat the free use of sulphuretted mineral waters has been found to be beneficial.

When the malady invades the larynx, and there is marked hoarseness of the voice, then other appropriate treatment will be called for, the nature of which will be indicated when the affections of the larynx are considered.

As a rule, though the treatment is tedious, the prognosis is usually favourable; but the throat will rarely recover itself entirely, or the voice regain its full power, so that all exciting causes must be constantly avoided. The exudative is more difficult to cure than the hypertrophic form of the disease.

Aphthæ.

This affection, frequently called *Thrush*, commences as a red papular eruption on the inside of the lips, cheeks, sides of the tongue, velum, and tonsils, and extends in some cases to the pharynx and even into the œsophagus. The papules soon become white, burst, frequently ulcerate, and coalesce into patches of variable size, which are covered by a pseudo-membrane, which is found on examination to consist mainly of a species of fungus, *Oidium albicans*. Sometimes these patches are of a white colour, causing the parts to look as though they had been dusted with flour; in others they are ash-coloured, dark or almost black they are apt to assume this latter appearance in cases of extreme prostration, although sometimes the discoloration may be explained in other ways, as by the medicines taken. In severe cases successive crops of papules continue to appear for some days, and even for two or three weeks, passing through their several stages of maturation, rupture, and ulceration.

French and German physicians distinguish two forms of this affection, the one characterized by the presence of the fungus, 'muguet,' the other by a non-parasitic membranous exudation, and to this the application of the word 'aphthæ' is confined. This affection is particularly apt to occur in infants and very young children; in the advanced stages of phthisis; after measles, and in old age with prostration of the vital powers. The chief determining cause would appear to be the same in all these cases, namely, debility, arising from defective nutrition.

The disease in infancy is characterized by pain and soreness of the mouth, and by difficulty in swallowing, these being in some cases so great that the child is unable to take the breast.

Usually in infancy the prognosis is favourable in the absence of any considerable diarrhœa or enteritis; but when the disease occurs in the advanced stages of phthisis, or after measles, or in old age, it is frequently the forerunner or harbinger of approaching dissolution.

In some exceptional cases a difficulty may be experienced in discriminating the fungoid patches of aphthæ from the false membrane of diphtheria, but ordinarily the distinction is easy. In aphthæ the patches are thin and easily broken up, while in diphtheria they are tough and like wash-leather; examination by the microscope will at once put an end to any doubt.

Treatment.—Our first endeavour should be to maintain the vital powers, especially by attention to the condition of the digestive organs, by the free administration of suitable nourishment, and by appropriate tonics, varied according to the age and condition of the sufferer. In infants, acidity must be corrected by the use of lime-water, and the diarrhœa checked. In advanced life, plenty of nourishment, with a moderate amount of stimulant, must be allowed, and bark and ammonia administered.

The local treatment is also important.

There are several remedies of approved efficacy in the milder and simpler cases of aphthæ.

One of these is alum, which may be applied as a powder directly to the spots, or it may be employed as a spray. Another good remedy is chlorate of potash, which may be used in a similar manner. An old and very serviceable application is borax with honey or glycerine.

For the destruction of the parasitic fungi, a solution of sulphite of soda, ½ to 1 drachm in an ounce of water, may be applied with a brush, or in the case of adults the spray may be used. Sir William Jenner has stated that this will remove the disease in 24 hours.

An affection which is sometimes confounded with the preceding, and to which the term 'thrush' is also sometimes applied, is stomatitis, or simple inflammation of the follicles of the mucous membrane of the cheeks and throat, probably of an herpetic character; the follicles become inflamed and exhibit white spots, which, however, do not usually pass into a state of ulceration. It occurs in young children and is sometimes associated with derangement of digestion or with the irritation of teething.

Relaxed Sore Throat.

This affection is the consequence of certain maladies of the throat and conditions of the system, and is not a special disease in itself.

It is apt to occur as the result of repeated attacks of catarrh, of chronic inflammation of the mucous membrane of the throat, independent of catarrh; in some forms of dyspepsia and in debilitated states of the system. A very frequent cause is over-indulgence in strong spirits; this gives rise to irritation, and even engenders chronic inflammation of the mucous membrane of the stomach.

The condition of the mucous membrane will vary in accordance with the cause. It will often be more or less congested, varicose, swollen, and œdematous, and in very chronic catarrhal cases it may be even permanently thickened; the surface of the membrane may be more or less coated with adherent mucus, or it may present an unnaturally pellucid appearance, but sometimes it is dry rather than moist. In cases dependent upon debility and anæmia the membrane may be pale and relaxed, either with or without œdema.

In the general relaxation of the throat the uvula of course is liable to participate, it becoming elongated and sometimes swollen and more or less œdematous; indeed the whole of the soft palate is frequently dependent.

The symptoms are allied, except in the absence of pain, very much to those of ordinary sore throat; there is a feeling of relaxation, of fulness and stiffness, with, in some cases, considerable irritation and cough. The irritation may be occasioned by the presence of secretion on the mucous membrane, but more frequently it arises from the elongated uvula, which, resting from time to time on the epiglottis or tongue, provokes sudden and severe paroxysmal fits of coughing.

Treatment.—This must be constitutional and local, and first the cause of the relaxation must be ascertained : if it be due to repeated catarrhs, then measures to avoid these must be taken, particularly, when practicable, residence in a dry and equable climate. If abuse of ardent spirits has led to this condition, then no treatment will be effectual until this indulgence has been abandoned. Again, if it be associated with other forms of dyspepsia, or with anæmia, or general debility, then the appropriate remedies for these conditions must be employed.

But still in most cases constitutional treatment will not be sufficient without the employment of local measures. The local treatment has for its object, when the relaxation constitutes the chief fault, the bracing up of the mucous membrane and the underlying muscular and other tissues.

This purpose is best accomplished by the inhalation of certain stimulant and astringent medicaments, as sprays of a strong solution of alum, of tannin and glycerine, or of perchloride of iron.

If the throat be simply relaxed, and without any obvious irritation or inflammation, a spray of tincture of capsicum may be inhaled, 10 minims of tincture in 1 ounce of water, or a solution of sulphate of zinc, 2 to 4 grains to the ounce; or, again, if with relaxation the mucous membrane be pale and flabby, a spray of tincture of cubebs, 1 drachm to 1 ounce, a little extra of spirit being added, or mucilage, in order to keep the oleo-resin of the tincture in solution.

It is very possible that in some cases a spray containing tincture of ergot combined with mucilage might prove serviceable by its action on the unstriped muscular fibre.

Now the local treatment by inhalation is, in general, at the same time more or less constitutional, since no inconsiderable portions of the sprays employed become swallowed and so reach the stomach.

It is not always easy by local means to restore a hypertrophied uvula to its normal size, so that it occasionally becomes necessary to abscind a portion of the organ. This operation, though very simple, is sometimes followed by a good deal of bleeding, rendering the application of strong styptics requisite, as well as by a considerable amount of irritation and pain, which, however, usually subside in the course of a day or two.

Rheumatic Sore Throat.

This form of sore throat occurs particularly in persons of a rheumatic diathesis, and who have suffered from other manifestations of rheumatism. It is often characterized by the suddenness of its onset and departure, it being sometimes followed by the occurrence of local rheumatism in other parts of the system, as the neck, shoulder, or lumbar region. There is usually acute pain in the throat, impeding, or perhaps preventing, deglutition, but this takes its departure in the course of a day or two as suddenly as it commenced; the throat is more or less red and the uvula somewhat œdematous.

This affection is also apt to occur in those who have suffered from repeated attacks of tonsillitis; in fact there would appear to be a close connection between the two complaints.

Treatment.—The constitutional treatment should consist of the administration of salicin or some of its preparations, and if the patient suffer from chronic rheumatism, and is not sufficiently relieved by the salicylic treatment, recourse may be had to iodide of potassium. The local treatment should consist mainly in the employment of warm emollient and anodyne sprays or gargles. A spray containing tincture of opium or pulvis ipecac. Co. is well calculated to afford relief to the pain. Sprays containing bicarbonate of potash or salicin will both be likely to prove beneficial.

Gouty Sore Throat.

A somewhat analogous affection to the preceding occasionally occurs in persons of a gouty diathesis. It is not distinguished by any special local symptoms, but its true nature is sometimes revealed by the suddenness of the attack and its equally sudden disappearance, followed promptly by symptoms of gout in the foot or in some other part of the system.

Ulcerated Sore Throat.

Ulcerations of the throat frequently occur in persons who have been exposed to unwholesome and perhaps infectious emanations, such as are encountered in dissecting rooms and in attendance on cases of scarlet fever and diphtheria, the health having become impaired thereby and a cachectic condition induced, due, it is generally supposed, to some form of septicæmia.

There will be in such cases swelling and soreness of the throat, with pain in deglutition; the fauces and tonsils may be congested and swollen, and there will be one or more round or oval and whitish ulcerations of variable size; the tongue is coated, the breath offensive, and there are other evidences of impaired health.

Treatment.—In affections of this nature, the constitutional treatment is of the first importance; suitable tonics must be administered, as bark and ammonia or quinine, and generous but not stimulating diet allowed; above all, change to the pure air of the seaside or country is indicated.

The treatment for the ulcerated condition, occurring in persons who are out of health, and due possibly to some slight infection, will depend on the state of the throat. If there be congestion with painful deglutition, ice or iced water will usually give relief, and, if this fails, recourse may be had to anodyne sprays, containing 5 to 10 drops of opium.

When the congestion has in a measure subsided, and if the breath be offensive, then sprays of certain antiseptics and deodorants should be employed, as of boracic or carbolic acid, chlorinated soda, or permanganate of potash.

Inflammation of the Larynx.

This affection is usually catarrhal, but it is also induced, like ordinary catarrhs, by other causes besides exposure to cold and damp, although this is its most frequent exciting cause.

It passes of course through two stages, the acute and chronic, although it may be chronic from the commencement, as is often the case when it is due to the descent of a cold from the head or throat. When brought on by cold it is seldom a dangerous affection, and its occurrence from other causes is exceptional.

Among the more special causes of Laryngitis the following may be mentioned: violent or over exertion of the voice, the inhalation of irritating vapours, or the extension of the disease from the pharynx in cases in which boiling water or corrosive substances, such as mineral acids, have been swallowed.

The acute stage of the affection is indicated by a feeling of soreness in the larynx, pricking, and a general feeling of uneasiness, hoarseness, alteration or loss of voice, cough, and perhaps some slight difficulty in swallowing. The cough is at first dry, as shown by the sound; a little later on secretion sets in, moist râles become detectable, but the expectoration is scanty, mucoid, or muco-purulent. The concomitant fever will be proportionate to the severity of the attack, but it soon subsides as the chronic stage is approached.

The condition of the laryngeal mucous membrane will of course vary with the intensity and stage of the malady. The membrane is, as a rule, congested and red, and certain parts may be more affected than others, as the epiglottis and vocal cords, and there is usually some amount of tumefaction or, more rarely, œdema.

The chronic stage of the malady is characterized by several of the symptoms above noticed, but the secretion is now increased, although it is never very abundant, and it is occasionally streaked with blood; the cough has lost its dry character, and there is considerable abatement of the fever.

INFLAMMATION OF THE LARYNX.

It has been stated that although acute laryngitis usually arises from catarrh, it is sometimes occasioned by other causes, and this is especially true of the chronic form, which is not unfrequently excited by the inhalation of irritating powders, as of pollen in hay fever, of glass and steel by glass cutters, knife and needle grinders.

The congestion of the mucous membrane in the chronic stage is less, although the membrane will still be swollen and sometimes, in very chronic cases, even thickened; it will be more or less coated with mucus, while superficial ulcerations, scarcely extending beyond the epithelial covering, are sometimes visible. The vocal cords are not unfrequently congested and thickened. It need scarcely be pointed out that, with a view to an accurate diagnosis and prognosis, a careful laryngeal examination must be made.

Treatment.—The acute and chronic stages of this affection should for the most part be treated on the same principles and by the same means as nasal and pharyngeal catarrh; in fact, these are often associated more or less with laryngeal catarrh.

In the acute stage, the patient must be confined to the house, compresses or warm linseed cataplasms should be applied round the neck and to the sternum, the feet placed in hot water, and some mild sudorific and aperient remedies taken. For the fever, aconite, either by the mouth or in spray, should be administered in the manner and in the doses already described. A very useful sedative and sudorific, when aconite is not given, is Dover's powder, 4 or 5 grains of which may be administered at first every 3 or 4 hours.

In this stage also warm, emollient, and anodyne sprays and vapours will be found of service.

To the above sprays, if the mucous membrane be dry, or the mucus be tenacious, chloride of sodium, carbonate of soda, or borate of soda may be added, with a view to moisten the membrane and dissolve the mucus, and so to facilitate expectoration. When the acute stage has in a measure subsided, the mildly stimulating fumes of chloride of ammonium may be inhaled.

In the chronic form of the malady, particularly when this has continued for a long time, there are two classes of remedies the employment of which is indicated, astringents and stimulants.

The astringents should be employed when it seems desirable

to lessen the irritation by checking secretion, which is, however, rarely excessive in this affection, and to brace up the weakened mucous membrane.

The most useful astringents are tannin and glycerine, alum, perchloride of iron, chloride of zinc, or acetate of lead. These may be applied by the brush, or preferably in the form of sprays, since no disturbance of the parts is thereby risked. When the brush is used the solutions must be stronger than when they are intended to be inhaled as sprays.

The stimulant remedies are chiefly required in those chronic cases in which the mucous membrane has become relaxed, and the most appropriate of these are creasote, the oils of cubebs, of eucalyptus, and of fir wood. The vapours of all these may be inhaled, provided a suitable apparatus be selected; but they can also be applied as sprays.

The inhalations in chronic cases should be repeated night and morning, and sometimes more frequently.

Remedies either containing chloride of sodium and carbonate of soda, or sulphur, usually in the form of sulphides, have long deservedly enjoyed a high reputation in the amelioration of many severe and persistent cases of chronic catarrh, as the waters of many alkaline or sulphuretted mineral springs.

The fact has been already noticed that in cases of chronic laryngitis the mucous membrane of the throat is often similarly affected, and when this is the case it must be included in the treatment.

Again, attention should be paid to the condition of the uvula, whether it be elongated, relaxed, œdematous, or hypertrophied.

Lastly, appropriate constitutional treatment will be necessary in some cases, while in all, rest of the vocal organs should be enjoined.

Catarrhal Croup.

Children are particularly liable to this form of catarrh—false croup, or, as t is now called, Laryngitis Stridulosa. It is in reality the same complaint as that previously described, namely, catarrhal inflammation of the mucous membrane of the larynx, but in children this presents one great peculiarity : while in adults the affection seldom gives rise to severe spasm, in children this is a frequent and alarming complication.

The reason assigned for their liability is the smaller size of the rima glottidis, which allows of its being more easily closed, as by mucus adhering to the vocal cords, and the presence of which in that situation is so apt to excite spasm.

Treatment.—It is important that the nature of these sudden attacks, which may recur from time to time on taking fresh cold, should be at once discriminated, with a view to the prompt employment of suitable means of alleviation.

Unfortunately these attacks usually come on in the night, so that some time is unavoidably lost before the measures requisite can be carried out. If the case be urgent, an emetic of sulphate of zinc or ipecacuanha may be given at once, and as soon as possible a hot cataplasm should be applied to the throat and sternum; the air of the room should be filled with steam by means of a 'croup kettle,' while the child should be made to inhale, as far as practicable, the vapour of hot water, which may be medicated with tincture of conium or hops. Another speedy means of relief is the inhalation of 3 or 4 drops of chloroform.

If by the above means the impending suffocation is relieved, then time is afforded for the treatment necessary to guard against a recurrence of the attack.

Usually there is more or less fever; this should be treated with $\frac{1}{2}$-minim doses of tincture of aconite, given for the first hour every 15 minutes and afterwards at longer intervals, till the frequency of the pulse is reduced, the temperature lowered, and the skin becomes moist.

The chronic stage having been reached, alkaline and expectorant remedies, either in the form of a palatable drink or in that of sprays when practicable, will help to dissolve the mucus and render it less irritating. Should the external air be very dry, then the moist state of the air of the room should be maintained for two or three days, until all active symptoms have disappeared. Should the child be out of health, the necessary constitutional treatment must now be adopted.

Aphonia.

Loss of voice, or aphonia, is not a special disease in itself; it arises from imperfect action of the vocal apparatus; the loss may be partial or complete.

The causes are various; it may be a nervous affection, as

hysterical aphonia, or it may be due to disease of the nerve centres, giving rise to paralysis of the cords; it may arise from simple debility, whereby the abductor muscles of the rima are unable to act properly; or the vocal cords may be relaxed; or it may depend upon anæmia, and sometimes that form of it which accompanies chlorosis. Very frequently it is due to congestion of the cords, the result simply of catarrh; or it may be caused by structural alterations affecting the vocal cords and other parts concerned in vocalization, arising from specific inflammations and diseases, as phthisis, scarlet fever, and diphtheria.

Rarer causes of loss of voice are the presence of foreign growths, including warts and polypi, in the larynx or on the cords.

The examination of the vocal organs by means of the laryngoscope, it is scarcely necessary to say, is usually indispensable in these cases, both for determining the cause of the voicelessness and the exact condition of the parts concerned.

Treatment.—This must depend on the causes, and must in most cases be both constitutional and local.

For the restoration of the voice in hysterical aphonia, the hysteria itself must be attacked; this is usually associated with some fault in the uterine functions, and which must be diagnosed and appropriately treated.

The loss of voice arising from debility, or from anæmia, must be treated by the tonics, especially nerve and blood tonics, which are adapted for those conditions, combined with nourishing food, plenty of fresh air, and moderate exercise.

The local treatment must also vary with the several causes and conditions which have brought about the loss of voice; if the cords be simply anæmic or relaxed, then stimulating applications should be employed, as nitrate of silver; a solution of this may be used in the form of a spray, or it may be applied to the larynx by means of the curved brush. For the spray 2 grains may be dissolved in 2 or 3 drachms of water, the patient closing the nostrils and inhaling deeply; with the brush a stronger solution of 20 grains to the ounce may be used.

But there are other sprays which may be employed with great advantage, and which sometimes are to be preferred to nitrate of silver, in consequence of their being less irritating, as perchloride of iron or carbolic acid, but the inhalation of the fumes of the acid in the dry state is preferable to employing it as a spray. Another

CATARRH OF THE TRACHEA. 267

remedy of approved value is iodine, which also may be inhaled either as a spray or in vapour.

Should these means fail, even when combined with the appropriate constitutional treatment, then recourse must be had to faradization of the larynx and vocal cords.

In the loss of voice from catarrh, with redness and congestion of the cords, the inhalations must be of a different kind. One of the best sprays which can be employed is a solution of ipecacuanha; this may be repeated night and morning, and it is not necessary that nausea should be produced, as secretion is promoted before this stage is reached. This remedy will prove equally serviceable in cases of hoarseness of the voice depending on congestion of the vocal cords.

But emollient and anodyne sprays will also be found most useful in a similar condition of the cords, and with these the wine of ipecacuanha may be combined.

The spraying of the throat and fauces is an easy matter enough, but when it is desired that the atomized liquid should reach the interior of the larynx it is necessary that certain precautions should be adopted, otherwise the greater part of it will be expended on the throat: one is that the delivery tube of the apparatus should be longer than that ordinarily used, and it should be curved. The advantage of this is that the spray reaches the throat in a more concentrated form, and hence can be inhaled with much less loss; another precaution is that the tongue should be drawn forward, either by the operator or the patient himself, with the finger and thumb. This makes the way to the larynx shorter and more direct, and fixes to some extent the epiglottis. Again, the patient should inspire deeply, while if the hand spray be used he should be careful to inhale at the moment the spray issues from the apparatus, otherwise but little of it will reach the larynx. When the spray is delivered into the mouth, the liquid should be cold.

Inflammation of the Trachea.

The most frequent exciting cause of Tracheitis is exposure to cold and damp. It seldom occurs as an independent disease, but is most frequently associated on the one hand with catarrhal laryngitis, especially of the subglottic region, or still more frequently with catarrhal bronchitis. Like other inflammations, it has two stages, the acute and the chronic.

In the acute form the disease is less severe than the corresponding affection of the larynx. There is less difficulty of breathing and less liability to spasm, owing to the large size of the trachea.

There is usually but little difficulty in distinguishing acute tracheitis. There will be a feeling of soreness and obstruction in the trachea, with some tenderness on pressure, irritation, and cough; this will be at first dry, but as the chronic stage is approached secretion returns, and is usually more abundant than in laryngitis, and especially if it be associated with chronic bronchitis. The presence and some idea of the amount of the secretion may readily be determined by auscultation.

Treatment.—For the most part this is the same as for laryngitis. In the acute stage it comprises the application of a compress or poultice externally, and the inhalation of the medicated vapour of hot water and of warm anodyne sprays; these will usually afford great relief.

In the chronic stage, in accordance with the symptoms, astringent or stimulant sprays may be inhaled; the latter are indicated when the chronic stage is of long duration and the mucous membrane is relaxed.

Inflammation of the Bronchi.

Bronchitis occurs in the acute and chronic forms; the latter in many cases is not simply due to a previous attack of the acute malady, but often arises independently from a variety of other causes.

Acute bronchitis is almost always due to sudden exposure to cold and damp, the young and the aged being the most liable to be attacked, but it may be brought on by other causes, as the continued inhalation of irritating vapours and dust.

The attack may affect the larger bronchial tubes only, or it may extend to the smaller tubes, or the bronchitis may be limited to the latter, constituting *capillary bronchitis.*

When the larger tubes are affected, the mucous membrane is at first dry, congested, and tumid, so that the calibre of the tubes is diminished, giving rise to some but not a considerable difficulty of breathing. After two or three days secretion sets in; this is at first frothy and mixed with air, afterwards more tenacious, and occasionally streaked with blood, but as resolution takes place it loses some of its tenacity, becomes more abundant, and may be for a time muco-purulent.

The capillary form of the disease is a much more dangerous affection; it is accompanied by more fever, the respiration is hurried, the difficulty of breathing great, the cough more paroxysmal and distressing, the expectoration difficult, the secretion being viscid, ropy, and sometimes purulent. In consequence of the extreme difficulty of breathing, the blood is imperfectly aërated, the countenance becomes livid, debility is extreme, cold sweats set in, the pulse is rapid and feeble, the breathing becomes shorter and shallower, and in fatal cases death from apnœa ensues, fibrinous clots being frequently found in the heart and large vessels. This last form of bronchitis is apt to occur in children after measles, in old people, when it has been described under the name of catarrhus senilis, and in emphysema.

Chronic bronchitis may be simply the sequence of the acute form of the disease, but it may arise from a variety of other causes and is not unfrequently a secondary affection, and may be due to emphysema, disease of the heart or kidneys, or to the gouty diathesis. While it may proceed from disease of the heart, in consequence of the impediments which exist to the circulation of the blood through the lungs, it may itself bring about disease of that organ, especially of the left side.

The prolonged continuance of chronic bronchitis tends to certain changes, not only in the mucous membrane itself, but in the adjacent structures; the submucous tissue becomes infiltrated and thickened, the fibrous and muscular coats hypertrophied, while ultimately in many cases emphysema becomes developed.

Three forms or varieties of chronic bronchitis have been recognized; the ordinary form, that attended with profuse secretion, distinguished by the name of bronchorrhœa, and dry bronchitis.

Chronic bronchitis in the usual form is at first a comparatively mild affection, occurring in the winter and altogether disappearing in the summer. After a time, if the existing causes continue in action, the attacks become more frequent and severe, especially if secondary and associated with some other disease. When advanced, the breathing is considerably affected, the cough severe and sometimes paroxysmal, while the expectoration may be scanty, viscid, and brought up with difficulty, or in cases of very long standing it may be discharged with ease.

Bronchorrhœa occurs chiefly in the old and feeble, and is sometimes associated with disease of the heart; in this form of the affection the secretion is profuse and consists of a thin, glairy fluid, or of a tenacious, ropy matter which has been compared to the white of egg.

The dry form of bronchitis is attended with an irritable cough, and if there be any expectoration it is transparent and viscid; the breathing is oppressed, and there is a feeling of tightness in the chest. This form is met with in the gouty, and is associated sometimes with emphysema.

Treatment.—With a view to treatment, a careful auscultatory examination must be made, to ascertain the extent of the mischief, whether one or both lungs are affected, the larger or smaller tubes; next whether the disease is primary or secondary, and depends on some other affection, emphysema, disease of the heart or kidneys, or on a gouty or rheumatic diathesis; then, lastly, the treatment must be modified according to the stage and the more important and pressing symptoms.

In the *acute* stage counter-irritants must be applied, as mustard cataplasms or turpentine, especially in the capillary form of the disease, while derivative remedies must be employed, as nauseants, sudorifics, diuretics, and suitable aperients, including such as promote the action of the liver. Care must be taken not to depress the patient and to weaken his vital powers by the treatment adopted, and this is especially necessary in the case of old people. For the relief of the congestion and inflammation, sprays of ipecacuanha in nauseant doses may be repeated at short intervals, to bring about as quickly as possible the restoration of secretion. When this is so abundant and tenacious as to give rise to dyspnœa, an emetic of ipecacuanha will often afford relief.

At the same time, and with the same object, the air of the room should be kept warm and moist by means of the bronchitis kettle. At this period also, if practicable, emollient and anodyne sprays should be inhaled. When secretion begins to be restored, though still scanty and viscid, the inhalations should be alkaline, and contain carbonate of soda or carbonate of potash. When it is thinner and more abundant, and the fever has considerably subsided, then the fumes of chloride of ammonium, with either ipecacuanha or squills, may be inhaled.

If the cough is very troublesome, anodynes may be combined with the alkaline and expectorant sprays, as conium hyoscyamus, or lettuce; but belladonna and opium should be rarely employed, because of their effect in checking secretion; chloral may be substituted for the opium in some cases with advantage, both for checking the cough and for procuring sleep.

If the case be secondary, and due to some other affection, this of course must be included in the treatment; if to a diseased or weak heart, digitalis must be administered, with occasionally efficient aperients. If it be associated with gout, this must be suitably treated with colchicum, citrate of lithia, and in some very chronic cases iodide of potassium.

In uncomplicated acute bronchitis, especially of the larger tubes, the prognosis is favourable, and the chronic stage usually lasts but a short time. In capillary bronchitis it is much less so, for this is not only a much severer disease, but it is often secondary.

In *chronic* bronchitis it makes all the difference whether the disease is primary or secondary; in the former case the prognosis is more or less favourable, while in the latter the treatment will be often merely palliative.

In this stage of bronchitis the same general principles of treatment must be acted upon, but they need not be pushed to the same extent as in the acute form, and they include the employment of antispasmodics, as chloroform, to relieve spasm and dyspnœa, of anodynes for the cough, of nauseants and solvents to promote and soften the secretion, and of expectorants, especially stimulant expectorants, to aid its discharge.

When the chronic bronchitis is of long standing, and the discharge profuse, then preparations containing tar in different forms should be inhaled.

The preference should generally be given to tar itself, rather than to any single constituent, however valuable. The analyses already given show that the composition of tar, whether wood or gas tar, is very complex, and it is highly probable that the beneficial action exerted by it is in part due to some of its more undetermined and less studied constituents. The success which has at all times attended the use of the tar compounds in chronic bronchitis has been remarkable, and it is only comparatively recently that they have been at all generally employed by the orthodox practitioner; formerly they were prescribed mainly by empirics, who obtained sometimes a wide reputation by the success they achieved.

For that form of chronic bronchitis which occurs in cold and damp weather in the early spring, and which is often distinguished by the name of winter cough, the tar preparations are even more effective, and exert in fact an influence which may be regarded as almost specific.

For the modes of inhaling the tar preparations the reader is referred to the previous chapter.

Other remedies possessing similar properties, and acting somewhat in the same manner, are benzoic acid, benzoin, ammoniacum, cubebs, and senega with ammonia. Although proving often very serviceable in cases of chronic bronchitis, they are on the whole less effective than the tar preparations.

Ammoniacum often affords very great relief in the bronchitis of old people, attended with much wheezing, while senega and ammonia are suited for those cases in which a tonic combined with a stimulant expectorant is needed.

A remedy of a different kind, also very valuable in many cases of winter cough, is ipecacuanha in the form of spray.

When the secretion shows any tendency to putrescence, then sprays of sulphurous acid may be used, as specially recommended by Dr. Dewar, or of creasote or oil of eucalyptus.

There is a special form or variety of capillary bronchitis, which occurs in old people, *catarrhus senilis*, and which has been sometimes described under the name of peri-pneumonia notha. It is usually a subacute affection having the symptoms of a catarrh, with an abundant secretion of an opaque frothy mucus, which sometimes even assumes a purulent character. In the latter stage of this affection, stimulant and balsamic inhalations will also do much good, combined with bark or senega, and either ammonia or squills.

Obviously, with the inhalation treatment other remedies, both external and internal, according to the cause of the bronchitis and the character of the symptoms, must be combined, as externally stimulating liniments and counter-irritants, and internally suitable tonic and supporting remedies.

Since in most cases of bronchitis the general health is much weakened, and since the disease is particularly prevalent in old age, care must be taken to maintain the strength. The diet should be light and nourishing, and stimulants should be allowed in both the acute and chronic forms, and even in capillary bronchitis; they promote expectoration and relieve the breathing. Tonics in most cases will be needed, combined with expectorants, as ammonia with bark or senega. When with debility there is profuse secretion, iron will often prove very effective, particularly in the form of ammonia citrate, ethereal acetate, or the tincture. If there be emaciation, cod-liver oil is indicated. The patient should at all times be warmly clad, and, if necessary, he should seek a warmer climate. As a rule dry climates suit the bronchitic the best, but occasionally the reverse is the case.

Dilatation of the Bronchi.

This disease is characterized by dilatation of one or more of the bronchial tubes, Bronchiectasis, and is usually a consequence of chronic bronchitis; this may consist of a general increase in the calibre of the affected tubes or of bead-like enlargements. It might at first be supposed that the dilatation was entirely due to mechanical causes, and that the persistent and more or less explosive cough caused the bronchi to give way, and this more particularly when they have become blocked up with solid pellets of mucus. It is found, however, that this explanation is not sufficient to account for all the phenomena as revealed in post-mortem examinations, and that other causes are at work.

Dr. Grainger Stuart, who specially investigated this subject, came to the conclusion, founded on the researches of Andral, Corrigan, Gairdner, and Stokes, as well as his own, that the essential condition in these cases is atrophy of the bronchial walls, which readily yield to internal pressure in the same way as an artery does when its middle coat is diseased. The contractile and expulsive power of the dilated and sometimes sacculated bronchi is lessened, secretion is apt to accumulate therein, expectoration becomes increasingly difficult, and the secretion is thus retained long enough to allow of decomposition setting in and the generation of offensive odours.

This decomposed secretion now begins to play its part in the consecutive changes: it excites irritation and chronic inflammation; the mucous membrane becomes thinned and ulcerated; the irritation now extends from within outwards, till various changes are produced in the surrounding lung tissues; the lungs may waste, or there may be increased growth of connective tissue, or abscess or gangrene may take place.

The expectoration in the advanced stage of this disease is apt to occur in paroxysms, is profuse, and often extremely offensive.

Treatment.—At first this will be the same as for chronic bronchitis; but when the dilatation of the bronchi has become considerable, the remedial measures adopted must be regarded mainly as palliative, although suitable inhalations will afford great and marked relief in several ways.

The decomposition and fœtor of the expectorated matter, and the consequent offensiveness of the breath, may be remedied by the efficient employment of carbolic or sulphurous acids. The carbolic acid may be inhaled either as fumes, by means of the 'globe' inhaler, aided by the water bath, or by the inhaler for concentrated vapours, or the acid may be inhaled as an atomized spray. The sulphurous acid should be used as a spray, the acid being properly diluted.

After the fœtor has been removed, another class of remedies may be resorted to with the best effect—namely, the tar preparations and the stimulant oleo-resinous balsams, under which turpentine may be included; this is often an efficient remedy, but its odour and taste are repugnant to many. Terebene is a much pleasanter remedy and probably quite as effective; the oils of eucalyptus and fir-wood may also be employed with advantage.

The above antiseptic and stimulant treatment will usually produce a great reduction both in the character and amount of the expectoration; but when the secretion is profuse, astringent remedies will often effect a great change for the better, as sprays of perchloride of iron, alum, tannin, and glycerine, or of acetate of lead.

The decomposition and fœtor being stopped and the amount of expectoration greatly reduced, the mucous membrane of the bronchi is brought into a healthier condition, and the irritation both within and without the tubes becomes greatly lessened.

The disappearance of the fœtor is of course a favourable sign, but too much stress must not be laid on this. A very slight decomposition will often give rise to an offensive odour.

Expiration into rarefied air would, I believe, afford great relief, by promoting expectoration whereby the bronchial tubes would become freed from the retained secretions.

Plastic Inflammation of the Bronchi.

This peculiar affection, Plastic Bronchitis, is characterized by the formation in the bronchi of fibrinous casts; these may be simple or branched, hollow or solid. Those from the larger tubes are laminated and more or less hollow, being denser externally than internally, while the casts from the smaller tubes are often solid. In the formation of these casts this variety of bronchitis would appear to stand in relation to croup.

It is a very chronic affection of rare occurrence, its causation being obscure. The first intimation usually given of its existence is the expectoration of portions of the casts.

The detachment of the casts is very difficult and is accompanied with much irritation, cough, dyspnœa, and not unfrequently hæmorrhage, which is usually slight, but may be profuse.

Treatment.—When the existence of this disease has been established, the treatment must be directed to the amelioration of the predominant symptoms, the lessening of the irritation and cough. An endeavour, however, must be made to bring about the softening and expectoration of the casts. With a view to effect this object the inhalation of atomized sprays containing emollient and alkaline remedies, such as those which have been already more than once referred to, should be employed. Sometimes an expectorant will prove useful, such as ipecacuanha, and if there be little or no irritation or hæmorrhage, carbonate of ammonia will aid the expectoration.

If there be hæmoptysis and this is but slight, it is better not to interfere with it, as it probably indicates detachment of a cast, of which it is in fact an almost necessary accompaniment. If the bleeding be profuse, however, recourse must be had to the inhalation of astringents, which must not be of an irritative character, as ergot or turpentine, or gallic acid with glycerine.

A remedy which is often very serviceable in these cases is iodide of potassium, which may be administered by inhalation in 5-grain doses twice a day.

It is affirmed that plastic bronchitis sometimes predisposes to phthisis.

Inflammation of the Lungs.

There are several distinct forms of Pneumonia; one of these is simply catarrhal, and is due to the same causes which give rise to catarrhs of the mucous membrane of the respiratory track; another form is true or acute pneumonia; a third is chronic or interstitial pneumonia, while in a fourth variety may be included those pneumonias which arise from some special or specific cause.

Acute lobar, or, as it is sometimes called, from its supposed relationship to croup, *croupous pneumonia*, is not a purely local affection, and although frequently excited by exposure to cold and damp, there is reason to believe that it is in reality a specific disease and due to a specific cause, probably a bacillus. The chief reasons for this belief are, that the fever does not stand in any definite relation to the condition of the lungs; it usually precedes the manifestation of local mischief and disappears generally before resolution has set in.

Until a comparatively recent period the belief was generally entertained that pneumonia afforded a typical example of a pure inflammation of a parenchymatous organ, the inflammation being modified by the extent of the disease and the age and constitution of the patient. On this view the heroic treatment which formerly prevailed was mainly based. As soon as more exact observations were made the soundness of the antiphlogistic treatment came to be doubted, and it was at length almost abandoned. For some time, however, no sufficient explanation was forthcoming of the change of treatment, which it was endeavoured to account for on the ground of either a change in the type of the disease or in the constitutions of those who became its subjects. Soon the unsatisfactory and insufficient character of these assumptions became apparent, and enquirers looked about for another explanation. Now, however, that the fact has been recognized that the disease is not local or purely inflammatory, but part of a general and specific malady, the change which this treatment has of late years undergone becomes intelligible.

The acute stage, though running a certain definite course and terminating in resolution, is seldom followed by any prolonged chronic stage; in fact, chronic pneumonia may generally be regarded as a separate disease, not ordinarily preceded by any well-marked acute stage.

The chronic form has been distinguished by the name of *interstitial* or *fibroid pneumonia*, and it affects principally the intercellular fibrous tissue of the lungs, which becomes thickened and contracted, giving rise to cirrhosis of one or both lungs, usually one only, with loss of elasticity and diminished capacity. This condition may lead ultimately to emphysema or dilatation of the bronchi, and it may occur either as a separate affection or associated with the inflammatory process in some other lung affections, as in fibroid phthisis. It is sometimes catarrhal in its origin, but occasionally it is induced by the bronchitis resulting from the continued inhalation of hard, irritating particles, as of stone, glass, or iron. It is a very chronic affection, lasting often for many years, but usually proving fatal in the end.

Then there are other forms of pneumonia, which have been discriminated of late years, and two of which depend on the presence of special organisms, *epidemic* or *endemic* and *intermittent pneumonia*.

Again, the pneumonia which is apt to occur in the course of certain infectious maladies, as measles, typhoid fever, &c., owes its origin, in part at least, to the specific poisons of the diseases with which it is associated.

Then, the pneumonia depending on obstructed circulation through the lungs forms another class.

The inflammatory process may affect the whole or part of a lobule, when it is known as *lobular*, or a whole lobe may be involved, *lobar pneumonia*. The disease is usually unilateral or single, but sometimes it is bilateral. When it is confined to part or the whole of a lobule the upper portions of the lungs are the most frequently affected, the mischief being in many cases simply local and catarrhal; when it is lobar, the bases of the lungs are usually implicated.

In very many cases of lobular pneumonia the pleura is more or less involved, constituting *pleuro-pneumonia*; in others the bronchial tubes are affected, *broncho-pneumonia*; this form is also usually distinct from true pneumonia, and is not unfrequently, like bronchitis itself, a purely catarrhal affection.

Occasionally lobular pneumonia terminates in the formation of an abscess, or in gangrene; but these events are apt to occur only in cases in which the inflammation is particularly intense or in which there is some special peculiarity, or in very depressed conditions of the system.

Of the two forms of pneumonia due to or associated with the presence of special organisms, one resembles the pneumonia of bovine animals, and is due to a similar cause. The presence of such an organism in certain cases of pneumonia occurring in the human subject was long suspected, and it was in part proved by Klebs and others, but it was reserved for Koch and particularly Friedländer to give the first accurate and complete description.

The micrococci or pneumonococci are of an elliptical form, joined together in pairs, diplococci, or even forming chains, made up of a succession of pairs; they are surrounded by a capsule, and this serves to distinguish them from other micrococci. They very closely resemble the organisms found in bovine pleuro-pneumonia, only that those occurring in the human subject are a little smaller. The micrococcus has been separately cultivated and the disease produced by the inoculation therewith of mice and guinea-pigs, and in the former by inhalation as well. It is introduced into the system from without, and it has been found and satisfactorily identified by Emmerich, of Munich, in some mortar taken from dormitories in which pneumonia was from time to time epidemic.

From more recent observations and experiments there is good reason to believe that croupous pneumonia is identical with the above, the same organism having been found in both.

The other form of pneumonia associated with a particular organism occurs in intermittent and remittent fevers, and hence has been termed intermittent pneumonia.

The mixed form of lobular pneumonia and bronchitis, *broncho-pneumonia*, may be either acute or chronic; it is apt to occur in childhood, in old age, and in debilitated states of the system. It also often occurs after measles in an acute form, or after whooping-cough, when it is less acute.

In the acute form it is a much more serious complaint than ordinary bronchitis, and is attended by considerable fever; in the more chronic form it is an insidious and dangerous malady. In old people it usually commences with the symptoms of an ordinary catarrh, as in *catarrhus senilis*.

Broncho-pneumonia, or, as it is sometimes called, catarrhal pneumonia, is generally catarrhal in its origin, the inflammation of the bronchi, or bronchitis, being antecedent to the pneumonic lesions.

The condition of the lobules varies much; they may be hyper-æmic, inflamed, consolidated, emphysematous, or even collapsed, these conditions being sometimes more or less associated.

The mucous membrane of the smaller bronchi is congested,

thickened, sometimes softened, and the tubules are more or less filled with tenacious or muco-purulent secretion.

The expectoration is very different from that of acute pneumonia; there is an absence of blood, but the cell elements abound, derived mainly from the bronchi, but in part also from the air cells. Now it is to catarrhal or broncho-pneumonia that the term desquamative pneumonia chiefly applies. Should the inflammation be greatly prolonged, an increased formation of fibroid tissue may be one of the results.

Treatment.--Under the head of Inflammation of the Lungs several distinct conditions and affections are included, each of which requires its own special treatment.

First, the treatment of the typical or croupous form of inflammation has to be considered; then that of catarrhal pneumonia or broncho-pneumonia, of chronic or interstitial pneumonia, of pneumonia arising from obstruction to the circulation through the lungs; and, lastly, of epidemic and intermittent pneumonia.

In these several forms the treatment is necessarily complex, and comprises constitutional means, the application of external derivatives, and the employment of various medicaments and medicated solutions in the form of sprays.

In acute pneumonia the first care must be to avoid all proceedings and remedies which tend to weaken, and especially those which exert a depressent action on the heart, since cardiac weakness constitutes the great danger in this disease. For this reason ipecacuanha is preferred by many to tartar emetic; it should be given in nauseant doses, repeated so as to keep up the nauseant effects, and there is no easier or more advantageous way of administering this than by sprays. It matters little whether the sprays are wholly inhaled or not, as the effect will be much the same even if a portion of them be swallowed, the object of their employment being to promote secretion and to lessen fever.

For the reduction of pyrexia, tincture of aconite may in most cases be employed in the usual manner, notwithstanding its somewhat depressent action, and it may be added to the sprays containing the ipecacuanha. Another remedy, which may be used on account of its derivative action, is digitalis, and this is not a cardiac depressant.

In Germany, when the temperature is very high, 106° or 107° F., physicians are in the habit of adopting one or other of the following proceedings with a view to its reduction: they either administer every 24 hours one large dose of quinine, some 20 or 30 grains, or they have recourse to the cold bath. Neither of these proceedings is in much favour in this country, although both are capable of rendering much service in severe and critical cases, in which alone one would be disposed to try them. Short of the cold bath, cold sponging with vinegar and water, or the sheet wetted with the same, may be employed; and if this plan be not adopted, then 10 grains of quinine may be given twice a day, one dose in anticipation of the maximum daily rise of temperature.

Very soon, and while the expectoration is still sanguineous, alkalies, as carbonates of soda or potash, for the sake of their solvent action, may be given in the form of a drink, and solutions of them may also be sprayed into the throat, whereby the local effect produced will be greater than when they are introduced into the stomach.

Then, as the sputa lose their rusty character, as resolution sets in and the expectoration becomes more abundant, the alkaline treatment should be combined, if the cough be troublesome, with such anodynes, as hyoscyamus and henbane, as do not dry the mucous membranes and retard secretion.

The stage of resolution in this affection is not usually much prolonged, but towards the close of the attack stimulant expectorants will sometimes be found of service, as chloride of ammonium or carbonate of ammonia, with, if there be much debility, bark or senega. The expectorants will be best administered in the form of sprays, it being always remembered that, in order to ensure their beneficial effects, it is not necessary that the quantity of liquid used should be great, or that the patient should make any special efforts with a view to their complete inhalation.

As mustard cataplasms are of value in the early stage, so stimulating liniments will be found equally so in the later stage.

The diet throughout should be nourishing, and brandy administered when the heart shows any sign of failing.

In catarrhal or broncho-pneumonia the inhalation treatment resolves itself very much into that which has already been described for acute and chronic bronchitis, and to which, to save repetition, the reader is referred. The fever, even in the acute cases, is usually not so considerable as in acute pneumonia, while in chronic cases it is very much less. It is when the capillary bronchi are affected that the fever is greatest and the difficulty of breathing most marked.

The constitutional, like the local, treatment must of course be adapted to the stage. In the acute stage, stimulant cataplasms or compresses must be applied, and, with a view to promote expectoration, sprays of ipecacuanha may be employed and the vapour of hot water inhaled; but when secretion has become so abundant as to interfere with respiration and to excite much cough of a spasmodic character, chloride of ammonium or carbonate of ammonia should be freely administered, both internally and as sprays; in the chronic stage the expectorants may be combined with bark and senega.

In chronic or fibroid pneumonia, the main objects of the treatment must be to remove as far as practicable the exciting cause of the pneumonia, usually either catarrhal bronchitis or that form of the same disease which arises from the continued inhalation of the irritating particles incidental to certain industrial occupations.

In the advanced stages of this disease, when there is confirmed bronchitis, dilatation of the bronchi, profuse and perhaps fœtid secretion, the treatment will be almost identical with that for bronchiectasis.

In the intermittent form of pneumonia the most important part of the treatment will consist in the administration of those remedies which have acquired the reputation of being almost specifics for intermittent and remittent fevers.

Should gangrene ensue, as it occasionally though rarely does, in the lobular form of pneumonia, in greatly debilitated subjects, in addition to the constitutional treatment, sprays containing carbolic acid, turpentine, or sulphurous acid, or the fumes of these or of iodine, should be freely inhaled. The preference should be given to turpentine or terebene, and these remedies may be taken internally at the same time with advantage.

Of course the powers of the patient must be sustained in every possible way by nourishment, stimulants, and suitable tonics.

Asthma.

A spasmodic affection of the unstriped muscular fibres of the bronchi, the attacks being paroxysmal, occurring at more or less definite intervals, and excited by a variety of causes, all of which act through the nervous system in the production of the spasm. One of the most distressing results of the spasm is dyspnœa, which, if long continued and frequently repeated, gives rise to serious structural effects in the organs of respiration and circulation.

The causes of asthma are various, but they all have this in common, that they give rise to the spasm through the medium of the nervous plexûs supplying the bronchi. The irritation may be communicated directly from the mucous membrane of the bronchi, from that of the stomach and bowels, or from the brain.

The most frequent, and therefore the most important, of all the causes of asthma is bronchitis, however induced, whether due to catarrh or caused by certain specific diseases, as measles or whooping-cough.

Other exciting causes acting directly on the mucous membrane, are irritating vapours and powders, whether organic or inorganic, as the powder of ipecacuanha, the pollen of certain plants, especially those belonging to the natural order *Gramineæ*, and various kinds of mineral substances. If these be applied for a short time, the effects are but temporary, but if the application be long continued, bronchitis may be established and permanent results ensue.

By some writers, climatic influences have been placed among the direct causes, these influences in some cases acting by aggravating the pre-existing bronchitis or by calling into action certain, and sometimes inherited, peculiarities. As a rule dry climates and well-drained towns and cities are best suited to most cases of asthma. It has been supposed that the increased amount of carbonic acid in the air of populous towns has its effect in warding off attacks of asthma, but this view is doubtful, seeing that the proportion of the acid, though increased, is still but small.

Amongst the more remote excito-motory causes of asthma are irritation arising from the stomach and bowels, or due to dentition.

Then occasionally an attack of asthma is brought about by emotional or centric causes, as by great mental excitement.

Certain forms of blood-poisoning also sometimes excite paroxysms of bronchial spasm, as of gout, Bright's disease, and of intermittent and remittent fevers.

Attacks of asthma may follow the disappearance or cure of some chronic and habitual skin affections; this also has been explained on the supposition of the presence in the blood of some morbific constituent.

Then, finally, heredity plays an important part in predisposing to asthma; in this case the liability is probably explained by the existence of certain susceptibilities and structural peculiarities, the nature of which is not understood.

When the asthma is associated with severe and long-continued bronchitis, then morbid structural effects are produced in the bronchi themselves, in the air cells of the lungs, and in the organs of circulation.

The muscular coat of the bronchi, particularly of the smaller branches, becomes abnormally developed; the mucous lining of the bronchi is more or less congested and thickened, a diminution in the calibre of the bronchi being thereby occasioned. In consequence of the difficulty of expiration the air cells are subjected to great internal pressure, so that they become gradually and at length permanently enlarged, and emphysema is established. The obstruction to the circulation through the lungs leads ultimately to dilatation of the right side of the heart, while the lungs themselves become congested.

Each attack of asthma usually terminates in expectoration; this in uncomplicated cases is usually thin and scanty, but if there be bronchitis it will be more abundant and thicker.

Treatment.—First, the cause of the asthma must be accurately determined, and, next, this as far as possible must be removed; in some cases the removal is easy, in others more difficult, if not impossible; but if the cause cannot be wholly removed its effects can usually be greatly mitigated, the severity of the attacks lessened, their duration shortened, and the intervals between them lengthened.

If the cause be mechanical, and depend on the inhalation of irritating dusts and powders, of which there are a great variety, these of course must be avoided; if the attack be associated with bronchitis, as in the more severe and permanent cases is commonly the case, the treatment, with a view to ensure lasting effects, must be directed mainly to this. Again, the fact must not be forgotten that the bronchitis is usually associated with considerable congestion and obstruction of the circulation through the lungs, and with engorgement of the right side of the heart; the bronchitis therefore will in many cases be best relieved by measures which tend to remove the congestion and obstruction, as particularly by digitalis and by the moderate use of saline aperients.

If the cause be peptic, then the condition of the organs of digestion will claim the first attention. The seizures which occasionally arise from the condition of the blood, although of an asthmatical nature, constitute a separate class of cases, differing in some respects from ordinary asthma.

The treatment of the bronchitis, whether general or by inhalation, must be conducted on the principles and in the manner already described.

For the relief of the spasmodic seizures a great variety of remedies have been employed, and many with considerable success.

During the seizures the patient has often great difficulty in inhaling even the most unirritating vapours, or he may be totally unable to smoke or swallow. In such cases, sprays may be employed with great advantage, the medicament being dissolved in only a small quantity of the menstruum. For the beneficial effect of these it is in general not necessary that the fluid should penetrate into the bronchi.

The remedies may for the most part be classified under one or other of the following headings :—

Antispasmodics, stimulants, anodynes, depressants, and nervine tonics. These several classes of remedies are capable of being effectively applied to the organs of respiration by means of inhalation, that is to say, to the parts chiefly involved, in place of reaching them by the more circuitous route of the stomach.

The chief *antispasmodic remedies* are chloroform and ether, separately or combined, and nitrite of amyl; they are inhaled chiefly in the form of vapour, either in anticipation, or for the relief, of the paroxysmal dyspnœa.

Dr. Talfourd Jones has found nitrite of amyl of great service in asthma, 5 to 10 drops being placed on a handkerchief and inhaled cautiously. Some persons are much more susceptible to its action than others, especially women, so that it is better to commence with the smaller doses. Dr. Ringer states that the relief afforded by this remedy lasts but a very short time, the dyspnœa quickly returns, and the remedy soon loses its effect. The same remark applies equally to chloroform and ether.

The *narcotics* act also as antispasmodics, and some, as stramonium and tobacco, exert a *depressent* effect as well.

The stramonium for smoking should consist of a mixture of the ripe seed vessels and leaves, as the former contain a much larger proportion of daturin, and hence are more active. The herb is sometimes smoked alone, either in pipes or as cigarettes, but is often mixed with datura tatula, lobelia, cannabis, and nitrate of potash.

Many nostrums, or so-called cures for asthma have been devised, the majority of which contain stramonium as one of their constituents. The chief of these have been already noticed. One is Himrod's cure, which consists mainly, if not entirely, of powdered stramonium leaves, aniseed in powder, together with a small quantity of the leaves of tobacco, the whole being rendered combustible by the presence of nitrate of potash. This mixture must be burned and the fumes inhaled.

Dr. Fothergill states that Himrod's cure contains also belladonna, but Mr. E. G. Clayton, F.C.S., who at my request very kindly undertook the analysis of the mixture, says that he has not been able to detect the presence of atropia, nor indeed of arsenic or theine, the presence of both of which has been suspected.

A mixture consisting of 2 parts of the powder of stramonium leaves, 1 part of aniseed, and 1 of nitrate of potash will be found to answer very much the same purpose as Himrod's cure.

Dr. W. H. Beverley, of Scarborough, gives the following formula for the preparation of a powder which he has found very useful in the relief of asthmatical attacks, including those of hay fever. Six drachms of each of the following: datura tatula, stramonium, cannabis, and lobelia, with 4 drachms of nitre and $\frac{1}{2}$ drachm of eucalyptus oil. A teaspoonful to be burnt occasionally and the fumes inhaled.

The cigarettes d'Espic contain extract of belladonna, henbane, thorn apple, water-fennel leaves, and extract of opium. This is very much Trousseau's formula for narcotic cigarettes, which, however, in place of fennel, contained cherry-laurel water.

Dr. J. Mortimer Granville has also lately published a formula, under the name of pulvis boracis Co., the composition of which has already been given, and which he states to be really curative of the asthma accompanying hay fever.

A very favourite remedy consists in the inhalation of the fumes arising from the burning of paper saturated with nitrate of potash, and it has this advantage, that the fumes can be inhaled when the patient is unable to smoke a cigarette. The paper should contain a known quantity of nitre, and should be burned in a special apparatus, so that the fumes may not be too much diluted by the air of the room. See Chapter V.

Another remedy for asthma, more or less useful, which has been long employed, is the so-called 'Ozone paper.' It is made by passing bibulous paper through a strong solution of nitrate of potash, chlorate of potash, and iodide of potassium. Cigarettes are made with paper similarly prepared, and may be procured of Mr. Huggins, Strand, London.

Occasionally tobacco is mixed with the other narcotics, but any such mixture should be used with great caution by non-smokers, especially when the heart is weak.

Lobelia is usually inhaled in the form of an ethereal tincture, and sometimes by means of an oro-nasal inhaler; this proceeding is, however, useless, as the alkaloid of lobelia is non-volatile. The ethereal tincture should therefore be inhaled as a spray, or, best of all, the herb should be smoked. The remedy must be freely used, and in considerable doses, to be effective. Ringer states that it is serviceable both in the peptic and bronchial forms of asthma, and he remarks, 'My experience leads me to esteem lobelia higher the more I try it, and I frequently hear it extolled by patients.' It is useless in the dyspnœa aising from heart-disease.

Of course the ethereal tincture of lobelia may be administered in the ordinary way by the mouth; in this case the safest plan is to commence with 10 drops every 10 or 15 minutes, till relief is obtained, but as much as a drachm may be given at longer intervals, say every hour. It is apt to produce great faintness and nausea, but, in the absence of disease of the heart, the depression is never dangerous.

Cannabis, or Indian hemp, though useful in some cases of asthma, is very inferior to lobelia. It being an excitant of the nervous system, it should not be given when there is congestion of the brain or lungs. The cigarettes should contain from $\frac{1}{4}$ to 1 grain of the extract, and the spray from 5 to 20 minims of the tincture in about $\frac{1}{2}$ ounce of thin mucilage.

Belladonna is another valuable remedy for the dyspnœa, and, if administered early enough, it will sometimes even prevent an attack, but it must be freely given, and if inhaled it should be in the form of spray, 10 minims of the tincture every ½ hour; but the effect must be watched, and when dilatation of the pupils occurs, and a disposition to drowsiness, the remedy must be suspended for a time. In some cases as much as ½ drachm of the tincture given hourly will be required before any sensible effect is produced.

Opium in some cases of asthma gives relief, but, strange to say, morphia will sometimes produce an asthmatical seizure, and it is well to bear this fact in mind. A spray of from 20 to 40 minims in ½ ounce of water of the ammoniated tincture of opium, or from 10 to 15 minims of the liquor opii sedativus, will often greatly relieve the cough of bronchitic asthma and so diminish or remove the dyspnœa.

Amongst the *stimulant remedies* which are often serviceable, either in warding off or relieving an attack of asthma, are assafœtida, coffee, or citrate of caffeine.

Assafœtida gives relief partly through its action on the nervous system and in part by its stimulant effect on the mucous membrane of the bronchi, causing an increased secretion of mucus; sometimes it exerts an aperient effect, and in large doses it causes nausea and even vomiting. It would appear, therefore, to be a remedy well adapted to afford relief in bronchitic and dyspeptic asthma. It is not a pleasant remedy, and ammoniacum would in some cases be better tolerated, although it must be remembered that it differs from assafœtida in the fact that it exerts but little action on the nervous system.

It has long been known that a cup of strong black coffee will often greatly relieve the breathing in a fit of asthma, this effect being due mainly to the caffeine which it contains. This fact has led Dr. Thorowgood to recommend and prescribe citrate of caffeine in 2-grain doses; this remedy may be employed effectually in the form of an atomized spray, 2 grains of the salt in ½ ounce of water.

The *depressants* act of course in a very different manner. Antimony in the form of tartar emetic, as already noticed, proves very serviceable in the quasi-asthma of children from 6 to 12 years of age, and which is so apt to come on in consequence of the slightest exposure. Ringer recommends that 1 grain of tartar emetic should be dissolved in ½ pint of water, and that a teaspoonful of the solution should be given every quarter of an hour for the first hour and afterwards hourly. He writes, 'If the wheezing comes on at night, it is sufficient to give the medicine at this time only. The good effects of the medicine become speedily evident, for on the very first night it often greatly benefits the child. So small a dose, it may be thought, must be inefficacious, but when first given it generally produces vomiting once or twice in the day, and, as it is not necessary to produce sickness, the dose in this case must be still smaller.'

Of course the tartar emetic may be administered in the spray form, 5 to 10 minims of vinum antimoniale in 1 or 2 drachms of water being used for each spray; the dose, however, must be regulated according to the age of the child and the effect produced.

The effects of the inhalation of sprays of ipecacuanha resemble those produced by tartar emetic; in bronchitic asthma, they increase the secretion of mucus, but occasionally, in place of relieving the breathing, ipecacuanha occasions a sense of constriction and tightness: in such cases of course the remedy must be abandoned. Their use is also contra-indicated when there is marked cardiac weakness.

Another well-known remedy for asthma is arsenious acid. This may be inhaled in cigarettes each of which should contain about $\frac{1}{12}$ to $\frac{1}{8}$ grain of arsenic; or from 3 to 6 minims of the liquor arsenicalis P. B. may be administered by the stomach, twice a day in the intervals of the attacks.

Arsenic is a valuable remedy in a class of cases of quasi-asthma, which has been specially described by Ringer, attended with periodical attacks of profuse running from the eyes and nose, occurring in the morning or two or three times a day, followed by difficulty of breathing at night: cases, in fact, which simulate true hay asthma, but are not really that disease, in which arsenic is but of little value.

There may be an interval of several days between these attacks. The causes are various and include cold, dust, and other mechanical irritants; in such cases from 2 to 3 drops of the solution of arsenic three times a day, administered in the usual way, is often serviceable. In a still severer form, which, like the first variety, may last even for years, the irritation may extend from the nose to the throat and thence to the lungs, producing difficulty of breathing, wheezing, and free expectoration. Children are liable to a similar affection, of which there may be several attacks in the course of the year, especially during the winter, and they may even end in true asthma.

These cases, Ringer remarks, appear to be related on the one hand to bronchitis and on the other to hay fever, from which, indeed, it is impossible to distinguish some of them, although the irritant exciting cause may be different.

Arsenic in many of these cases is of much service, although some days may elapse before its beneficial effect is apparent in relieving the paroxysmal sneezing and coryza. Where arsenic fails, the inhalation of the fumes of iodine will often prove of service in arresting the coryza.

There are still several other valuable remedies for the relief of asthma which remain to be mentioned.

One of these is the internal administration of the iodide of potassium. Dr. Tanner was of opinion that this was the most valuable of all remedies in asthma. It is more especially in bronchial asthma that its beneficial action is most apparent, 5 grains being administered two or three times a day. There can be little doubt that the remedy would have the same effect if inhaled in a spray. The iodide should be continued for some time.

Another remedy which, in the hands of Dr. Begbie, has proved very beneficial in some cases of asthma is bromide of potassium. This acts in a different way from the iodide, and mainly by its sedative effect on the nerves.

In the same class of cases hydrate of chloral also frequently proves of much service.

One of the latest remedies employed in the treatment of asthma is Grindelia robusta, which is much esteemed in America and is of undoubted value. It may be obtained of Mr. Martindale, of New Cavendish Street, London, and is used with a view of either anticipating and so preventing an attack of asthma, or of cutting it short. The dose of the solid extract is from 2 to 3 grains two or three times a day as a preventive. and from 20 to 30 minims of the liquid extract every half-hour or hour at the onset of the dyspnœa. The liquid extract may be inhaled as a spray.

Lastly, sulphurous acid has been recommended in asthma by Dr. Dewar, as has been already stated.

Emphysema.

When Emphysema involves only a part or the whole of a lobule it is called *lobular*, but when one or more lobes of the lungs are affected it is termed *lobar*. So great are the differences between these two forms that they may almost be regarded as distinct affections.

The *lobular* or partial form of the disease is brought about by unusually strong expiratory efforts, necessitated by some impediment in the bronchi to easy and natural expiration, as that occasioned by the irritable, congested, and swollen mucous membrane, with often excess of secretion, arising from chronic bronchitis. The cough causes the lung to contract on the air confined in the lobules behind the seat of obstruction, and this, not being able to escape readily, produces distension of the air cells, which, at first temporary, ultimately becomes permanent. This form of emphysema is usually limited to one lung and is not hereditary.

The second or *lobar* form of the disease is often hereditary, and it generally affects both lungs; at first there may be no cough or bronchitis and no mechanical obstruction in the bronchi. It usually commences in early or middle life, is very insidious, the first symptom which attracts attention being increasing breathlessness on slight exertion. Lastly, the cause of the enlargement of the air cells is different, it being produced by unusual inspiratory efforts, occasioning enlargement of the air cells, owing to a want of power in the elastic intercellular tissues, and this again is regarded as an evidence of mal-nutrition.

At first the air cells are simply enlarged, but as the disease progresses the enlargement leads to pressure on the intervesicular vessels, diminishing their blood supply; as it advances still further the vesicles become more or less ruptured, so that now many vesicles unite to form larger cells. The result of these changes is great diminution of the aërating surface of the lungs, destruction of the capillary vessels, diminished blood supply, lessened gas exchange, great impairment of lung elasticity, consequent inability to empty the lungs, and in very advanced cases distressing breathlessness, augmented on the slightest exertion.

EMPHYSEMA.

The lungs are large, pale, non-crepitant, and undergo but little alteration of size either during inspiration or expiration, and hence the movements of the thorax are confined to narrow limits.

At first lobar emphysema may be uncomplicated with any other disease, but when of long standing secondary affections always arise. One of the most frequent complications is bronchitis; this is rather congestive than inflammatory, involves the smaller tubes, which sometimes become dilated, is often attended with profuse secretion, which, in consequence of the diminished resiliency of the lungs, is expectorated only with difficulty and in some cases not at all, death by apnœa being the result. Another complication is hypertrophy of both the right and left sides of the heart, with sometimes even valvular disease, while, in consequence of the difficulty of circulation through the lungs, fibrinous clots are apt to form in the heart and large vessels. The liability to valvular disease is explained on the supposition that lobar emphysema is a disease of mal-nutrition, as gout is said to be; indeed, this is often associated with emphysema. A third consequence is asthma; this, as in most other cases, occurs at night, owing, it is believed, to the hypostatic congestion which take place in consequence of a continuance for some hours in the recumbent position.

In the lobar form of the disease, in the absence of confirmed bronchitis, the act of inspiration is performed with ease, although it is shallow; but that of expiration, in consequence of the impaired elasticity of the lungs, is slow, difficult, accompanied by wheezing, and the lungs themselves are only partially emptied.

Owing to the rupture and destruction of the air cells, obliteration of capillary vessels, and to the incomplete manner in which the lungs are filled and emptied, the blood is imperfectly aërated and the skin assumes a congested and dusky appearance.

Treatment.—Before proceeding to treatment, the first thing to ascertain is, whether the case be one of lobular or lobar emphysema, and next the nature of any complication which may exist.

In most cases bronchitis will be found to be present; this, owing to the difficulty of circulation through the lungs, is usually, especially in the lobar form, congestive rather than inflammatory; the small bronchial tubes are frequently affected, and there is copious secretion, which it is often extremely difficult to bring up from the want of resiliency in the lungs.

With a view to relieve the congestion the several emunctories must be kept in action, especially the kidneys. Digitalis, from its diuretic effect and power of strengthening the heart, will often prove of value, and it may be as easily inhaled in the form of a spray as given by the stomach.

Another indication is the promotion of expectoration; for this purpose there is no remedy so efficacious as carbonate of ammonia, with which squills may be combined. In the lobar form care must be taken to avoid the use of all remedies which dry up or hinder secretion, for this would increase congestion.

If the secretion be thick and tenacious, then the solvent carbonates of soda or potash may be combined with the expectorant ammonia salt. The effects of these remedies are better obtained by sprays than through the medium of the stomach.

Anodynes will not often be required for the relief of the cough, and unless these are very judiciously employed they may do harm, particularly when the cough is due to excess of secretion in the bronchial tubes.

With a view to relieve the vascular system the treatment must be in part derivative.

The asthmatical seizures will require very much the same treatment as asthma from other causes, and it comprises especially the inhalation of antispasmodics.

The inhalation of the fumes of arsenic are often of much service in relieving the asthma so frequently attendant on emphysema. It is also beneficial in the case of persons who are more or less emphysematous, and who, on taking cold, are subject to wheezing, with some difficulty of breathing, but not amounting to asthma. Arsenic is not, however, efficacious in the paroxysms of true asthma, or when this is associated with severe bronchitis.

Another remedy often of value in cases similar to the above, with wheezing and shortness of breath, is chloral; this must not, however, be administered if there be obstruction of the circulation, as evidenced by lividity, and which would be increased thereby.

Again, lobelia is stated to relieve the dyspnœa in emphysema with capillary bronchitis. Here caution is also required when there is disease of the heart.

The lobar form of the disease being of constitutional origin, all those means must be adopted which are calculated to improve the general health and to give tone to the system.

If the emphysema be associated with gout, this also must be treated; in such cases, as also in some which are not gouty, iodide of potassium proves beneficial, it aiding in the relief of the bronchitis, as well as being serviceable in other ways.

A very valuable method of treatment, much employed abroad, is by the inhalation of compressed air, with sometimes expiration into rarefied air. The latter ensures the complete discharge of the residual air from the lungs, and so allows the over-stretched elastic tissues gradually to regain their power. The inhalation of compressed air is, however, in many advanced cases not advisable, and it is better to limit the treatment to expiration into rarefied air, which in most cases will prove beneficial and is not likely to do harm. For the carrying out of this treatment only a very simple and inexpensive apparatus is requisite.

Another method of treatment which should be adopted in emphysema with defective aëration of the blood, and which can scarcely fail to afford beneficial results, is the free and continued inhalation of oxygen gas. The inhalation should be repeated twice a day, as much as 30 or 40 litres being inhaled each time. If it be diluted with atmospheric air the quantity of the mixture inhaled must be proportionately increased.

Lastly, change of climate is often very beneficial in cases of emphysema, especially the lobar form, but those suffering from the disease should be cautioned against any extreme exertion and the climbing of hills or mountains.

Pulmonary Consumption.

We have now arrived at the consideration of the treatment of lung consumption, more particularly by means of inhalation, this being the most important subject of the whole work. A strong feeling in favour of this treatment has long possessed the minds of many medical men, although it must be confessed that the trials of this method hitherto made have disappointed the expectations generally entertained. Some few medical men have, indeed, lauded the plan beyond measure, and have attributed to it surprising results. That the results of inhalation in phthisis should be so often disappointing is doubtless in the main due to the very imperfect manner in which the inhalation has, until very recently, been carried out. This part of the subject has already been treated of, and it is not necessary to enlarge upon it now.

The treatment of phthisis by inhalation embraces a variety of important objects, as the following :—

The alleviation of the cough, the diminution of expectoration, alteration of its quality, the destruction of the bacteria, the healing of the ulcers and cavities, the lessening or removal of the night sweats and diarrhœa, and the arrest of the hæmoptysis which is so apt to occur, either at the commencement or in the course of the disease.

Consumption may be defined to be an infective disease, running sometimes an acute but more frequently a chronic course, having its seat primarily in the lungs and tending to the formation and deposition of a characteristic puriform material, termed tubercle, which in its progress undergoes a series of degenerative changes, causes lesions of the normal tissues, and the effusion of various morbid products and secretions, and ending, in the advanced stages, in ulceration, followed usually, when the disease is situated in the lungs, by excavation and the formation of cavities.

Although the disease ordinarily commences in the apices, often of both lungs, but sometimes of one only, yet it usually extends from the part first attacked, involving at length more or less of the structure of the lungs and spreading thence to other portions of the respiratory mucous track, this tendency to spread being not even limited to the lungs.

The infectivity of the disease is shown by the results of experiments by inoculation with the sputa of phthisis.

It has been abundantly established that these sputa, in what may be termed true or undoubted phthisis, contain one invariable element, namely, the tubercle bacillus, and there is much probability that this organism is the infective element of the sputa, seeing that when such animals as rabbits and guinea-pigs are inoculated with the pure cultivated bacillus they become affected with a malady not distinguishable from phthisis in the human subject, the bacillus being invariably present in the tubercular formations.

Now, although the disease is usually first manifested in the lungs themselves, as just stated, it shows a remarkable liability to spread and to infect, first, other parts of the lungs, including the mucous membrane of the respiratory track, as the bronchi, larynx, &c.; secondly, other parts outside the lungs, but which are within easy reach of infection by the sputa, as the mucous membrane of the small and large intestines and the liver; and, thirdly, parts and organs still more remote and which are only to be reached through the absorbent and vascular systems, as the mesenteric, bronchial, and cervical glands, and the membranes of the brain. The stomach generally escapes infection; at all events it resists it for a long time, and this comparative immunity is probably due to its power of digesting the sputa and of destroying the bacillus. No doubt some portion of the sputa, becoming mixed up with the food, escapes digestion and passes on into the bowels to work its mischief there; but, since the digestive process is constantly going on in the stomach, it is very difficult for the sputa or the bacilli to survive the solvent action of the gastric juice and so to obtain a permanent footing there.

Consumption, then, is not a simple local or inflammatory malady, although it is more or less localized at first, and different names have been given to the disease, according to the parts chiefly implicated, as *pulmonary, bronchial, laryngeal*, and *pharyngeal* consumption; and, quitting the organs of respiration, *intestinal* consumption may without impropriety be spoken of. The pharyngeal and even *buccal* complications, or rather extensions, of the disease have found a separate place and been specially described in more than one work treating of phthisis.

The following figures represent the comparative liability in cases of pulmonary phthisis of other organs and parts to become affected, usually secondarily, with tuberculosis. Out of 1226 cases of pulmonary phthisis recorded by Heinze, the statistics of which are given by Morell Mackenzie, tuberculosis was found in the pleuræ in 137 cases; bronchi, 15; glands (bronchial and cervical?), 106; trachea, 99; larynx, 376; pharynx, 14; tonsils, 8; intestines, 630; mesenteric glands, 106; peritoneum, 95; liver, 286; spleen, 120; kidneys, 150; and in the membranes of the brain, 43.

Willigk's figures differ somewhat from the preceding; he found that out of 1317 cases of tuberculosis, in 656 cases the intestines were affected, in 237 the mesenteric glands, in 182 the larynx, and in 242 cases other parts and organs.

The difference between these two sets of figures, so far as the mesenteric glands are concerned, is very great and not easily to be explained. It might have been supposed that these glands would be particularly liable to become affected, more so than the bronchial glands ; the few cases in which the glands were found to be affected by Heinze is remarkable.

Several different forms or varieties of lung consumption have been described, most of which have one source or origin, notwithstanding considerable diversity in the pathology and symptoms. The chief of these, not including the mere extensions of the disease adverted to above, are the following, they being divisible into the acute and chronic.

The principal *acute* forms include acute tuberculosis, acute phthisis, and acute tuberculo-pneumonic phthisis.

In *Acute Tuberculosis*, that is, in the acute *miliary* variety of the disease, the tubercles are more or less transparent, grey or dark-coloured granulations, of about the size of hemp-seeds, and which are usually widely distributed ; this form of consumption is prone to attack serous membranes, as that of the pleura and brain, giving rise in the latter case to fatal *cerebral* symptoms. This variety runs a rapid course, often terminating in a few weeks and occasionally even in a few days; it is apt to occur in those who have inherited a predisposition to phthisis.

Of *Acute Phthisis*, or, as it has been termed, *scrofulous pneumonia*, not a very appropriate name, the characteristics are thus summed up by Dr. C. Theodore Williams : 1, ' the acuteness of the disorganizing processes, excavation quickly succeeding consolidation ; 2, the inflammatory nature of the lesions and the rarity of miliary tubercle ; 3, the occurrence of pneumo-thorax ; and 4, the freedom of other organs from tuberculosis.'

Acute Tuberculo-Pneumonic Phthisis presents somewhat the characters of the two previous forms. It agrees with the first in the rapidity with which tuberculization takes place in the lungs, and it often extends to the intestines, and with the second in presenting consolidations of a pneumonic character, while it differs from them both in the tendency of the tubercles to caseate, break down, and to form cavities.

The more important *chronic* varieties have been described under the following names : catarrhal phthisis ; fibroid phthisis ; scrofulous phthisis ; hæmorrhagic phthisis, and chronic tubercular or ordinary phthisis.

Catarrhal Phthisis.—This form is described as having its origin in chronic bronchitis, which may have existed for years, but which has gradually passed into catarrhal pneumonia. It is stated to be more common among the young than the old, and to arise frequently from whooping-cough and measles.

Fibroid Phthisis.—The term fibroid phthisis originated with Sir Andrew Clark, and it well expresses the condition of the lungs in this form of the disease. While in nearly every form of phthisis more or less of fibrosis exists, this special form frequently takes its origin in pleuro-pneumonia and chronic interstitial pneumonia, which give rise to a greatly increased development of the interstitial fibroid tissues. The presence of this increased amount of fibrous tissue renders the course of the malady very chronic, because by its dense character it acts as a barrier to the breaking down of other tissues and the extension of the disease. The adventitious fibrous tissue undergoes a process of gradual contraction, whereby the lung is made to shrink and is reduced in size; this shrinking, supposing the disease to affect one lung only, which is usually the case at first, though ultimately both lungs become involved, causes a falling in of the chest on the affected side and a dragging over and displacement to the same side of the opposite lung and the heart. When the contraction is very considerable it gives rise to dyspnœa, obstruction to the circulation through the lungs, dropsy, and albuminuria. Again, the lung having lost its elasticity, expectoration is difficult and the secretion may be retained, causing dilatation of the bronchi; it even becoming sometimes fœtid and so giving rise to the danger of blood poisoning. This is a particularly chronic form of phthisis; it may last for many years and is attended with but little or no fever. In some cases doubtless the lung is affected by fibrosis without the subsequent development of phthisis; again, although it is liable to attack those who are predisposed to phthisis by heredity or family history, it yet may occur independent of any hereditary tendency.

All observers, including particularly Sir Andrew Clark, concur in the statement that no tubercle bacilli are present in the lungs in cases of fibroid phthisis. This is a very important fact, and would seem to point to some essential difference between this and some of the other forms of phthisis, a further distinction being the chronicity of fibroid phthisis, which may last 20 years or even longer. Since more or less fibrosis exists in many cases of phthisis in which bacilli are present, it seems difficult to distinguish in all cases the one form from the other. If only a small amount of fibrosis exists, unless the bacilli were also absent, it would not be correct to regard it as a case of fibroid phthisis; it is only when the fibroid element is in great amount that the affection has been denominated 'fibroid,' and thus until quite recently without reference to the presence or absence of the bacillus.

Scrofulous Phthisis.—This variety occurs usually in early life, in cases in which the scrofulous diathesis is well marked and in which frequently the disease is actually progressing in other parts and organs of the body. It is particularly apt to become developed in the lungs on the cessation of scrofulous discharges elsewhere; it is specially characterized by the disposition to the implication of the lymphatic glands, as those of the mesentery, bronchi, and neck; and its progress is usually slow.

Hæmorrhagic Phthisis.—This term is applied by Dr. C. T. Williams, Dr. Hughes Bennett, and others to a form of phthisis marked by the occurrence of considerable and repeated hæmorrhages, there being at the same time but a small amount of detectable mischief. It occurs more frequently amongst men than women, and usually in those past the middle period of life; in most cases the disease does not extend beyond consolidation, and the patient may live for many years, although in fatal cases, on examination after death, the pathological conditions indicative of advanced phthisis are met with.

Chronic Tubercular Phthisis.—In this, the ordinary form of the disease and the last to be noticed, all the elements of phthisis are to be found in the lungs: the small miliary tubercles, the white and yellow large tubercles, identical with the former, only that they are in a more advanced stage and tending to disintegration, caseation, and softening; larger caseous masses and tubercular infiltration; ulceration and excavation; more or less pneumonia and excess of fibrous tissue.

A few brief remarks may now be made concerning the temperature, pulse, the night perspirations, and the diarrhœa of phthisis.

Temperature.—This varies considerably, according to the type of the disease and the condition of the lungs. The temperature is of course much above normal in the acute forms of the disease, and also when any inflammatory complications arise in chronic forms of the malady.

In ordinary cases the temperature, even in the course of 24 hours, is both above and below the normal standard. The increase usually commences early in the afternoon and continues until 4 or

5 o'clock in the morning, by which time the temperature has become subnormal. In the chronic forms the increase may not amount to more than a degree or two, except when some inflammatory complication exists, but in acute cases the temperature may rise to 106° and 107 °F., while the decrease in both acute and chronic cases is sometimes equally considerable; usually the temperature falls to 95° or 96° F., but it may descend even 3 or 4 degrees lower. The afternoon exacerbation is to be regarded as an index of the type and progress of the malady, and the morning depression as an evidence of the greater or less exhaustion of the vital powers; it must be remembered, however, that the disease may make slow progress even while the temperature is normal or subnormal. Normal temperatures are usually met with at about 10 or 11 A.M.

The temperature is reduced usually by the night sweats, when these depend on pyrexia; as also by the diarrhœa, unless in cases in which there is extensive ulceration of the intestines.

Night Sweats.—These occur in two forms; in acute and sub-acute cases they are indicative of the temporary subsidence of the fever, while in other cases, in the absence of much fever, they are to be regarded rather as a sign of debility and a want of vaso-motor power.

Pulse.—This will of course vary greatly, in accordance with the fever and the temperature; when these are high, the pulse will be quick and may reach 130 to 140, while in the absence of fever it may even, in advanced stages of the disease, when this is in a quiescent condition, be but little above normal; in nearly all cases it is small and feeble.

Diarrhœa.—In the early stages of phthisis there is often some amount of constipation, but when the disease has made considerable progress there is frequently more or less diarrhœa; this at first may arise from acidity, or some other form of dyspepsia, or it may be due to ulceration of the solitary and aggregated glands of the intestines, and if the ulceration be extensive it may involve the jejunum, ilium, and even the large intestines, although it is usually in the lower part of the ilium and cæcum that the ulceration is most marked. This form of diarrhœa constitutes of course a serious complication.

Another grave cause of obstinate diarrhœa is lardaceous degeneration of the mucous membrane of the small intetsines.

Lastly, in some cases diarrhœa is to be regarded, like the night sweats, as a kind of flux, due to want of vital power.

The diarrhœa dependent on ulceration is doubly serious, first because it indicates the almost hopeless extension of the tuberculization, and next because it rapidly destroys the little remaining strength of the patient.

Associated with the changes due mainly to inflammation, tubercular deposit, and ulceration are other important changes, namely, lardaceous and fatty degenerations. The first form of degeneration is very widespread; it not only affects the bowels, giving rise to diarrhœa, but it involves other organs, as the liver, and if there have been albuminuria the kidneys as well; while fatty degeneration is found in many cases in the liver, kidneys, and, as Dr. Quain has shown, in the muscular tissue of the heart itself.

Treatment.—This comprises a great many objects and purposes, and may be considered under the following heads : the prophylactic or preventive treatment, which is in part hygienic, the general or constitutional, and the local treatment. Under each of these divisions inhalation will be found to play an important part.

The object of the *prophylactic* treatment is twofold : first, to strengthen the vital powers and to promote health in every possible way; secondly, to increase the capacity of the lungs and augment the force both of inspiration and expiration by improving the power of the elastic tissue of the lungs.

With a view to strengthen the vital powers the organs of digestion must be kept in good condition; the food must be light and nutritive, so that the body may be well nourished; the blood must be well aërated and measures adopted to guard against anæmia, as by the respiration day and night of an abundance of fresh and frequently changed air.

All depressing influences must be avoided. The climate should be dry, the soil porous and not clayey, and the situation elevated. The body should be warmly clad, but not heavily, for over-clothing is as great an evil as insufficient clothing, since it tends to make the skin so sensitive that even a breath of the cool, fresh, outer air gives rise to a sensation of chilliness and discomfort.

For the development of the lungs and thorax, the increase of their capacity and the strengthening of their elastic force, certain mechanical measures must be resorted to, and which in their results produce also chemical and vital effects There are several of such means, but some of them are not resorted to in this country as early or as frequently as is to be desired. One of these is a well-devised system of *gymnastics*, having for its object the development and strengthening of the muscles of respiration, the expansion of the lungs, and the consumption of a greater amount of oxygen. This method therefore is partly one of inhalation treatment. One of the first physicians to recognize the necessity of, and to practically carry out, though in an imperfect manner, the forcible exercise of the respiratory muscles and of the lungs was the late Dr. Ramadge.

Another means of carrying out the same object, is by *wind instruments* of a suitable kind and carefully used. Unlike many other remedial measures, these may be made a source of amusement as well as benefit.

Another means which acts partly mechanically and partly chemically, much employed abroad, but looked upon in this country with little favour, mainly because it is somewhat out of the way of ordinary routine practice, is the inspiration of *compressed air*, combined in some cases with expiration into ordinary, or *rarefied air*. The effects of this are to strengthen the respiratory muscles, to increase the capacity and elasticity of the lungs, to augment the consumption of oxygen and increase gas exchange, and so promote sanguification and the depuration of the blood. Now there is no difficulty worth mentioning in carrying out this system of inhalation : it is very easy and simple ; it makes very little demand on the time and attention of the medical man ; the necessary apparatus is by no means costly, and it can be readily used and managed by the patient himself.

The same purposes can, in the main, be accomplished by a prolonged stay in high *mountain regions* ; but mountains are few and far between, they are difficult to reach, and it is not always convenient to leave home for the purpose.

Now one or other of the methods of exercise and inhalation just indicated is not only suitable as a prophylactic, but is imperative in many cases of insipient phthisis in which congestion and consolidation have begun to manifest themselves in the apices of the lungs.

In those cases in which there is an insufficient supply of air and defective aëration, the inhalation of *oxygen gas*, more or less diluted with air, will often be found serviceable. The employment of this gas is extremely simple and involves scarcely any trouble.

The *general* and *constitutional* treatment of phthisis should be on the same principles as the prophylactic treatment; the physical exercises, however, for the expansion of the lungs and the strengthening of the muscles of respiration will require in some cases to be modified and in others they may be unsuitable. The diet must be nourishing and of easy digestibility, while cod-liver oil should, when practicable, form part of it. The necessity for an abundance of fresh air is even greater, as well as for a dry climate.

The nature of the climate will depend very much upon the type and extent of the disease and the condition of the patient. If the case be in an early stage, the digestion good, and the vital powers considerable, then dry, cold, mountainous regions may be the most suitable; but if the case be advanced, digestion weak and the powers generally feeble, then a dry, temperate climate should be selected.

A great deal of nonsense has been written to the effect, that consumptive patients in general will do far better by remaining in England amongst their friends, surrounded by 'home comforts,' these including, of course, good food, than by coming abroad. But what are the facts? The climate of England in winter is damp, cold, and variable, with but little sun. There are nearly 200 days in the year on which more or less rain falls; during the winter the sky is seldom seen, and the sun is only at rare intervals capable of penetrating the canopy of clouds which veils it from sight. Again, diseases of the organs of respiration are the scourge of this climate; about 51,000 deaths occur annually in England and Wales from consumption alone, and 63,000 from bronchitis. Are we to say that looking out of the window day after day and seeing mist, fog, and rain, and vainly hoping from week to week to get a glimpse of the blue sky and to feel the warmth of a ray of sunshine, are to be reckoned amongst the home comforts? Contrast for a moment with such a climate that of the Riviera, in which on an average there are but 50 days on which rain falls, where fogs are unknown, and where out of the 180 days constituting the winter season the sun shines on not less than 160 days, and on an average for 8 hours each day. Add to this the fact, that in a south room one can sit with the window open for most days of the winter without a fire, and that flowers abound throughout that season.

Then as to the food. In all the well-known health resorts of Italy this includes excellent meat and an abundance of poultry and game of all kinds, at prices more moderate than in England. While there is no lack of sea fish, this on the whole is, no doubt, better in England; and though exception may in some cases be taken to the cooking, with a little pains this may soon be set right. So far, therefore, from there being any want of variety or excellence in the food, the invalid, if he lives in any one of the many well-appointed hotels, is more likely to suffer from indigestion on account of the number of dishes provided, than from any deficiency in nutritious and suitable food.

Then as to another point often urged with a view to keep patients in England—namely, that they would be amongst their friends : It is no doubt very pleasant to see one's friends, but it must be remembered that many invalids are usually unfitted for much society, and that they are generally accompanied on going abroad by one or more of their nearest relatives. What the medical man abroad has often to complain of is, that usually there is too much attraction in the way of society, and that the young are in consequence frequently tempted to overstep the bounds of prudence.

Under the above circumstances therefore, to detain a consumptive patient in England, unless the case be very advanced indeed and almost hopeless, is but little short of cruelty.

With regard to cod-liver oil, the taking of this is so important that its use ought not to be readily abandoned, even should it disagree at first. It should be tried in all forms, as an emulsion or with extract of malt. Among the best substitutes for cod-liver oil are butter and cream.

Tonics are usually indicated in ordinary cases of phthisis, particularly quinine, the hypophosphites, and arsenic. Each of these fulfils a different object. Quinine is not only a tonic, but it is an antipyretic, and therefore should be given in pyrexia; the hypophosphites are nervine stimulants, and best suited for those apyrexial cases in which the nervous powers of the patient are exhausted; while arsenic is tonic, alterative, and aids nutrition.

In acute forms of phthisis the treatment during the pyrexial period must be directed to lessening the fever by salines, by quinine, and when there is congestion and inflammation by the employment of counter-irritants, such as iodine. In chronic cases with intercurrent or interstitial pneumonia the treatment must be somewhat similar, though in such cases there is usually much less fever. Although the treatment must not be depressent, yet benefit will often be derived from sprays of ipecacuanha.

In advanced cases of phthisis with extensive ulceration and suppuration, the fever is of a hectic character and requires entirely different treatment; this should be directed specially to the diminution of the suppuration, the night sweats, and diarrhœa which usually prevail. In such cases, the pulse being rapid, digitalis will often prove of service, and this does not depress.

Aconite reduces greatly the frequency of the beats of the heart, and renders the respiration slower, but it is also a powerful depressant; it is therefore only suited to the early stages of phthisis in which there is active congestion, or to more advanced cases with concurrent pneumonia.

Digitalis likewise not only reduces the frequency of the pulse, lowers temperature, and lessens fever, but, unlike aconite, it strengthens the heart and does not increase the cutaneous secretion.

The *symptomatic* treatment is directed to the accomplishment of a variety of purposes.

If the disease be in the early stage, with congestion only or consolidation of one or both apices, and there be no breaking down of tissue, the inhalation of the vapour of iodine, or of iodine and iodide of potassium combined, are indicated, and the use of counter-irritants, as liniment of iodine or even a small blister; the general health being at the same time well maintained.

If, however, the disease has already advanced to the stage of softening, expectoration, and excavation, then other proceedings will have to be adopted, having for their object the diminution of the expectoration and the arrest of any putrefactive changes which it may have undergone. Now for these purposes there are few more potent or suitable remedies than carbolic acid. For its effective employment, however, it is essential that its volatilization should be aided by a moderate heat, that is to say, by a temperature several, and in some cases many, degrees above that of the ordinary air. This augmented volatilization may now be readily and simply accomplished by means of the author's 'globe' inhaler and water bath, figs. 16 and 17, by the chamber inhaler No. 2, fig. 15, as well as by the apparatus for concentrated vapours, fig. 19. I have already conclusively shown that the oral and oro-nasal inhalers, which have hitherto been so generally, and I might say almost exclusively, used in cases of consumption, are most ineffective, for several reasons, to two of which only I will here simply allude. One of these is, that the quantities of carbolic acid hitherto employed in these inhalers is far too small; another, that even of this small quantity the greater part, four-fifths, is recoverable from the sponge of the inhaler at the completion of the inhalation.

For the successful employment of creasote, tar, thymol, and many other medicaments, it is equally necessary that the vaporization should be aided by a moderate elevation of temperature.

Creasote and tar, as well as carbolic acid, have both been frequently used with advantage in lung consumption, although it is in chronic bronchitis particularly, that the best results are obtained.

Benzoic acid inhaled in the fumes of boiling water has also proved very useful in chronic phthisis, it lessening expectoration and so easing the cough.

Contrary to what I at first supposed, I now find it is necessary that the temperature should be considerably raised, even for the effective inhalation of some of those substances which are usually regarded as among the more volatile, as the oils of eucalyptus and juniper, and fir-wood oil. This is clearly shown by the table on p. 23, from which it appears, that at a temperature of 60° F. the rate of evaporation per hour in each case amounted to only 3·7, 3·9, and 3·7 per cent.; so that it is useless to employ ordinary oronasal inhalers for these oils. With chloroform, ether, and carbonate of ammonia, the rate of evaporation is sufficient, but not so with alcohol, unless a considerable quantity be taken.

Now whether the bacillus of Koch be or be not the cause, or a prime factor in the production of phthisis, yet carbolic acid, creasote, tar, thymol, and perhaps to a less extent fir-wood oil and eucalyptol, efficiently inhaled, fulfil a variety of important purposes, even if they do not destroy the bacillus. The effect of inhalation on this, must for the present remain an open question.

Thus these antiseptics retard and often prevent fermentative and putrefactive changes in the secretions, they improve their condition and render them less irritating, and so afford relief to the cough; they also diminish the quantity of expectoration, while carbolic acid lowers temperature and so helps to reduce irritative fever; points all of the very first importance.

It has already been shown how beneficial these remedies are in many cases of chronic bronchitis, with which disease lung phthisis is often associated.

Quinine is another remedy which is not only antiseptic, but exerts a marked effect in lowering temperature. It is stated that it destroys germs and arrests putrefaction more thoroughly than even arsenic or creasote.

Salicylic acid is another substance which possesses considerable antiseptic and germicide power. So far as I am aware it has never been employed in inhalation, nor in consequence of the largeness of the dose required is it well adapted for the purpose.

Carbolic acid, creasote, and remedies containing these, as different kinds of tar, are of special value when the expectoration is profuse or offensive. Their effect in arresting putrefaction, when efficiently employed, is shown in many cases by the removal of the fœtor where this does undoubtedly proceed from the lungs. It is only in this way, that the disappearance of the offensive odour is to be explained, since they are non-oxydizing agents and have no direct or chemical power of destroying such odours when once formed.

For the destruction of the fœtor when once formed, other agents must be resorted to, as permanganate of potash.

In all cases in which the breath is offensive, an endeavour should be made to determine whether the odour proceeds from the sputa or not. Not unfrequently the cause of the fœtor will be found in the condition of the mucous membrane of the nostrils, throat, or pharynx, and not in that of the lungs at all. While the secretion in bronchitis is often offensive, it is not very frequently so in simple lung phthisis. It is surprising how readily the odour disappears in some cases on the employment of an antiseptic.

In lung phthisis not only is it necessary that the inhalation of carbolic acid or some similar antiseptic should be accomplished in an effective manner, but it must be repeated usually two or three times a day. Nor is this all; if a successful issue is to be assured, the air of the room, especially the bedroom occupied by the patient, should be impregnated at night with the medicament employed, an object which can now be successfully carried out by means of the Chamber Inhaler No. 1.

Moreover, it is desirable, when practicable, that the patient should spend one or two hours each day in an Inhalation Chamber such as has been already described in this book. It is also advisable in some cases that the antiseptic should be administered by the stomach, the object being not merely to bring the medicament into contact with the part affected in as concentrated a form and for as long a time as possible, but also that as much of it should be introduced into the system as will produce constitutional effects.

I will now pass on to make a few remarks relative to the remedies to be employed for the relief of the *Cough* of phthisis. These consist chiefly of anodynes and antispasmodics, combined sometimes with solvents and expectorants.

Before proceeding to prescribe, we must first ascertain the precise exciting cause of the cough. Sometimes this is high up in the throat and is found in an elongated uvula, in a red and irritable condition of the pharyngeal mucous membrane, or in a similar condition of that of the epiglottis, or it may be more deeply seated, as in the larynx or bronchi.

Sometimes the cough is kept up by the character of the secretion. This may be quickly formed and abundant, so as to occasion almost constant irritation; or it may be very scanty, requiring considerable effort to bring it up; or again it may be tenacious and adherent, or it may be in the form of pellets and plugs in the bronchi. Now these several conditions necessitate different treatment. When the secretion is profuse and thin, a mineral acid combined with an expectorant may be used. Half to one drachm of the compound tincture of benzoin suspended in mucilage and used as a spray, often gives great relief in the chronic cough of phthisis, when there is relaxation of the mucous membranes of the throat, larynx, and bronchi.

When stimulant expectorants are indicated recourse may be had to chloride of ammonium, or carbonate of ammonia, with or without squills.

If the sputa be tenacious and brought up with difficulty, then alkalies must be employed, with sometimes expectorant and anodyne remedies.

Again, in cases in which the secretion is scanty and difficult to bring up, or in which there is emphysematous wheezing, sprays of ipecacuanha, employed so as to avoid the nauseating effects of the remedy, will often be followed by considerable relief.

An effective prescription for the cough of phthisis is 20 to 30 minims of spirits of chloroform with 20 minims of the liquor morphiæ hydrochloratis in ½ ounce of distilled water. This may be used as a spray on going to bed. Another useful receipt consists of 30 minims of spirits of chloroform and 40 minims of tincture of conium in ½ ounce of almond mixture.

In the cough of chronic phthisis attended with ulceration, no preparations give greater relief than do those which contain opium.

A spray of chloral again, in small doses, may be used with benefit when the cough comes on in fits and is of a spasmodic character.

A remedy which recently has been much used in the treatment of cough, and praised for its effects, is gelsemium. It diminishes secretion, increases the power of expectoration, and lessens nervous sensibility and excitability. Gelsemium should at first be very cautiously used as a spray, containing from 10 to 20 minims of a tincture made with one part of the root to four of spirit.

When the cough depends upon or is aggravated by a relaxed or elongated uvula, then astringent sprays of alum, or tannin and glycerine, will afford much relief.

A very effective method of treating the cough is undoubtedly by inhalation, and it is not even necessary, although of course it is desirable, that the remedies employed should really reach the seat of the mischief. It is sufficient in many cases that the spray be thrown upon the mucous membrane of the throat: some of it will become quickly absorbed, while a portion will also be swallowed, and the effects will be produced more readily and completely, and with less disturbance of digestion, than if the remedies, partly anodynes, were administered entirely by the usual channel, the stomach.

A few remarks may next be made on the remedies used for checking or preventing the persistent and exhausting *Night Perspirations* or sweats attendant on advanced cases of phthisis. These may proceed from two very different causes : they may be merely an indication of extreme debility or they may result from pyrexia.

It may here be observed, that whatever lessens the fever and expectoration will increase the strength, and thus diminish much more effectually the night sweats than the employment of astringent remedies. Still there are several remedies which are more or less efficacious in lessening the perspiration of phthisis, and all of which may be inhaled as sprays.

First, there are the acid remedies, all of which tend to check excessive cutaneous secretion, particularly when this is due to debility; but those which are chiefly employed for the purpose are acetic and sulphuric acids. One or two drachms of acidum aceticum dilutum P.B. in a wineglassful of water may be taken at bed-time, or the formula of my former teacher, the late Dr. Graves of Dublin, may be employed. This was as follows: distilled vinegar 2 ounces, cherry laurel water 2 drachms, syrup 2 drachms, made up to 8 ounces with water, 1 or 2 ounces to be taken every 3 hours.

A spray of dilute sulphuric acid may be inhaled, containing 10 to 15 minims of the acid in 6 drachms of water, and either with or without 10 to 15 minims of liquor morphiæ hydrochloratis.

Other acids which sometimes may be usefully employed for the arrest of the sweating are tannic and gallic acids; these are best administered by the stomach.

A very effective spray consists of 2 to 3 grains of sulphate of zinc in 1 ounce of water; or 3 to 6 grains of oxide of zinc may be given in the ordinary way by the stomach, by the acids of which it is readily dissolved.

Atropia has been used with great success in checking the sweating of phthisis by Drs. Fothergill, Ringer, and Murrell; it may be hypodermically injected, given in pill, or inhaled as a spray. If injected, from $\frac{1}{200}$ to $\frac{1}{100}$ grain is sufficient; for the pill $\frac{1}{70}$ to $\frac{1}{50}$ grain, and for the spray $\frac{1}{200}$ to $\frac{1}{100}$ of a grain.

Dr. Ringer states that 'sometimes its effects are delayed; thus, if administered at bed-time, it may not check sweating till the following night; or its beneficial influence may extend over several nights, then gradually wear off so that each night the perspiration returns a little earlier. In a few cases it permanently checks the sweating.' The atropia sometimes produces a disagreeable dryness of the throat, but Dr. Fothergill, who uses the remedy in the form of pill, states that this neither dries the throat nor affects the brain. In some cases he has found it necessary to increase the dose to $\frac{1}{30}$ grain. Each drachm of the liquor atropiæ P. B. contains $\frac{1}{2}$ grain of atropia, and the liquor atropiæ sulphatis 4 grains to the ounce: 1 minim of the latter is equal to $\frac{1}{120}$ grain of atropia. Tincture of belladonna may be used for the same purpose, but its strength is of course subject to some variation.

Dr. John E. Howe, in a paper in the *Lancet* of the 6th of September, 1884, gives the results of treatment with strychnia, of the night sweats of phthisis, in the City of London Hospital for Diseases of the Chest. This remedy was recommended in 1879 by Dr. Lauder Brunton of St. Bartholomew's Hospital, he having successfully treated several cases with doses of from 5 to 30 minims of tincture of nux vomica. From 6 to 12 minims of liquor arsenicalis were given to the patients at night, the larger dose being seldom required. It succeeded, Dr. Howe states, admirably in most cases, it also promoting sleep, and no ill effects followed even when the 12 minims were administered.

Quinine also helps by its tonic and antipyretic properties to check the perspiration; it may be given with dilute sulphuric acid; a large dose will often succeed when a small one fails.

Opium and morphia, with or without sulphuric acid, will often produce the same effect; while, strange to say, ipecacuanha will in some cases relieve the sweating when zinc and other remedies have failed.

It is thus apparent that there is no lack of remedies for alleviating the sweating of phthisis, but there are still others which should be mentioned; one of these is jaborandi or pilocarpine. From $\frac{1}{2}$ to $\frac{3}{4}$ grain of the alkaloid may be administered by the stomach thrice daily. Pilocarpine is believed to act on the peripheral nerve apparatus; Ringer says it is a most efficient remedy in the perspiration of phthisis.

Dr. Howe also gives the results of treatment in the City of London Hospital with nitrate of pilocarpine in doses varying from $\frac{1}{20}$ to $\frac{1}{4}$ grain, $\frac{1}{12}$ being the usual dose. 'In the majority of the large number of cases in which it was given, the result was good'—'it seldom produced any of its physiological action, and even if it did so it was very slight sweating for half an hour after the draught, except in one case,' when it continued for an hour. It occasionally happened, however, that after the lapse of 5 or 6 hours the sweating would return. Sometimes the nitrate failed altogether or lost its power after a time.

The action of both pilocarpine and strychnine may last for several nights or may even be permanent, so that these remedies may sometimes be discontinued for a time.

On the whole, Dr. Howe regards pilocarpine and strychnine as nearly as effective as atropia, but with the advantage that they do not, like it, dry the throat.

Another remedy is picrotoxine, a non-nitrogenous crystalline and neutral principle obtained from *cocculus indicus*. Dr. Murrell has shown that this also is very effectual. It acts on the sweat centres of the cord: the usual dose is $\frac{1}{120}$ to $\frac{1}{60}$ grain.

Another important indication to be fulfilled is the reduction of *Temperature* and the consequent lessening of fever. The chief remedies employed for this purpose are carbolic acid, quinine, aconite, and digitalis. The antipyretic action of these remedies has already been noticed in the chapter which treats of the medicaments employed in inhalation.

Another complication of chronic phthisis is *Diarrhœa*, which may arise from different causes. It may be due merely to acidity or other fault of digestion, or it may be more serious and arise from an adenoid condition of the mucous membrane of the intestines, or more commonly from tubercular deposit and ulceration. If from derangement of digestion, the diarrhœa will be relieved by the measures usually adopted in such cases.

In treating the primary affection in the lungs by the inhalation of carbolic acid and other similar remedies, it must not be forgotten that we are really treating at the same time the malady in the intestines. It cannot be doubted that much of the carbolic acid inhaled in the form of fumes or spray never enters the lungs at all, but finds its way into the stomach, and thence into the system generally.

Even much of that which enters the lungs is in the end absorbed, and it is in this way that the acid is found, first in the blood and ultimately in the urine. It is probably through its presence in the blood, that the lowering of the temperature is produced.

Supposing the inhalation of carbolic acid to be effectually carried out, it is still desirable, especially if there be tubercular deposit in, and ulceration of, the mucous membrane of the bowels, to give the antiseptic in the form of pills as well. The pills should contain from 1 to 3 grains each of the acid. Whether the diarrhœa proceeds from acidity or flatulency, or even from ulceration, the sulpho-carbolates, especially the soda salt, will be found very valuable. They will correct the flatulency due to decomposition, while the specific effects of the carbolic acid will still be produced; the sulpho-carbolates are far less irritating than the uncombined acid, and a larger quantity of the acid can by their means be introduced into the system. The dose of sulpho-carbolate of soda is from 10 to 15 grains. When the diarrhœa is dependent on ulceration, sulpho-carbolate of zinc would probably be still more effective, in doses from 2 to 3 grains.

The question may be asked, how is it that more or less ulceration of Payer's glands so commonly occurs in cases of advanced phthisis? Is not one reason of this the injurious habit which some patients acquire of swallowing more or less of the sputa, these sputa being capable, when they find a favourable nidus, of infecting the mucous surface with which they are brought into contact? That the stomach and upper portion of the small intestines are not liable to be similarly affected may be due, as already suggested, to the destructive agency of the digestive ferments, which have of course all disappeared by the time the ilium is reached.

The effectual treatment of the bowel complication of advanced phthisis is attended with many difficulties. The portions of intestine chiefly involved are so deeply seated that it is not easy for the remedies to reach them so as to exert their usual local action. In the case of diarrhœa from tubercular ulceration, the remedies should be given by the stomach or rectum, as they are thus more likely to come into contact with the surfaces, on which their effects are to be exerted.

A very simple and efficacious remedy for checking the diarrhœa dependent upon simple irritation of the mucous membrane of the stomach and bowels, is the sub-nitrate of bismuth, in doses of from 1 to 2 scruples; it may be given in milk and repeated three or four times a day as may be required, combined if necessary with opium. This will sometimes succeed when other remedies fail.

The mineral acids, especially dilute sulphuric acid, are often very beneficial in cases of diarrhœa in which the evacuations are abundant and watery, with a tendency to decomposition. Sulphuric acid will not only prove very serviceable in checking this diarrhœa, but in controlling the night perspirations as well.

Another valuable remedy is acetate of lead, usually combined with opium in the form of pill; it should be given in 5-grain doses, and may be continued, it is stated, for weeks and months without giving rise to any symptoms of lead-poisoning. Its administration is indicated particularly in those cases in which there is reason to believe that ulceration of the intestines already exists, and it acts both by its powerfully astringent properties, and by

PULMONARY CONSUMPTION. 313

forming an insoluble compound of albumen on any abraded or ulcerated surfaces with which it may be brought into contact.

Other remedies which act somewhat in the same way, mainly by their astringency, are nitrate of silver and sulphate of copper. These should be given in the form of pills, $\frac{1}{4}$ to $\frac{1}{2}$ grain of the nitrate, or 1 to 2 grains of the sulphate in each pill, the pill being repeated two or three times a day.

Bichloride of mercury is another remedy which sometimes proves effective. The doses should be very small, from as little as $\frac{1}{100}$ to $\frac{1}{50}$ grain, every three or four hours; its use is indicated when the motions are pale and watery.

Another valuable remedy in chronic diarrhœa, even when due to bowel ulceration, is arsenic, in doses of from 1 to 2 minims given shortly after food.

In some severe cases the treatment of the diarrhœa may be materially aided by enemata of some of the above-named remedies, as of sub-nitrate of bismuth, sulphate of copper or acetate of lead, combined with opium. Given in this way there will be less risk of the remedies disordering the stomach.

The next complication of phthisis to be treated of is *Hæmoptysis*. Hæmorrhage of the lungs, which in its several forms is usually termed hæmoptysis, may be active or passive; that is to say, with or without fever, a point of practical importance in relation to treatment and one easily ascertained by the thermometer. The hæmorrhage may occur in different forms and in any stage of phthisis.

Hæmoptysis is very often one of the first outward or visible symptoms of threatened or impending phthisis, and is then usually indicative of congestion, or if the case be more advanced, of commencing softening. The bleeding is seldom profuse in these cases, but the sputa often consist for a time wholly of blood.

Sometimes the sputa are more or less deeply tinged throughout, and approach in character those of pneumonia. This condition is also an evidence of congestion, of inflammation, or even of the solution or breaking down of a portion of consolidated lung. If a cavity exist, the bleeding will probably indicate congestion of the lining membrane.

If the sputa be simply streaked from time to time, and a cavity exist, this would indicate a rupture of some minute vessels on the surface, and if the streaking were considerable and persistent, that the destructive or ulcerative action was progressing.

If the bleeding be somewhat abundant, and the blood be gulped up in mouthfuls, then it is evident that some small vessels must have given way, possibly in connection with tubercular deposit.

If still more profuse, and the blood be rapidly brought up, some even escaping by the nostrils, then some large and unsupported vessel has become ruptured, in consequence probably of the extension of ulcerative action, probably in a cavity.

When, in a case of advanced phthisis, repeated attacks of somewhat profuse hæmorrhage take place, the rapid progress of the malady is usually indicated, fresh portions of the lungs becoming affected, and the area of the disease enlarged. Some of the effused blood often makes its way into more distant parts of the lung and even into the other lung, these effusions acting as fresh foci for the disease.

In the hæmorrhagic form of phthisis, the periodical hæmorrhages which are so apt to occur are not usually followed by an extension of the disease, a fact which is probably to be explained by the absence, or the small quantity, of tubercle usually existing in such cases, at all events in the earlier stage.

Sometimes the expectorated blood is fluid, bright-coloured, and frothy; this shows that it has been very recently effused, and that it is well aërated : at others it is liquid but dark-coloured, proving recent effusion and a want of aëration. Again it may be coagulated and dark, almost black; this proves absence of aëration, and that the blood must have been effused a sufficiently long time to allow of its coagulation.

The above various conditions and appearances of the blood brought up, afford in many cases useful practical information.

The treatment of hæmorrhage of the lungs by inhalation possesses in some cases considerable advantages over that by the stomach, since the remedies are brought into more or less immediate contact with the parts from which the bleeding proceeds, and the local effects are exerted in most cases much more strongly and quickly than when the remedies reach the part through the general circulation.

The choice of the remedies to be used, of which there are many, will depend very much upon the presence or absence of fever, the state of the circulation, and the character and quantity of the hæmorrhage; their mode of action varies in different cases.

The first class to be noticed are the acid remedies, including sulphuric acid and some of its salts, as sulphate of alumina, sulphate of copper, solution of perchloride of iron, tannic and gallic acids.

Sulphuric acid, even when much diluted, is very astringent and has the property of coagulating the blood. Administered in the usual way by the stomach, and in the ordinary doses of from 10 to 15 minims of the dilute acid every 3 or 4 hours, it becomes in its passage through the system so greatly weakened that it can scarcely be capable of producing any marked effects in arresting hæmoptysis; inhaled in the form of spray it is doubtless much more effective.

The salts of sulphuric acid retain not only a good deal of the astringency belonging to the acid, but this, in the case of sulphate of alum and sulphate of copper, is increased by its union with the bases themselves.

One of the most generally used of these salts is sulphate of alumina and ammonia; its astringent properties, however, are not so great as some of the other remedies employed. Alum is often prescribed with excess of sulphuric acid and with tincture of opium, also with tannic and gallic acids; but when tannic acid is employed, the addition of the sulphuric acid is undesirable.

Sulphate of iron and sulphate of copper are much more powerful astringents, but they are not very often employed for the arrest of hæmorrhage, on account of their irritant properties. If prescribed, it is better that they should be administered in the form of pill, with a little extract of belladonna to obviate constipation. Of the iron salt 3 to 5 grains may be given, and of the copper salt $\frac{1}{2}$ to 1 grain.

The most effective preparation of iron for checking hæmorrhage, especially from the smaller vessels, is the solution of perchloride of iron; from 10 to 30 drops of the liquor ferri perchloridi P.B. should be added to not more than 1 ounce of water and used as a spray, this being repeated at intervals as may be necessary. It is best suited for passive hæmorrhage, and care should be taken that the solution be neutral. A preparation of iron regarded by some as superior to the perchloride, is acetate of iron, which also may be conveniently inhaled as a spray. The watery solution of the acetate is preferable to the tincture, and if made of a strength corresponding to the liquor ferri perchloridi, from 10 to 30 minims in about $\frac{1}{2}$ to 1 ounce of water should be used for each spray.

Another useful remedy is acetate of lead, which is best suited for hæmorrhage from the smaller vessels. From 3 to 5 grains should be dissolved in not more than 1 ounce of distilled water, and inhaled, the inhalation being repeated according to the necessities of the case. The addition of 10 minims of tincture of opium will increase its effect.

A valuable hæmostatic in cases in which the loss is but moderate, is tannic acid, dissolved in glycerine and water and inhaled as an atomized spray. Part of the tannin is no doubt swallowed, and of this a portion becomes slowly dissolved in the stomach and so enters the circulation. It becomes transformed, however, in the blood into glucose and gallic acid, and it is to this acid probably that its remote effects are due. Although more soluble in water than gallic acid, tannic acid is yet, owing to its extreme astringency, absorbed by the mucous surfaces with much greater difficulty.

The remedy which is more to be relied upon, when the action of a remote astringent is desired, is gallic acid, it being more readily and quickly absorbed; as a local astringent it is, however, far inferior to tannin. It has been already said that the latter in passing through the system undergoes certain chemical changes, and is converted in part into gallic acid: taking therefore equal quantities of the two acids there is reason to believe that gallic acid would prove the more efficient remote astringent. That gallic acid is readily absorbed, quickly reaches the lungs, and penetrates the sputa, is shown by the dark green or blackish hue that these are apt to acquire some time after being expectorated. Gallic acid is best inhaled as a spray containing from 4 to 10 grains in 1 ounce of water and a small quantity of glycerine.

Another famous remedy for hæmorrhage from the lungs is Ruspini's styptic; this is stated to consist of gallic acid dissolved in spirits of wine and scented with attar of roses.

A remedy of great repute in America for hæmorrhage, even when this is profuse, is *Hamamelis Virginica*, or witch-hazel. It may be inhaled as a spray, from 1 to 2 minims of the tincture being added to 1 ounce of water, the inhalation being repeated every 2 or 3 hours if necessary. Large doses are to be avoided, as they are apt to produce certain disagreeable symptoms, as pain in the head. It is best suited for passive hæmorrhages, and has been found very effective when hypodermically injected.

Two of the most valuable of all known remedies for controlling and arresting profuse hæmorrhage, even when arising from the rupture of vessels of considerable size, may next be alluded to, namely, ergot and turpentine.

Ergot acts chiefly by its power of producing contraction of the blood-vessels throughout the system, and which is effected mainly through the unstriped muscular tissue which forms one of their coats. This remedy may also be advantageously inhaled as an atomized spray. There is no reason why its administration in this way should not be quite as effective as when it is hypodermically injected. If, however, the loss of blood is so rapid and profuse as to interfere with the inhalation of the spray, then the ergot must be subcutaneously injected.

Turpentine also, when emulsified, may be inhaled as a spray. If the hæmoptysis be slight, 10-minim doses repeated according to the urgency of the symptoms may be sufficient; but when the hæmorrhage is serious, then 1 or more drachm doses, at intervals of about 3 hours, should be inhaled. The larger doses may occasion sickness, diarrhœa, and even the appearance of blood in the urine, but these symptoms quickly pass away on the discontinuance of the remedy.

There are still a few other remedies to notice; one of these is digitalis. The employment of this is indicated since it lowers temperature, quickly reduces the frequency of the pulse, and acts as a diuretic. Digitalis was highly esteemed in hæmorrhage by my former colleague at the Royal Free Hospital, the late Dr. Brinton. The best test of its action is the retardation of the pulse. It must, however, be given in considerable doses : 1 ounce of the infusion or from 30 to 40 minims of the tincture in water, either of these being inhaled as a spray, and repeated as may be necessary.

Another medicine which acts in a different manner from the foregoing, namely, by its nauseant and depressent effects, is ipecacuanha. It has proved very effective in some cases. The doses must be sufficient to produce a decided feeling of nausea : 20 to 40 drops of the wine as a spray in a little water, and repeated as required ; or 5 to 15 grains of the powder divided into three parts, one being given every 5 or 10 minutes.

Opium or some of its preparations is also frequently employed in hæmorrhage from the lungs, usually, however, in combination with alum, tannin, and other medicaments. How it acts is not very clear, possibly in part by its drying effect on the mucous membranes. Dr. Braithwaite has found small hypodermic injections of morphia to be very useful in hæmoptysis.

Lastly, quinine proves beneficial in some cases of passive hæmorrhage.

A smart, quickly acting aperient will sometimes of itself stop bleeding from the lungs, such as a full dose of sulphate of magnesia. It may be administered in obstinate cases of hæmorrhage in aid of the other remedies employed.

At the same time that remedies are administered in the form of sprays, they should be given by the stomach as well, if the hæmoptysis be profuse or obstinate.

Another very distressing condition which so frequently occurs in advanced stages of phthisis is *Vomiting* after meals. This, of course, is very exhausting and interferes greatly with nutrition. The food distends the stomach, and this pressing upon the lungs excites paroxysms of cough and expectoration, as a consequence of which vomiting occurs. The treatment for this condition is to strengthen the stomach and to diminish its reflex sensibility; this may frequently be insured by the judicious use of either quinine, arsenic, strychnia, or bismuth, with a few drops of chloroform.

Another troublesome condition which is apt to occur in chronic cases of phthisis, is the occurrence on lying down of a 'click' or crepitation, due to the presence of a small amount of secretion; this is sometimes high up and seems almost in the throat. It is in some cases so disturbing that it prevents the patient from sleeping for hours. It is best treated by astringent sprays of acetate of lead, or tannin and glycerine, or a solution of nitrate of silver may be applied to the throat by means of a brush, or gargles may be employed.

A few remarks may now be made respecting the action of certain other medicines which possess a special value in the treatment of phthisis, namely, the hypophosphites, arsenic, and iodoform.

The hypophosphites of soda and lime were first brought into prominent notice by Dr. Churchill, and of their efficacy in cases marked by nervous exhaustion no doubt can be entertained. They act as nervine tonics, increase appetite, and promote digestion and nutrition. They are most serviceable in the first and second stages of phthisis, and it is affirmed that they not only ameliorate all the symptoms, but that they will sometimes even effect a cure. They may be inhaled in the form of spray, but it is better to administer them by the stomach in the usual way. Large doses are to be avoided, as they sometimes produce great weakness and other undesirable symptoms. Dr. Thorowgood has employed the hypophosphites with great success.

Arsenic acts as an alterative and tonic, and there can be no doubt that in many cases it promotes appetite, aids digestion, and improves nutrition. It also reduces temperature in phthisis: the decline may begin soon after commencing to take the remedy, or it may be delayed for several days. Arsenic is not only serviceable in improving appetite and digestion, but it seems to exert a special effect in cases in which there is a difficulty in emptying the lungs, as in emphysema, and also, but from a different cause, in fibroid phthisis.

Dr. R. Shingleton Smith, of Bristol, at the International Medical Congress held in Copenhagen in August 1884, narrated in a paper on the use of iodoform in the treatment of phthisis the results in 46 cases: in 29 there was an absolute gain in body weight, amounting in one case to 32 lbs., and in another to 33 lbs. Other indications of amendment were fall of temperature, diminution of cough and expectoration, cessation of night sweats, and improved appetite. The drug was given in doses varying from 1 to 6 grains, 3 times daily, and was continued for several months. Toxic symptoms of a mild character were observed in some of the cases. Dr. Smith also referred to the local application of iodoform in diseases of the larynx.

Dr. J. Sormani, of Pavia, has also obtained favourable results in the treatment of phthisis by iodoform, while Dr. G. Hunter Mackenzie has found that the remedy, even when pushed to such an extent as to give rise to mental excitement, did not exert any effect on the bacilli even in a single case.

The subject of the inhalation of oxygen gas in the early stages of phthisis, or where there is defective aëration, has already been noticed; the remedy has however been employed at different times in the more advanced and suppurative stages of the disease, the reported results being somewhat contradictory, and not on the whole particularly favourable. Dr. de Tymowski, of San Remo, informs me that he has treated several of his patients by the inhalation of oxygen with very beneficial results. The fever was at first increased, but after the lapse of two or three weeks it was in all cases reduced, the expectoration diminished, the number of bacilli was lessened, and in some instances they disappeared entirely. The appetite, sleep, strength, and the general health all improved.

The patients inhaled at intervals from 30 to 60 litres of oxygen daily, by means of Limousin's apparatus, and took from 5 to 15 drops of turpentine internally after each inhalation. Tymowski believes that the turpentine, in conjunction with the inhalation of oxygen, aids the ozonizing of the oxygen, and states that the tubercle bacillus cannot exist in blood which is rich in oxygen.

Should these results be confirmed on extended trial, it is evident that the inhalation should not be discontinued on account of a temporary increase in the temperature.

One of the consequences of inflammation of the lungs, of pleuro-pneumonia, broncho-pneumonia, and of congestion or consolidation of the apices of the lungs, is more or less encroachment on or obstruction of the air cells, entailing diminished lung capacity and a lessening of the aëration of the blood. With a view to the restoration of the capacity of the lungs in such cases, treatment by the inhalation of compressed air is indicated, and will often prove very serviceable. This inhalation may in many cases be combined with expiration into rarefied air, whereby the lungs are enabled more completely to empty themselves. By the inhalation of the condensed air, cells which are simply compressed become gradually expanded, while absorption is promoted in those which are more or less occupied by products effused as the results of inflammatory action.

A remedy of which a good deal has been heard lately is the Great Mullein, a species of Verbascum, of which a notice appears elsewhere in this work. It is stated to ease the cough, arrest diarrhœa, and in particular to improve nutrition, whereby patients often increase greatly in weight. It is usually given in milk, and it has not yet been clearly proved whether these good effects are due to the mullein or to the milk. Certainly the latter must be credited with some part of the improvement in weight.

Laryngeal Consumption.

Laryngeal phthisis, so far as yet ascertained, is never a primary affection, but is consecutive on tubercular disease of the lungs, although this bears no definite relation to the extent of the disease in the larynx; in fact, in many cases the lung affection at the outset is of a limited character.

The mucous membrane of the larynx is at first unusually pale, but after a time becomes more or less pink or congested in parts; the membrane covering the vocal cords, the under surface of the epiglottis, and the ary-epiglottidean folds being specially liable to implication. One or both of these last swell up and assume a pyriform shape, which is almost diagnostic of the disease in its early stage.

As the disease progresses, the mucous membrane of the larynx, including that of the parts just named, becomes thickened, infiltrated, and studded with minute tubercles; these are chiefly situated in the mucous membrane beneath the epithelial layer, which, it is stated, is at first often entire. Gradually the tubercles soften successively here and there, giving rise to small discrete ulcers, which, however, in the more advanced cases coalesce, forming patches of considerable size; these at length involve and destroy nearly the whole of the mucous membrane of the larynx. The parts most liable to infiltration, œdema, and ulceration are the epiglottis and the ary-epiglottidean folds, which are sometimes so much swollen as to preclude entirely all view, even by means of the laryngoscope, of the parts below; finally, more or less of the epiglottis may be eaten away, the ulcerative action not even sparing its cartilage, which becomes denuded, and at last in part destroyed; the ligaments and cartilages elsewhere in the larynx being also liable to become involved and similarly affected. At first the membrane covering the cartilages becomes diseased, perichondritis; then the cartilages themselves, chondritis; or these, deprived of their nutritive investing membrane or perichondrium, may even be necrosed.

The symptoms at first resemble those of chronic laryngitis; there is more or less hoarseness, some degree of aphonia, soreness of throat and cough, with, after a time, shortness of breath, pain and difficulty in swallowing. Of these symptoms, the cough and dysphagia are usually the most urgent. The cough, when the disease has made much progress, is dry, husky, rasping, and so peculiar that by any one accustomed to the sound it must be at once recognized. The act of deglutition becomes increasingly difficult, until at length the pain is so great as seriously to interfere with and limit the amount of nourishment taken. Dr. Morell Mackenzie points out, and the distinction is important, that the difficulty in swallowing may arise from one of three causes. 'In the early stage it generally partakes of the character of odynphagia, being due to pain in swallowing. Later on there is often obstruction from the enlarged epiglottis and the swollen ary-epiglottic folds; whilst at a still more advanced period the difficulty of swallowing is due to the imperfect closure of the larynx, and the consequent passage into that tube of the ingesta.' The escape of particles of food into the larynx gives rise to violent fits of coughing, and altogether the distress is so great that the patient dreads and often avoids the act of swallowing as much as possible.

The shortness of breath depends very much on the condition of the lungs and the state of the epiglottis. If there be extensive disease in the lungs, or the epiglottis be seriously involved, in either case the breathing will be short and hurried, and if the epiglottis be chiefly affected there will be spasmodic attacks of dyspnœa.

The amount of expectoration will vary according to the extent of the disease in the lungs, and whether the bronchi are implicated or not; in the latter case it will of course be much more abundant.

Again, the irritative or hectic fever will depend on the stage of the disease and the amount of lung implication.

When the disease is fully established, there is in most cases but little difficulty, from the history and general symptoms, in distinguishing it, but should any doubt exist, this may always be set at rest by a microscopical examination of the sputa. If the tubercle bacillus be found, then no room for conjecture remains as to the nature of the case.

Treatment.—This must be based upon the same general principles as that for *Lung Consumption*, and it must be remembered that we have not simply the laryngeal affection to treat, but the disease in the lungs as well. This complication of course adds greatly to the gravity of the case, and were it not for this circumstance the treatment would be more promising, since the remedies admit of being more effectively applied to the larynx than to the lungs.

The treatment must be both constitutional and local. The first has for its object the maintenance of the nutrition of the body and of the vital powers by all available means, dietetic and hygienic. The purposes to be fulfilled by the local treatment are various : to promote absorption and so lessen infiltration, to heal the ulcers, to diminish the expectoration, to alter the character of the sputa, to disinfect them and, if possible, to destroy the bacilli—for, whatever may be their import, our plain course is to spare no effort to bring about their destruction; to assuage the more urgent symptoms, particularly the cough, pain, difficulty of swallowing, and the vomiting.

The remedies to be employed for these several purposes include disinfectants, astringents, anodynes, and anæsthetics.

It will be remembered that the mucous membrane in laryngeal phthisis is at first thickened and hypertrophied from effusion, tuberculous, and at length ulcerated. Now antiseptic, astringent, and absorbent remedies are equally necessary in these several conditions, which are all consequent on one and the same cause; they include resorcin, terebene, iodide of ethyl, and iodoform, as amongst the most likely to prove beneficial. At the same time that these are inhaled, the same or similar remedies should be administered internally, especially either sulpho-carbolate of soda or terebene. The two most effective astringent remedies are acetate of lead and perchloride of iron.

One or more of the preceding remedies should be employed in succession. Several observers have recently borne strong testimony to the efficacy of iodoform and terebene.

When the disease is far advanced, then the more urgent symptoms will claim attention, and, fortunately, much may be done for their relief. While the necessary means are taken with this object, the absorbent and disinfectant remedies must still be continued, and these as a rule should be pushed to such an extent as to produce constitutional effects, and to cause their appearance in the renal excretion. Among the disinfectant remedies from which beneficial results have been obtained, tar, or the preparations containing it, occupy a prominent place.

A remedy frequently employed in the ulcerative stage of laryngeal phthisis is nitrate of silver; 2 grains of this diluted with 5 grains of finely powdered sugar may be occasionally insufflated, or a solution containing 20 or even 30 grains to the ounce of water may be applied with a curved brush. But it appears to me that this remedy is too irritating, especially when a concentrated solution is used, to be very frequently employed, although it doubtless gives temporary relief to the cough and pain in swallowing by its effect in lessening for a time the sensibility of the mucous membrane.

For the relief of the pain, ice may be used in the early stage, but when the disease is more advanced greater relief is obtained by the inhalation of the vapour of hot water, medicated with conium or lupuline; but the employment of this is not to be encouraged, as it may probably have a tendency to promote tuberculization and ulceration. One of the most effectual remedies for the pain, including that attending deglutition, is morphia; this may be in solution and inhaled as a spray, but most frequently the morphia is mixed with some soluble substance, as sugar, and is insufflated, the quantity varying according to the case. The administration should be so timed that food may be taken as soon as the full effect of the anodyne is obtained, as the act of deglutition will then be attended with far less pain.

Another means of relieving the distressing pain in swallowing has recently been placed in our hands, namely, cocaine. The mucous membrane of the pharynx and larynx should be brushed over with a 10 to 20 per cent. solution of the alkaloid, and the application of this should also precede the taking of food, the anæsthetic effects being produced in a few minutes, in place of in the course of an hour, as with opium.

For the relief of the vomiting due to reflex action and spasm, brought on sometimes by severe paroxysms of cough, bromide of potassium will be found very useful.

To the œdematous epiglottis two classes of remedies may be applied by means of the brush: astringents, as perchloride of iron, tannin and glycerine, or acetate of lead, and remedies which relieve œdema by their exosmotic action; one of these is carbonate of soda, and another, less effectual probably, glycerine. The frequent application to the throat and epiglottis of olive or almond oil, or almond emulsion, will be found to have a very soothing effect, as will also the introduction of these oils into the larynx in the form of sprays. Dr. Marcet recommends that 20 grains of iodine, with 5 grains of iodide of potassium, should be dissolved in an ounce of olive oil, and applied over the larynx occasionally.

Should these means be ineffectual for the relief of the œdema, then scarifications may be resorted to in those cases in which the œdematous condition is dependent on fluid and not on tubercular infiltration; in the latter case they would probably be hurtful.

There is, therefore, much to be done in the general and local treatment of laryngeal phthisis, and which requires great perseverance on the part of both the medical man and the patient.

Much discrimination is required in the selection of the food. None is suitable which is dry, or in particles, or crumbs, as, should any of these enter the throat, they would give rise to a severe paroxysmal cough. The food should consist of strong soups, milk and eggs, well cooked wheat-flour with milk, jelly, and it should be in a semi-solid form, so that it can be drunk in considerable amount, only one act of deglutition being required.

Should the epiglottis be so much destroyed by ulceration as to be unable to guard the passage to the lungs, it may become necessary to administer the food through the œsophageal tube, and this method of feeding may be supplemented by nutritive enemata.

Tuberculosis of the Pharynx and Nose.

Within the last few years *Tuberculosis of the Pharynx* has been noticed by several observers, and described by Isambert and Fränkel. It is generally, if not always, associated with tubercle in the lungs. In most cases the larynx has been found to be similarly affected, but whether it became so by extension of the disease from the pharynx, or whether the larynx was first affected, does not appear to have been determined. When the pharynx is the seat of the malady, there is always soreness of the throat and great pain in swallowing; this seriously interferes with the amount of nourishment taken, and contributes to bring about the early and invariably fatal termination.

The mucous membrane of the pharynx, including that covering the uvula, is more or less swollen, infiltrated, and ulcerated, and should the epiglottis be involved in the ulcerative process, it may become almost entirely eaten away. The cervical glands are always enlarged.

Several cases have also in recent years been recorded of *Tuberculosis of the Nose*; these, like the pharyngeal and laryngeal affections, being in most, if not in all, cases preceded by tubercular disease in the lungs. In one only of the cases was the pharynx found to be similarly affected. In suspected tubercle of the nose the diagnosis should be confirmed by microscopical examination, and the discovery of the tubercle bacillus.

Now that attention has been called to the subject of tuberculosis of the nose and pharynx, doubtless many fresh cases will ere long be recorded. I believe that the throat is more frequently affected than is generally supposed.

Treatment.—This will not differ in any very essential particulars from that already described under the head of laryngeal phthisis.

Influenza.

This specific disease usually prevails in an epidemic form, and hence is sometimes called Epidemic Catarrh. It affects particularly the mucous membrane of the air passages, and is attended with considerable debility.

It occurs in two forms, simple and complicated.

It is ushered in generally with marked febrile symptoms, chilliness, or even rigors, alternating with flushes of heat and severe headache. The mucous membrane of the nose is at first sore, there is a feeling of tightness in the chest, and a dry cough. Soon secretion sets in, and a thin, watery liquid runs from the nose; the expectoration is scanty and glairy, the chest feels sore, there is some degree of difficulty of breathing, and the respirations are accelerated. During inspiration the sounds are at first dry and harsh, but soon moist and sibillant rhonchi are heard, particularly at the base of the lungs. Throughout, sensations of chilliness, flushes of heat, and severe headache continue, with tenderness of one or both eyeballs. At night all the symptoms become aggravated. The tongue is usually moist, and covered with a creamy or brownish fur, but the tip and edges are often red; the pulse is more or less accelerated, but always compressible and usually feeble. In mild cases the attack generally lasts some four or five days, but in severe ones much longer. The cough may remain for some time after, and there is always considerable debility. Relapses are common, and towards the termination of the attack pains and affections of a rheumatic character frequently set in, and these are often remittent or intermittent, so that the frontal headache assumes the form of neuralgia.

The preceding description has reference to the simpler form

of the disease, in which the organs of respiration are only affected in a slight degree; but the severer cases are often complicated with capillary bronchitis and more or less pneumonia. When this is the case all the symptoms will of course be greatly aggravated, respiration will be difficult and hurried, the face will become pale or even livid, and the debility may amount to prostration; such cases not unfrequently ending fatally. The expectoration is at first scanty, and consists of small tenacious pieces of mucus: after a time the secretion increases in quantity, is still viscid, but often becomes more or less muco-purulent and is occasionally streaked with blood.

In some epidemics there are marked gastric symptoms, sickness, bilious vomiting, and a jaundiced skin, while the pains in the forehead, back, and limbs present not only a rheumatic, but also a remittent character, and are always much worse at night.

It has been observed that in those years in which there have been epidemics of influenza, the other specific febrile diseases have been more than usually prevalent.

Treatment.—This disease being due to a specific poison, it is marked by more or less debility or prostration throughout, so that the treatment adopted must not be of a lowering or depressing character.

As is the case with so many of the specific fevers, the force of the disease is mainly expended on the organs of respiration, so that the condition of these requires to be carefully noted throughout.

The treatment at first should have for its object the free action of the skin. The patient should be kept warm and in a room of a temperature of about 60° F. Citrate of ammonia should be freely administered, with in some cases small doses of ipecacuanha; these may be given with more effect in the form of sprays than by the stomach. If at the outset there be evidence of prostration, the ipecacuanha may be omitted and carbonate of ammonia substituted for the citrate, or spirit of nitrous ether. At the same time counter-irritants, such as mustard cataplasms and turpentine, should be freely applied to the chest, especially over the base of the lungs, and this although the signs of lung implication are but slight. The foot-bath is another good derivative remedy, and this should be occasionally repeated; if elimination by the kidneys be defective, then non-depressent diuretics should be employed, as digitalis or spirit of juniper.

If from the stethoscopic sounds and other symptoms there are evidences of capillary bronchitis, then a further object will be to lessen congestion by the promotion of secretion, this being at first scanty and difficult of expectoration. For this purpose alkaline sprays of carbonate of soda with ipecacuanha should be inhaled, and if there be great debility, citrate or carbonate of ammonia may be substituted for the soda salt. At the same time the vapour of hot water should be freely inhaled. The inhalation of the fumes of chloride of ammonium would be likely to prove beneficial in this stage; effective counter-irritation must be perseveringly carried out. As secretion increases, the alkaline and expectorant treatment must be continued.

At the same time the powers of the system must be sustained, because if there be lung complication, the prostration will be greater, and the duration of the illness much prolonged. Not only must the diet be light and nutritious, but stimulants must be administered.

When secretion is freely established, bark and ammonia should be given with or without squills, while for the rheumatic symptoms and the brow headache, especially if this be intermittent, quinine is the remedy indicated. Counter-irritants should still be applied to the chest, either in the form of the liniment of iodine, or of turpentine and acetic acid.

When the rheumatic complication is marked, salycin or some of its compounds would probably prove very serviceable, and might take the place of quinine.

When the liver is at fault, as indicated by a feeling of weight in the right hypochondrium and the jaundiced skin, some mild mercurial aperient should be administered.

Actea racemosa, it is said, has been given with much success in nfluenza accompanied with headache, pains in the back and limbs, and symptoms of rheumatism. Since its action is comparable to that of aconite, and it depresses both the force and frequency of the pulse, it should not be given when there is considerable prostration.

Sulphurous acid has been recommended by Dr. Dewar, but it is to be feared that in some cases this would be too irritating.

Whooping-cough.

Whooping-cough, or *Pertussis*, is an infectious disease manifesting itself principally in the organs of respiration, and chiefly attacking children. It occurs usually once only, but relapses are frequent, and it often prevails epidemically.

The precise nature of the infective cause has not yet been demonstrated, but the secretions thrown off by the mucous membrane abound with bacteria and micrococci, none of which, however, so far as yet ascertained, are special to the disease; these are also found in the leucocytes of the blood and tissues. Whatever its nature, it may be conveyed by the clothes of unaffected persons.

Although due to a specific contagion, the liability to the disease is increased by several causes, especially those which give rise to general or even local susceptibility, as a weakly constitution, as shown particularly by rickets, or by the irritable condition of the respiratory mucous membrane, due to catarrh, measles, &c.

The disease usually attacks children under eight years of age, and in three-fourths of the cases in those under two years; the ordinary mortality from the disease amounts to about $2\frac{1}{2}$ per cent., the danger being greater the younger the child.

It occurs chiefly in the spring, and commences with coryza, oppression, feverishness, and the usual symptoms of catarrh; these are sometimes accompanied, but more frequently after some days are followed, as secretion sets in, by a convulsive cough.

The incubative stage varies usually between 4 and 10, but may even extend to 14 days; the catarrhal stage lasts for about a week, and the convulsive or spasmodic from 6 to 8 weeks in the absence of complications; after this time the disease ceases to be infectious, although the cough remains and still retains its periodic and spasmodic character, in some cases for months.

The cough is excited mainly by the tenacious secretion, and the fits come on with varying frequency; they may occur several times a day, or even three or four times an hour. They are usually worse at night, and they gradually become less frequent and severe as the disease gets better.

The periodical and convulsive cough is not only dependent on the irritating character of the secretions, but on an undue sensibility or hyperæsthesia of the mucous membrane of the respiratory track, especially that of the larynx and bronchi. Ordinarily the irritation in uncomplicated cases does not amount to active inflammation, as shown by the temperature, which although increased somewhat during the catarrhal period is never very high.

The mucous membrane of the larynx, trachea, and particularly of the bronchial tubes, throws out a ropy and tenacious mucus, to get rid of which, violent expiratory efforts have to be made, whereby the lungs become thoroughly emptied, and the patient almost asphyxiated and exhausted. At this juncture a strong inspiratory effort is made, and the in-rushing air, in passing through the narrow rima glottidis, produces the peculiar or whooping noise so characteristic of the disease. At last, to the great and almost complete relief of the child, some ropy mucus is expelled and the fit is ended, to be, however, unfortunately speedily renewed.

Owing to the violence of the paroxysms the face becomes turgid and red, the eyes seem as if they would start out of their sockets, and vomiting is sometimes induced, as well as bleeding from the nose.

Another serious consequence of the violent expiratory efforts made is that one or more of the lobules of the lungs may be so completely emptied of air that they become collapsed, a result aided by the great elasticity of the chest walls and of the diaphragm in childhood. This untoward event is more apt to occur if the child be weak and delicate, and therefore unable to make the vigorous inspirations necessary to bring about the re-expansion of the collapsed portion of lung. This complication is not seldom the cause of a sudden and fatal termination of the case.

During the catarrhal and more active stage of the affection, more or less redness of the respiratory track will be found to exist from the pharynx downwards, even to the smaller bronchi. This redness is conspicuous in the larynx, but the trachea, and especially the bronchi, furnish the greater portion of the secretion. The bronchial glands soon become implicated, tender, and swollen, whether owing to the presence in them of the poison of the disease, which is probable, or from simple irritation, is uncertain. They sometimes become softened.

Other complications which are apt to arise are collapse of one or more of the lobules of the lungs, lobular pneumonia, limited emphysema, and sometimes dilatation of the smaller bronchial tubes, those not strengthened by annular cartilages; all these complications add greatly to the gravity of the disease. The occurrence of convulsions is particularly serious. The child may die from apnœa in one of the convulsive seizures, or death may result from capillary bronchitis or pneumonia.

It is stated that whooping-cough, especially when attended with lung implication, may, in those predisposed to the disease, lay the foundation of and be followed by tuberculosis of the lungs.

Treatment.—This of course must be adapted to the stage of the disease. During the catarrhal stage the patient should be kept to the house, and indeed to one airy room, the temperature of which should be maintained at about 60° F., and febrifuge remedies administered. When cough has set in, and particularly if the breathing be at all impeded or the temperature elevated, and there be any signs of capillary bronchitis or lung implication, then counter-irritants should be applied, as mustard cataplasms, turpentine, or the iodine liniment; also antispasmodics, anodynes, solvents, and expectorants should now be employed. In the chronic stage astringents will be useful, and during convalescence tonics, particularly those of a nervine and antispasmodic character.

In the first stage, if the child be old enough, a spray may be used of acetate of ammonia and tincture of aconite until the feverishness and temperature are reduced; 20 to 40 minims of the liquor ammoniæ acetatis and $\frac{1}{2}$ to 1 drop of tincture of aconite in $\frac{1}{2}$ ounce of water, repeated as may be requisite, and if the spray cannot be employed then a mixture or drink sweetened with sugar, containing the same ingredients in the same doses, may be substituted.

If the breathing should be embarrassed, an emetic of ipecacuanha will be found of much service, and this may be repeated daily with a view to bring away the tenacious and abundant secretion which sets in after the first few days, and which is so provocative of spasm. It is often easier to administer the emetic in the spray form than by the stomach, and in order to insure the effect it is not necessary that the whole of it should be made to enter the lungs.

When the feverishness has been in part relieved, the solvent alkalies and non-stimulant expectorants, as carbonate of potash, chloride of ammonium with, in some cases, small doses of ipecacuanha should be continuously employed, both in the form of sprays and administered by the stomach.

In the convulsive paroxysms a few drops of chloroform or ether will usually give speedy relief; but if the child be very young, then it is safer to have recourse to bromide of potassium or ammonium, or chloral hydrate. Should the attacks not be very frequent, these remedies may be administered in the intervals. The smallest dose of bromide to be used is about 3 grains, and of the chloral 1 grain, the doses being increased in the proportion of 2 grains in the former case and 1 grain in the latter for each year of the child's age.

In a still later stage, when fever is absent, secretion abundant, and there is no special lung complication, morphia or belladonna, either separately or combined, or united sometimes with ipecacuanha, will often greatly lessen the spasm; the effects, however, must be watched and the doses carefully regulated. The dose of liquor morphiæ hydrochloratis will usually range between 2 and 5 minims, and of the tincture of belladonna, of which children are very tolerant, between 8 and 12 minims, repeated in urgent cases every hour or two. The opium and belladonna should be pushed to the extent of producing constitutional effects, as evidenced by slight drowsiness. These remedies will also prove serviceable in checking the vomiting consequent on the spasmodic seizures, and which so often leads to rejection of the food. It is usually much easier, as well as more effective, to administer the anodynes as sprays than by the stomach.

Another very serviceable anodyne is conium; this does not, like opium and belladonna, check secretion, but reduces reflex spinal irritation, and hence a strong liniment of the same applied to the spine may likewise be found useful.

Amongst the antispasmodic remedies, lobelia and cannabis must not be forgotten.

In some epidemics of the uncomplicated disease lobelia affords great relief in the antispasmodic stage, greatly diminishing in two or three days the frequency and force of the attacks. Ringer recommends 10 minims of the tincture every hour, and an extra dose when the paroxysm gives signs of its approach. The tincture may of course be readily inhaled as a spray.

Cannabis has been serviceable in some cases of whooping-cough, but there does not appear to be any strong evidence in its favour.

I must now mention another class of remedies which have

proved of value in many cases of whooping-cough, namely, the antiseptic class, and foremost in this stand carbolic acid, creasote, and tar. They may be administered with a view to lessen secretion, diminish irritation, and to act as germicides. It is of course difficult to administer these remedies in so concentrated a form as directly to kill the bacteria and micrococci; still, by altering and improving the character of the secretions, the multiplication of these germs may possibly be hindered and their destruction ultimately brought about. For any of the purposes above referred to, it is necessary that the remedies should not only be effectually inhaled, but also administered by the stomach, so that the system may be brought to some extent under their influence. For the direct introduction of either the acid or creasote into the lungs, a spray producer or the Chamber Inhaler No. 2 may be used; by this last the atmosphere of the room around the patient is charged with the vapour. For administration by the stomach, sulpho-carbolate of soda, in the form of powder mixed with sugar, should be given by the stomach three or more times a day, in doses varying according to age.

The inhalation of tar, usually that derived from wood, in whooping-cough, has been much extolled, and it no doubt proves effective in many cases. Gas tar is used for the same purpose. Dr. John Brown, in a short letter to the 'British Medical Journal' of January 19, 1883, states that in the west of England the parents of children suffering from whooping-cough take them to the gasworks; but in Lancashire the children are commonly made to inhale the tar at home. The efficacy of these inhalations is in great part, but not wholly, dependent on the carbolic acid contained in the coal, and the creasote in the wood tar. Some of the other constituents contained in the gas tar which contribute to the good effects experienced are ammonia, sulphides, and benzol. But it has been stated that the inhalation of the illuminating gas, carburetted hydrogen, is also in some cases attended with marked benefit, due, there can be little doubt, not to the gas itself, but to the presence, owing to imperfect purification, of more or less of the substances just mentioned.

As soon as the chronic stage has been reached, then astringent applications and sprays will often have an excellent effect in warding off or relieving the spasmodic attacks, as of alum or tannin and glycerine, or the fauces and throat may be brushed over with a strong solution of nitrate of silver. This last proceeding, however, sometimes gives rise to so much cough and spasm, that it is necessary to be very cautious in the application, and in some cases it would be better to abandon it altogether for other and milder means.

Dr. Warfwinge, of Stockholm, has treated 40 cases of whooping-cough with about 20 grains of alum per day. He found it to be almost a specific provided it was early employed, the intensity and frequency of the attacks being lessened and the duration of the illness shortened. In these cases no doubt the remedy was administered by the stomach, but it would be at least as effective when inhaled in solution as a spray. Letzerich has recommended and obtained good results in whooping-cough by the use of inhalations of sprays of a solution of quinine.

Red clover is another remedy which has been recommended for whooping-cough by Mr. Foster, of Huntingdon, and Dr. Sargent, of Boston, U.S. The latter gives a wineglassful during the day of an infusion made with 2 ounces of the dried flowers to a pint of boiling water; a syrup is also prepared of which a teaspoonful is given 3 or 4 times a day.

In the chronic stage, and with a view to promote convalescence, tonics, and especially antispasmodics, are indicated, as tincture of valerian, valerianate of zinc, oxide and sulphate of zinc.

Other remedies sometimes employed, and which may be prescribed according to the condition of the patient, are quinine, iron, and cod-liver oil. The diet should be light and nutritious. Finally, to complete the cure, change of air will often in this stage exert a most beneficial effect, the cough and spasm quickly disappearing.

Diphtheria.

Diphtheria is a general and contagious malady, depending on the presence of a specific poison, and attended with an almost constant disposition to the formation of false membranes on the mucous surfaces of the several outlets of the body, but more particularly that of the pharynx and of the respiratory track, including the larynx, trachea, and bronchi; it prevails in an epidemic or an endemic form, and, owing to the depressing effects of the poison, it always assumes a more or less adynamic form.

The tendency to the formation of a false membrane is not confined to the mucous membranes, but seems to be an almost universal disposition. Thus, if the skin be thin and delicate, or the epidermis abraded or removed as by a blister, the membrane is still liable to make its appearance.

The nature of the poisonous material has not yet been conclusively demonstrated, but Oertel, particularly, contends that it is due to the round sporules of a species of micrococcus, M. diphtheriæ, while Letzerich has described another fungus which he believes to be the cause of the disease, namely, Zygodesmus fuscus. Senator, however, believes that the spores which Oertel has described are those of Leptothrix buccalis, and Dr. Mackenzie has succeeded in finding this fungus in 5 out of 7 cases which he examined, it being in each instance situated in the superficial layer of the lymph. Oertel states that the spores which he has described are to be found abundantly in the early stage of the formation of the pseudo-membrane, and that after inoculation of animals with a diphtheritic exudation, they have been found in great number in the blood and lymph vessels, in the muscles and even the kidneys.

The predisposing causes of the disease are various, but the principal are childhood, the great majority of cases occurring before 10 years of age, and family susceptibility. Other causes are impairment of the health, however brought about, as by imperfect nutrition and living under insanitary conditions. Previous illnesses strongly predispose to diphtheria, not merely, probably, by their lowering effect on the general health, but by their rendering the mucous membrane more susceptible to the invasion of the poison; thus diphtheria is apt to occur after ordinary catarrhal affections, but especially after scarlet fever, and also to a less extent after measles, whooping-cough, typhoid, and even phthisis.

It has been shown that, whatever its nature, the poison is contained in the secretions, excretions, and exhalations, and from these it is liable to pass into the air breathed and the water drank. Again, it has been proved that it may lay dormant for years, and yet retain vitality and virulence. It may occur at any period of the year, but usually in the spring, and it is more prevalent in rural than in urban districts, and one attack affords but slight protection against a return of the malady. The period of incubation is very short, usually two or three days, but in very severe cases it may prove fatal in 24 hours.

Several different degrees or varieties of the disease have been described—the *catarrhal* or *benignant*, the *typical*, the *inflammatory*, the *malignant*, and the *chronic*.

In an ordinary or typical case, after a short period of incubation the temperature rapidly rises to 103° or 104° F.; soon there is a feeling of stiffness in the neck, soreness of the throat, and pain in swallowing. If, now, an examination be made, the mucous membrane of the pharynx, tonsils, soft palate, and uvula will be found red and swollen; and a little later on, even in a few hours, the first indications of the formation of a false membrane will become apparent in the form of transparent yellowish exudations, somewhat elevated above the general surface; these gradually become opaque and change from yellow to grey, the patches increasing in size and ultimately coalescing, so as to form a membrane; this at the same time increases in thickness by successive additions to the under surface, so that it often exhibits, when fully formed, a laminated structure, and presents an appearance which has been compared to that of wash-leather or wet parchment. At this stage the membrane may be artificially removed, and still later it may become spontaneously detached; in the former case the underlying mucous membrane will be red, raw, and ecchymosed; in the latter it will be covered with a puriform material. If forcibly removed it will be more likely to re-form than when it is thrown off naturally. The pharyngeal secretions very soon undergo decomposition, and emit a highly offensive odour.

Contemporaneously with these changes, the salivary and cervical glands become swollen and tender.

In favourable cases the temperature usually falls considerably on the appearance of the adventitious membrane, but it may, when the glands are much affected, subsequently rise again. This is sometimes ascribed to a secondary absorption of the poison, but in those cases where the secretions are much decomposed it may be due to ordinary or septic toxæmia.

If the case be severe, the pulse will be very compressible, usually rapid, but sometimes slow, and the debility will be extreme. If the urine be examined it will often be found to contain albumen.

But very frequently in the more serious cases the pseudo-membrane is not limited to the pharynx and adjacent parts, but extends downwards to the larynx, trachea, bronchi, and even the lungs themselves, giving rise in some cases to lobular pneumonia or lobular collapse, the danger being of course augmented in a corresponding degree. It is not possible to mistake the symptoms which indicate this extension—there will be distressing dyspnœa, and death may ensue from asphyxia or coma.

The extension may take the upward direction and pass into the nares, through the lachrymal duct to the conjunctiva, or it may invade the Eustachian tube. The implication of the pituitary membrane will be shown by, amongst other symptoms, the discharge of a sanious fœtid liquid, and sometimes by alarming hæmorrhage.

The prognosis will depend not merely on the extent of the local manifestations of the disease, but upon the degree to which the constitutional powers are affected.

If the case is to terminate favourably, at the end of about a week there will be an amendment in the symptoms; portions of the membrane may be thrown off, and it will not be re-formed; the temperature will be reduced and the pulse gradually regain its power. Sometimes, however, there is great prostration and weakness of the heart, entailing the danger of death from syncope.

It must not be forgotten that relapses frequently occur, possibly from reinfection, but more probably from septicæmia; also that temporary paralysis may supervene. If the progress be unfavourable the symptoms of blood-poisoning will increase, the prostration will be more profound, the heart weaker, and death may occur from syncope or coma.

In the benignant form of the disease, the cases usually recover in the course of three or four days; there is but little elevation of temperature, and though yellow spots of transparent secretion may appear in the throat, the false membrane is not formed. It occasionally happens, however, that these mild cases suddenly become aggravated.

In the inflammatory variety the disease usually runs a rapid course, and the local symptoms are intensified, the membrane being very quickly formed. Sir William Jenner has pointed out that in these cases the joints are apt to become affected, to be painful, swollen, or even inflamed.

The malignant form is ushered in with many of the same symptoms which announce the presence in the system of some of the other grave infectious maladies, as scarlet fever; evidences of blood-poisoning rapidly show themselves, hæmorrhages are apt to occur, and typhoid symptoms quickly supervene.

The chronic form of diphtheria is of infrequent occurrence, but cases of it have been recorded by Barthez, Isambert, and particularly by Morell Mackenzie. These cases are distinguished by the occasional formation of false membranes in the pharynx or nares, the diphtheria continuing for weeks and even months.

A few other particulars in reference to diphtheria may now be briefly noted. 1. The false membrane rarely invades the œsophagus, probably owing to the disturbance of the parts in deglutition. 2. The albuminuria is less persistent than in scarlet fever, and is seldom followed by anasarca. 3. In fatal cases, according to Dr. Faralli, unless respiration is impeded, the temperature remains high until the end. 4. A cutaneous eruption sometimes appears, as in scarlet fever; but it is stated that it is not followed by desquamation.

One of the most frequent and trying sequelæ of diphtheria is paralysis; it may follow even mild cases, and it occurs usually two or three weeks after recovery, but it may be earlier or later. The paralysis comes on gradually, and affects most frequently the muscles of the pharynx, then those of the eye, or of the lower extremities; the muscles of the larynx, neck, and trunk are more rarely affected, and whichever set of muscles is involved, one side is more affected than the other. With the paralysis there is also diminished sensibility; if the pharynx be affected there will be difficulty of deglutition, and a regurgitation of liquids through the nose; and if the larynx, alteration of voice and a liability to the entrance of particles of food; while in other cases expectoration will be impeded, and hence there is often accumulation of the secretions.

Diphtheria sometimes presents considerable difficulties of diagnosis, particularly in the benignant and malignant forms of the disease, in both of which the distinctive membrane may be absent. The difficulty of diagnosis is further enhanced by the occasional presence in the severer cases of a rash, which, with other symptoms, doubtless sometimes causes cases of diphtheria to be mistaken for scarlet fever. In both there is albuminuria, while the only distinctive feature of the rash is that it is not followed in diphtheria by desquamation.

The adventitious or false membranes are elastic, tear readily,

DIPHTHERIA.

swell up, and become transparent when treated with acetic acid, while by alkalies they are dissolved. Under the microscope, according to the most recent authorities, they consist mainly of cells which present the appearance of being more or less united together by fusion, but with channels or interspaces running between them, there being no fibroid basis or intercellular substance. The cells forming the upper layers of the membrane are two or three times larger than the lower cells. When naturally thrown off, the membranes undergo both fatty degeneration as well as ordinary decomposition, whereby they become more or less disintegrated, the breath being rendered extremely offensive. Ulceration, and more rarely gangrene, occurs; this latter may occasion loss of a portion of one or both tonsils, or of the uvula, and if, after extensive ulceration, the throat be examined on recovery, cicatrices will still be visible.

In many cases not only are the larynx, trachea, large and small bronchi involved, but even the lungs themselves. In the bronchi effusion will often be present; this may be more or less plastic, membranous or purulent; the air cells themselves may contain similar secretions, while some of the lobules may be affected with pneumonia, and even be found collapsed.

Treatment.—The principal indications are, to support the vital powers, to prevent or lessen decomposing changes in the blood, to limit the formation of the false membrane, to promote its separation or solution, to disinfect the secretions, and to guard against syncope.

The room occupied by the patient should be light, airy, and the uniform temperature of about 60° F. should be maintained.

The diet from the first must be light and nourishing and not too abundant, as the power of assimilation is at first but limited; nourishment should be given at short intervals, and continued during the night. It should be chiefly in the liquid form, and amongst the most suitable articles are milk, eggs, beef-tea, meat juice, Valentin's extract and peptonized foods, while as an aid to digestion pepsine with dilute hydrochloric acid should be given. Vomiting and diarrhœa are very apt to occur in the early stages of diphtheria, and both are probably to be viewed as natural efforts for the elimination of the poison from the system; still, if persistent, they so seriously interfere with the nutrition of the patient, that it is necessary they should be controlled. For this purpose ice should be constantly sucked, and, if necessary, pills each containing ½ grain of creasote or from 1 to 2 grains of oxalate of cerium should be administered every three or four hours, or powders of the subnitrate of bismuth either with or without hydrochlorate of morphia, while warm cataplasms or turpentine may be applied over the stomach. Should these means fail, nutrient and medicated enema may be resorted to, and in some cases it may even become necessary to employ the œsophageal tube.

At the same time, and probably from the very first, stimulants, chiefly brandy, should be freely given according to the condition and age of the patient, it being remembered that alcohol does not in such cases produce its usual effects, and that hence there is a great tolerance of the remedy. Another stimulant which may often be employed at the outset is carbonate of ammonia; this will also do good by its diaphoretic and expectorant action.

For the purpose of checking the tendency of the blood to decomposition, there are no remedies so effectual as the astringent preparations of iron, including particularly the tincture of the perchloride of iron. For an adult as much as 30 minims should be given every 3 hours; for a child the dose must be proportionate to its age; but the remedy should be freely administered. If the child be old enough the iron may also be employed in the form of a spray. Other remedies which may be employed for the same purpose are sulpho-carbolate of soda and chlorate of potash; the sulpho-carbolate is a very effective preparation, and should be given in 10-grain doses every 3 or 4 hours. Chlorate of potash is credited with antiseptic properties, but in what way it acts is not very apparent, seeing that nearly if not the whole of the salt escapes from the system in an undecomposed state, and may be recovered from the urine. Again quinine and salicin and its compounds have been employed for the same object, and no doubt in some cases with beneficial results.

The special liability of the mucous membrane of the throat and air passages to implication is probably to be explained by the delicacy of its structure and its accessibility to the outer air, and it seems to me that derivative treatment might possibly be carried out with beneficial results; treatment in fact designed to divert the malady from vital to other less dangerous parts. With a view to accomplish this purpose the skin should be freely acted on; acetate of ammonia and spirit of nitrous ether should be freely given so as to secure their diaphoretic action; blood should also be determined to the skin by very brief but repeated immersions in a hot bath, the duration of these being only a minute or two; or by the wet sheet; or stimulating applications should be applied on a large scale to the skin, as cataplasms of mustard and linseed, or flannel sprinkled with turpentine.

In mild or benignant cases of diphtheria, the mucous membrane may be only red and swollen with merely diphtheritic specks of secretion, which do not spread or go on to the formation of a false membrane. In such cases beneficial effects have been derived from the internal administration of cubebs and copaiba. These remedies might also be applied to the throat by means of the brush, or to the larynx in the spray form.

When the false membrane has been formed, any efforts made for its forcible removal can only result in further mischief; the attempt will increase the irritation, occasion pain, some amount of hæmorrhage, and is almost certain to be followed by the re-formation of the membrane; except, therefore, it be already partially detached, it is better to adopt one of two courses, either to be content with taking measures to promote its separation or to effect its solution. To bring about its separation, the vapour of hot water should be continuously inhaled, and if the patient be too young or too ill to inhale the vapour directly, the bed should be surrounded by curtains, and the space thus enclosed charged with the vapour by means of some suitable apparatus, such as a croup kettle. The air and vapour may at the same time be charged with some volatile antiseptic, such as oil of eucalyptol, of fir wood or of cubebs, or with carbolic acid or creasote.

With the object of facilitating the detachment of the adventitious membrane, an infusion or tincture of the leaves of jaborandi or a subcutaneous injection of its alkaloid, pilocarpine, has been suggested. These remedies speedily produce a greatly increased flow of saliva, sweating, and augmented bronchial secretion. Of the infusion, as much as contains the extract of 60 gs. of the leaves should be administered. Of the alkaloid from $\frac{1}{2}$ to $\frac{1}{3}$ grain, but $\frac{1}{4}$ to $\frac{1}{2}$ grain is sufficient when hypodermically injected. One would certainly hesitate to employ these remedies in the case of young children. The infusion and the tincture are apt to produce some very disagreeable effects, as nausea and vomiting; these effects it is said do not follow the employment of the alkaloid. Again, jaborandi is powerfully depressent, and in some cases would be more likely on this account to do harm than good.

With a view to the solution of the membrane, a great variety of remedies have been recommended and employed. For particulars concerning many of these the reader is referred to the chapter on the Materia Medica of Inhalation, but so far as is yet known, no remedies are more effective than lime-water, solution of potash, acetic and lactic acids. Ordinary lime-water is scarcely strong enough for the purpose, but liquor calcis saccharatus is 12 times as strong, and this may be applied to the membrane by means of a brush undiluted, but if used as a spray 1 or 2 drachms of the solution made up to $\frac{1}{2}$ ounce with water will be sufficient. The presence of the sugar lessens the causticity of the solution and modifies its effects; in like manner the presence of glycerine in the glycerides of carbolic acid and tannin materially affects the action of those medicaments.

Foremost among the acid remedies is acetic acid; this applied to many animal tissues produces, as is well known, some remarkable effects; the tissues, and the epithelial and granular cells contained in them, swell, become transparent, gelatinous-looking, and more or less dissolved. This is particularly the case with any corpuscular structures which may be present. The acid may be applied as a spray, but it is better to use the brush, as then the application can be strictly limited to the affected surfaces.

The difficulty is to apply solvent remedies to all the parts affected and not beyond them, and this is greatly enhanced when the membrane has been formed on the surface of the larynx or trachea; still we must do the best we are able, and when the larynx is invaded by the disease, it must be reached, where practicable, by means of atomized liquids.

In those cases in which medicaments are applied to the throat itself, especially when they are used for their solvent or antiseptic properties, the mucous membrane should as far as possible be freed from secretion and dried prior to their application. This purpose can generally be effected by the employment of some absorbent pad of lint, medicated wool, or charpie, securely fastened to a suitable handle, care being taken not to use the same pad more than once. Blotting-paper has been employed for the purpose, but several folds of this would be required.

During and even prior to the detachment of the membrane more or less decomposition sets in, giving rise to great fœtor; to obviate this occurrence and to stay the decomposition antiseptic remedies are now specially indicated, although it is well that the air of the room occupied by the patient should from the first be charged therewith. The remedies most suitable for this purpose are carbolic acid, creasote, thymol, and permanganate of potash. Chlorine and sulphurous acid have been strongly recommended, but unless very carefully employed they will prove too irritating. If carbolic acid be used it should be in the proportion of 5 to 10 grains to the ounce, while the most suitable chlorinated preparation is liquor sodæ chloratæ, 60 minims to the ounce if applied with a brush, and 20 if used as a spray. The effects of chloral have been highly spoken of by medical men abroad and at home, and especially by Messrs. Hemming. This remedy is not usually credited with antiseptic properties, but Mr. Hughes Hemming states that it rapidly gets rid of the fœtor, and promotes the separation of the membrane, leaving a healthy surface beneath. Mr. Hemming uses the syrup of chloral, 25 grains to the drachm, and directs that it should be applied every hour or two.

Astringent remedies are not unfrequently employed, as perchloride of iron, tannin, and alum; all these act of course more or less antiseptically, but, with the exception of the astringent preparations of iron, they are all inferior to the antiseptics proper, especially those above named. It is scarcely necessary to point out that the most potent of these antiseptics act also more or less as germicides. Where it is possible, the antiseptic remedies should not only be applied directly to the throat by means of a brush, but should be used also in the form of sprays; the quantities of the medicaments being of course much less in the latter case.

The employment of caustics as solvents was formerly recommended on high authority, but in most cases they are too irritating, and their use has therefore been abandoned.

There is another class of remedies, the effects of which, when applied to the throat in diphtheria, so far as I am aware, have not yet been observed. I allude to certain fixed oils, as olive, almond, or cod-liver oil. The effects produced by the application of these are chiefly mechanical; but they are so important, and they so alter the condition of the structures and organisms entering into the composition of the false membrane, that highly beneficial results may reasonably be anticipated. These oils not only lessen irritation by the exclusion of the air, but they arrest decomposition, and by the exosmotic action set up they exert the most marked effects on all the corpuscular elements contained in the tissues, as also on any organisms which may be present therein; effects, indeed, incompatible with their vitality. Now it is not difficult to apply these oils either by the brush or as sprays to the false membrane of diphtheria, whether this be in the pharynx or larynx; the one condition, in the case of the pharynx, essential to success, is that the parts should be first dried as far as possible prior to the application. The application should be repeated several times a day, and however much of the oil becomes inhaled, passes into the air passages, or is swallowed, can only have a beneficial effect. The idea of excluding the air from the throat has occurred to Dr. Morell Mackenzie, who employs a spirituous solution of a gum resin; but the oil is a more effective application, and its operation is altogether different.

During the period of convalescence it will still be specially necessary to sustain the vital powers, and particularly to guard against syncope, by light nourishing diet, stimulants, and suitable tonics, such as perchloride of iron, quinine, bark, and ammonia, or senega and ammonia. Should paralysis ensue, nervine tonics must be administered, as the hypophosphites or strychnia, while, if necessary, the galvanic battery should be employed.

Croup.

The definition already given of diphtheria is equally applicable to membranous croup, *Cynanche trachealis*. This disease was formerly regarded as an acute and purely inflammatory affection, demanding the most prompt and vigorous antiphlogistic treatment. As soon, however, as English medical men became acquainted with diphtheria, the points of resemblance between the two diseases became apparent, and now the belief is general that they are identical, modified only in their symptoms, course, and termination by the parts specially involved. This view has led of course to a revolution in the treatment.

When, as in diphtheria, the membrane is formed in the pharynx, it is usually thicker and more adherent than when, as in croup, it is situated in the larynx; these differences being due to the diversity in the anatomical characters of the membrane in the two situations, but no essential difference can be detected by the microscope in the structure of the adventitious membrane.

Other differences arise out of and are dependent on the two localities; thus, when the disease attacks the throat, the specific poison is in close communication with the salivary and cervical glands, and with abundant absorbent vessels, through which septicæmia sometimes arises; again, in this situation, the nervous communications are especially numerous. In the larynx the anatomical surroundings are entirely different, and there are no salivary and but few other glands which lie in the way of infection. Then the presence in the larynx of the croupous membrane gives rise to a special train of symptoms, commencing with extreme difficulty of breathing.

A careful examination of the pharynx at the commencement of an attack of croup has revealed the fact that in the great majority of cases isolated specks of deposit are to be detected, although these do not go on to the formation of a coherent membrane; this, in the affection to which the term croup is usually applied, is limited to the larynx, but sometimes extends to the trachea and bronchi.

Croup is generally ushered in, like most other diseases of the same class, by symptoms of catarrh and fever. The voice becomes hoarse, or it may be almost lost, a short, dry, and shrill cough sets in, and very soon the inspirations become stridulous. If the pharynx and larynx be examined in this, the very early or first stage, the mucous membrane will be found red and congested, but in children it is not possible to make a laryngeal inspection.

Gradually, in what may be termed the second stage, the breathing becomes more or less difficult, mainly from the occurrence of sudden spasmodic attacks. During the spasm inspiration is hurried and difficult, the nostrils are alternately dilated and contracted, and the muscles of inspiration are called into strong action in order to overcome the obstruction occasioned mainly by the spasm, but also in part by the presence of more or less mucus. Should now an emetic be given, and this have the desired effect and the throat be cleared of the mucus, the breathing will be greatly relieved and may for a time be almost natural, so that the child falls asleep.

As the disease progresses and enters on the third stage, the hoarseness increases, the voice sinks to a whisper, and the breathing becomes permanently difficult, but little relief being experienced during the intervals of the spasms; this condition plainly indicating the formation of the adventitious membrane. The obstruction to the entrance of the air is now so great that suffocation is threatened and may even occur. The entrance of the air into the lungs is limited, gas exchange lessened, the blood is not sufficiently decarbonized, and congestion of the lungs and other organs may take place; the lips become blue and the countenance livid. The efforts now made to breathe are painful to witness; the little sufferer thrusts his fingers into his mouth in the vain endeavour to remove the obstruction, of which he is conscious; ultimately death ensues from coma, syncope, or exhaustion.

At the onset, it is not possible to distinguish true from catarrhal croup, or Laryngitis stridulosa; it is only when the graver symptoms make their appearance that the distinction can be made. Again, spasmodic croup, or Laryngismus stridulus, may be mistaken for true croup, although the former is distinguished by the absence of fever; further, true croup occurs usually at night, whereas the catarrhal form often commences in the daytime.

It is obvious from all that has been stated that croup is a most serious and often fatal affection: it is estimated that not more than 10 per cent. of the children attacked recover, unless the operation of tracheotomy has been resorted to, when the recoveries, which vary however in different epidemics, may reach 30 and even more per cent. The fatal termination usually occurs within a week.

Treatment.—The treatment, both local and constitutional, is very much the same as that for diphtheria.

Notwithstanding the belief generally entertained in the identity of the poison of the two diseases, croup does not usually present such marked evidences of blood disorganisation and general prostration as diphtheria, and it proves fatal in many cases rather by its effects on the respiration than by the extent or virulence of the poison.

The remedies to be employed at first should be such as counteract the fever and are calculated to prevent the formation of the false membrane. For the first purpose aconite or digitalis and acetate of ammonia, with sometimes small and nauseating doses of ipecacuanha, should be administered as sprays or by the stomach; preferably if practicable by the former means. At the same time other derivative measures should be adopted, as short immersions in the hot bath, or packing with the wet sheet, or by stimulating applications to the skin and by friction. The application of hot sponges or flannels to the throat, neck, and sternum will often help to give relief.

As a precautionary measure, and before any evidences have arisen of the formation of a false membrane, solvent sprays may be used; these will do good even if they enter the larynx only in part, but they must not be too strong lest they prove irritating. When the membrane is already formed there will be the less necessity for caution in this respect. Among the many solvent remedies used, preference should be given to lime-water, liquor potassæ, lactic or acetic acids. These are best applied by the brush, or if as sprays, care should be taken to use only a small amount of liquid.

At the same time that solvent remedies are employed, efforts must be made to bring about the separation and ejection of the membrane by the use of emetics of ipecacuanha, sulphate of zinc, or of alum.

Dr. Meigs prefers alum as an emetic for children; he gives a dose of 1 drachm with syrup or honey, and repeats the dose if necessary in a few minutes until free vomiting is produced. If the case is a severe one, he repeats the emetic sometimes as often as 3 or 4 times a day.

Emetics frequently afford very great relief, even when they do not cause the expulsion of any part of the adventitious membrane, by enabling the patient to get rid of the irritating secretion which is so often the cause of the paroxysmal attacks of difficulty of breathing, and for this reason emetics are indicated in the second stage and before the membrane has been thrown out. The detachment of the membrane should also be aided by the inhalation of warm steam; if the patient is old enough to inhale the steam directly, all the better, but if not, the croup kettle may be made to discharge its steam within the curtains around the bed.

With a view to lessen the spasm, whenever this may occur, a few drops of chloroform or ether may be inhaled, according to the age and condition of the patient, and such soothing anodyne remedies may be given as neither depress the system nor lessen secretion, as conium, hyoscyamus, or their alkaloids, but opium and belladonna should for the most part be avoided.

If the secretions become offensive, then antiseptic, or if profuse, astringent sprays should be used. Trousseau employed with success a spray containing 5 per cent of tannin, which he repeated several times a day.

In cases in which the blood is only imperfectly decarbonized and asphyxia threatened, oxygen may be inhaled either with or without an admixture of air, or doses of a solution of hydrogen peroxide may be administered.

The constitutional treatment will depend very much on the condition of the patient, but nothing should be done to weaken him. If there be great depression, carbonate of ammonia is indicated, and if there be a tendency to dyscrasia, tincture of the perchloride of iron. For further details as to treatment the reader is referred to the chapter on 'Diphtheria.'

If all the means adopted fail to give the desired relief, and the symptoms progress from bad to worse, then the operation of tracheotomy must be performed, and this even when there is reason to believe that the trachea has been invaded. It is the only chance left, and one greatly diminished by any undue delay in carrying it into effect. It is folly to wait until the child is *in extremis*.

Treatment by inhalation will still be needed after the operation, with a view to soften any membrane which may be within reach, to check decomposition, and to correct fœtor.

Scarlet Fever.

This disease finds a place in this work for the same reason as diphtheria—with which affection, indeed, it is often associated—namely, in consequence of the liability to throat and lung complications.

There is, in fact, much in common between scarlet fever and diphtheria. Both are due to a special poison, both are prone to implicate the pharynx and adjacent parts, in both there is a rash and kidney implication, as shown by the albuminuria, while, lastly, diphtheria is more liable to occur in connection with scarlet fever than with any other disease.

As, according to Oertel and others, a micrococcus is the cause of diphtheria, so, according to some, an organism of the same character constitutes the infective matter of scarlet fever. This has been found in the blood, in the blood corpuscles themselves, and in the tubules of the kidneys.

It would appear that one effect of the poison of scarlatina is to stimulate cell growth, not only of the skin, but of the mucous membrane, salivary and other secreting glands. It is in this way that the proneness to glandular and kidney complications in this disease is in part explained.

Scarlet fever occurs in varying degrees of severity from a mild affection unattended with danger to the malignant form which often proves fatal. Three degrees or varieties of the complaint have been specially distinguished and described, Scarlatina simplex or benigna, S. anginosa, and S. maligna.

In the first, or mildest form, in addition to the usual feverish symptoms, there will be some degree of redness of the pharynx, tonsils, and uvula, swelling of the same, and the characteristic rash, which is, however, usually but slight and quickly disappears. There may also be tenderness of the salivary and particularly of the parotid glands.

In the second form there is great redness of the mucous membrane of the pharynx and adjacent parts, which may assume even a bluish hue; the throat is much swollen, and the membrane more or less covered with a thick and tenacious secretion; there may be hoarseness, the neck is stiff, the salivary glands enlarged and painful; the inflammation continuing, suppuration takes place and abscesses form either in the cellular tissue or in the glands; these usually break externally near the angle of the jaw. Ordinarily, at the same time, the mucous membrane of the nose is involved, as shown by the occurrence of an acrid discharge. In this form the fever is considerable, as indicated by the high temperature. The abscesses often continue to discharge for a long period, giving rise to great exhaustion. Occasionally the epiglottis and larynx become involved, and if the former be much affected there will be difficulty in swallowing and regurgitation of liquids through the nose; usually the breathing is not much affected.

In the malignant form the fever is high and the symptoms progress with great rapidity. There is marked dyscrasia, the disease presents a thoroughly adynamic form, exhaustion rapidly sets in, and death sometimes ensues in the course of a few hours, often even before the rash has had time to appear.

In the third form there is great difficulty of deglutition and impediment to respiration, owing to the extent of the swelling and to the abundant formation in the throat of viscid phlegm. The mucous membrane of the throat is dusky red and coated with a dark incrustation of exuded lymph, soon to be followed by ulceration and extensive sloughing; hæmorrhage at this stage not unfrequently occurs; the breath is fœtid, and the lips and gums are covered with sordes.

The disposition in scarlet fever to laryngeal and bronchial complications is much less than in diphtheria.

When diphtheria occurs in the course of scarlet fever, it sometimes follows a mild attack, and does not show itself for several days and until the eruption has disappeared and the fever subsided. The diphtheritic membrane appears on the fauces and tonsils, a sanious discharge takes place from the nose, and the breath becomes exceedingly offensive; occasionally only, the membranous formation extends to the larynx, giving rise to hoarseness and some, but rarely serious, difficulty of breathing. Scarlatina is more particularly apt to be complicated with diphtheria when epidemics of this disease are prevailing at the time.

In malignant cases of scarlet fever, gangrene of the pharynx is apt to occur, followed by ulceration, suppuration, and, not unfrequently, by profuse hæmorrhage.

But scarlet fever is not the only disease which is liable to be complicated with diphtheria and with pharyngeal and laryngeal mischief. Thus, in *Small-pox* the mucous membrane of the pharynx and larynx is apt to become congested and red, and pustules to form on it, while in some cases the diphtheritic membrane also makes its appearance. The occurrence however of diphtheria is much less frequent as a sequel of small-pox than of scarlet fever.

The same parts are often implicated in *Typhoid Fever*. The mucous membrane may be simply of a dusky red colour, or it may be attacked with diphtheria; when this occurs in the larynx there is a great tendency to destructive ulceration, leading to denudation of the cartilages. The diphtheria may not make its appearance until two or three weeks after the commencement of the fever.

Treatment.—This of course must be adapted to the character of the attack, whether mild or severe, the stage, and the predominant symptoms.

In the milder form, with moderate fever and soreness of the throat, the patient should be put to bed and kept warm, suck ice or drink iced water, while chlorate, citrate or carbonate of ammonia should be regularly administered, with a view to promote the action of the skin and the development of the rash. Carbonate of ammonia is one of the most valuable and efficacious of remedies in this disease, and extensive trials have been made of it by several observers—by Dr. Peart, the late Dr. Williamson, Mr. Charles Witt, and Dr. Ringer—all of whom have highly praised its effects. The earlier it is given the better; by the diaphoretic action which it exerts, it helps the development of the rash and the elimination of the poison. It may be administered by the stomach or as a spray. Acids and all substances containing them, as fruits, should be avoided. Liquor ammoniæ acetatis is a useful remedy, but when there is much depression the carbonate is to be preferred. When the tongue is much coated and there is other evidence of stomach derangement, an emetic of ipecacuanha, which may be given in the spray form, will generally have a beneficial effect.

In the more severe forms of the disease, in which the fever is high and the throat specially involved, the following treatment should be adopted. With a view to relieve the throat, on which the force of the attack is about to be expended, the skin must be called into free action, and to some extent the kidneys as well. Carbonate of ammonia should be regularly administered, and recourse had to the air bath, warm bath, or wet pack. If the urine is scanty, digitalis should be given; this while it acts as a diuretic will also strengthen the already weakened heart. In this stage, if the neck be painful and the throat much swollen, the inhalation of the vapour of hot water will prove very soothing, as will also warm linseed poultices; but these applications must not be pushed too far, since they may do much harm by promoting suppuration, which might otherwise be avoided.

A remedy of great efficacy when the tonsils are turgid and inflamed, giving rise to much difficulty in swallowing, is hydrargyrum c. cretâ; $\frac{1}{2}$ grain of this should be given for several consecutive hours. It is affirmed, that it will quickly reduce the swelling, and that when an abscess is inevitable it will promote the maturation and evacuation of the pus. A remedy which has been recommended as a diaphoretic in scarlet fever is jaborandi, but it is much more likely to do harm than good. It depresses, often excites vomiting, and acts more on the salivary glands than on the skin, so that by increasing determination to those glands it may intensify the throat mischief. Aconite as a sudorific and antipyretic is far less objectionable, and in the anginous form of the complaint may be safely administered, and when the fever is high with beneficial results. With a view to diminish fever, quinine also may be given in efficient doses.

Citrate and carbonate of ammonia and aconite may either be inhaled as sprays or given by the stomach, the former method being preferable.

The inunction of the skin with oil and other greasy preparations, as often advised, is I believe very objectionable, especially in the early stage of the disease, on account of its impeding elimination by the skin; even during desquamation, so far as the patient is concerned, it is still unadvisable, although the practice may have some effect in diminishing the risk of infection to those in attendance. In place of blocking up the pores of the skin with the oil, and so preventing transpiration, the body should be sponged during the period of desquamation with tepid water containing aromatic vinegar or permanganate of potash. Later on the sponging may be followed by warm baths and the free use of soap; these will remove any loose cuticle, promote the action of the skin and relieve the congested kidneys, while they are probably quite as effective against infection as the inunction of oils.

Presuming that the throat affection has gone on to suppuration, ulceration, or sloughing, remedies of a different class must be brought into operation, as sprays of sulphurous acid, carbolic acid, of chlorine or permanganate of potash.

When sloughing has taken place one of the best applications is strong nitric acid, which must be very carefully applied by means of a brush.

The fœtor of the breath, so extreme when sloughing has occurred, will be corrected by any one of the disinfectants above mentioned, but particularly by the sulphurous acid and permanganate of potash.

For the rheumatic complications which sometimes arise during convalescence, salicin and its compounds are sometimes recommended; but in those cases in which there is kidney implication, as shown by albuminuria, these remedies may do mischief by their irritating action.

Lastly, the air of the room, especially near the patient, should, be disinfected throughout the illness, an object which may be accomplished either by Chamber Inhaler No. 1 or No. 2, or by a large-sized 'Queen Mab' vaporizer, which may be obtained of Messrs. Williams and Co., of Birmingham.

A strong belief was at one time entertained in the prophylactic power of belladonna; it is extremely doubtful, however, whether it really exerts any protective influence. I believe that arsenic is deserving of more confidence. Dr. Walter G. Walford, in the 'British Medical Journal' of July 5, 1884, praises highly the protective effects of arsenic in minute doses.

When scarlatina assumes the malignant character, the constitutional treatment required will be similar to that for diphtheria, since now there will be a marked disposition to disintegrating changes of the blood and other secretions, and great prostration. To counteract this condition the vital powers must be sustained by nourishment, by stimulants, including ammonia, and particularly by the free administration of perchloride of iron.

Syphilitic Affections of the Organs of Respiration.

The mucous membrane of the air-passages and subjacent structures, from the nose downwards to the pharynx, larynx, and even the lungs, are particularly liable to be attacked with secondary and tertiary forms of syphilis, either hereditary or acquired. When syphilis attacks the larynx or lungs it may simulate phthisis, to which, indeed, it is asserted that it sometimes gives rise. It is of the utmost importance that syphilitic affections of any portion of the air-passages should be early discriminated, in order that they may be treated on those principles which can alone arrest the destructive course of the disease and effect a cure.

The *Nose* is less frequently affected with lues in the present day than formerly, owing probably to the earlier discovery of the disease and its more effective treatment.

New-born infants and children not unfrequently suffer from inherited syphilis of the nose, which usually manifests itself under the guise of a nasal discharge or ozæna, and which is apt to be regarded as catarrhal. The discharge, which is muco-purulent and often very offensive, is attended with swelling of the pituitary membrane, on the surface of which mucous patches or condylomata may sometimes be detected, and with more or less narrowing and obstruction of the nasal passages. If the affection be not treated it will continue for a long time, and ultimately end in ulceration and all the evil consequences of tertiary syphilis.

In non-hereditary syphilis of the nose the secondary manifestations follow the original infection in the course of a few months, but the tertiary form not until after some years. The former are seldom visible, probably owing to the difficulty of making a complete inspection, but they usually take the form of condylomata, which on microscopical examination are found to be chiefly composed of epithelium; these readily soften, giving rise to superficial ulcerations which may quickly disappear, and they are generally situated at the external angles of the nostrils, just inside the anterior or posterior nares or on the septum. At the same time, condylomata are usually to be found on some portion of the cutaneous surface.

The tertiary form is characterized by deep and destructive ulcerations, which may occasion perforation of the septum or necrosis of the bones of the nose or base of the skull. If the cartilage of the septum be eaten away, the tip of the nose will fall in, and if the bones of the arch be destroyed, the nose will be flattened and great disfigurement ensue, or the whole nose may be destroyed by ulceration.

The necrosed bones may either be partly denuded or covered with a layer of blackish secretion, and there will be a dark-coloured and dreadfully offensive discharge, the smell of which no treatment succeeds wholly in removing. The bones often exhibit a blackish tinge, due probably to the formation of a sulphide, occasioned by the decomposition of the disintegrated tissues.

With the tertiary symptoms in the nose, the skin and periosteum will also be found to be more or less affected; there may be copper-coloured blotches on the skin and nodes on some of the bones, while the pharynx and larynx may likewise be implicated.

Syphilis of the *Pharynx* is almost invariably secondary or tertiary. The secondary affection consists, first, in an erythematous, dusky redness of the mucous membrane, especially characterized by the abrupt line of demarcation which separates it from the sound skin, and secondly, by the presence of mucous tubercles or condylomata. These tubercles are usually symmetrical, somewhat flat, and they generally ulcerate and disappear in the course of about two months. At the same time a papular eruption is usually found on the skin.

SYPHILITIC AFFECTIONS.

The tertiary conditions consist either in 'ulcerations, which may be superficial, but which are usually deep and destructive, or in gummata, in which in fact the deeper ulcerations originate. The ulcerations may involve the destruction of the uvula, or the adhesion of the velum to the posterior wall of the pharynx, and when extensive they may lead to narrowing and contraction of the upper part of the throat, and render swallowing very difficult.

The gummata are usually situated below the mucous membrane, and arise in some cases from the periosteum; they are placed mostly in the posterior wall of the pharynx; they slowly soften and ulcerate, giving rise to deep and perforating ulcers. When situated on the soft palate, a communication may be established through the ulcerative process between the mouth and nose, causing during deglutition an escape of liquids by the latter. Gummata may be found at the same time, as also rupia on different parts of the surface of the body.

Amongst the worst consequences of tertiary syphilis of the pharynx are the gradual and unavoidable contractions which take place as the result of the extensive ulceration.

The ulceration may extend to the deeper parts, and involve both cartilages and bones; the epiglottis may be in part destroyed, and even the vertebræ become necrosed.

The syphilitic affections of the *Larynx*, making allowance for differences of situation, correspond for the most part with those of the pharynx; the secondary affections consist chiefly of condylomata and the tertiary of gummata, and of superficial or deep ulcerations; the latter often situated on the epiglottis, leading to the destruction of cartilages, to necrosis, to contractions, and to stenosis.

The tertiary conditions ensue usually in three or four years after infection, but their appearance may be delayed for many years.

Lastly, in the *Lungs* themselves gummatous deposits and syphilitic infiltrations may take place, which may even be mistaken for tubercle.

It is a characteristic of syphilitic affections of the throat and larynx that they are not very sensitive, and give rise to but little irritation or pain, unless the lesions be in such situations as cause them to interfere with the act of deglutition.

In the tertiary form of syphilis, the health is apt to suffer and there is more or less cachexia and fever.

Treatment.—Secondary syphilitic affections usually disappear of themselves, and the opinion is held by some that no special treatment of them is required, the risk of tertiary syphilis not being increased by the omission. Still, it is in all cases safest to put the patient under some alterative or mild mercurial treatment as a precautionary measure, and to guard against the approach of tertiary symptoms.

In secondary syphilis of the *Nose*, or syphilitic coryza, since the discharge is usually offensive, antiseptic sprays should be injected up the nares, or antiseptic and astringent powders should be insufflated for the purpose of correcting the fœtor. Among the more recent devices for the insufflation of medicated powders into the nose, throat, and organs of respiration, is Kabiersky's Powder-Blower, which may be procured of Messrs. Krohne and Sesemann; it seems to be well adapted for the purpose and is furnished with nozzles or tubes of various lengths and shapes. If any condylomata are visible, these should be touched with tincture of iodine, or the powder of iodoform may be insufflated. In the case of infants, it is not easy to employ sprays, but still the nostrils must be cleared of secretion as far as possible, powders of bismuth, oxide of zinc, alum, or iodoform being afterwards insufflated.

In the tertiary lesions of the nose, iodide of potassium must be administered for a long time, both internally and in the form of sprays. Should this not succeed, then mercury must be given, either as grey powder, or the fumes of calomel may be inhaled through the nostrils. In some cases iodoform and mercury may be employed at the same time, the iodoform being insufflated especially where ulceration exists. If the ulcers be within reach they may be touched with nitrate of silver, or with the ointment of nitrate of mercury.

The secondary lesions of the *Pharynx* require for the local treatment, simply the occasional application of iodine or nitrate of silver, but should they not quickly disappear, then recourse may be had to iodoform, or to sprays of black or yellow wash, or of perchloride of mercury, or corrosive sublimate.

In the tertiary conditions, the chief remedy to be employed is iodide of potassium. This may be given as an atomized liquid; it should be continued for a long time, and should it not succeed, then hydrargyrum c. cretâ may be administered or the fumes of calomel inhaled. The application of nitrate of silver, or sulphate of copper to the ulcerated surfaces will help to bring about a healthier action and arrest the destructive tendency. The compound decoction of sarsaparilla, with either bichloride of mercury or iodide of potassium, were formerly frequently prescribed, and proved in many cases very effective.

The tertiary affections of the *Larynx* must be treated with iodide of potassium and iodoform, and should these fail, by the inhalation of the fumes of calomel. It will only rarely be necessary to bring the system under the influence of mercury by means of inunction. Sprays of a solution of bichloride of mercury, 1 in 1000 or in 500 parts, have been strongly commended as of great efficacy in obstinate tertiary syphilis of the larynx. Iodide of potassium will also prove very useful where there is much œdema.

The best correctors of the horrible fœtor arising from necrosis are, in the case of the pharynx, gargles and sprays of a solution of chlorinated soda or of permanganate of potash. Should the fœtor proceed from the larynx, then sprays of permanganate of potash or of carbolic acid may be employed.

For the cachexia, often so marked in tertiary syphilis, suitable tonics and cod-liver oil must be prescribed, such as bark, with one of the mineral acids, or carbonate of ammonia, or iodide of iron.

Further experiments in Evaporation by Mr. Clayton, F.C.S.

Name	Quantity taken	Percentage proportion by weight lost during one hour's evaporation at					
		50° F.	60° F.	70° F.	80° F.	90° F.	100° F.
Ol. terebinth...	grains 40	3·9	4·3	5·0	6·3	7·7	9·8
Ol. cubebæ...	,,	1·0	2·1	2·9	4·1	5·2	7·7
Terebene...	,,	2·2	2·8	3·7	4·4	5·9	8·0
Camphor...	,,	0·5	0·9	1·4	2·0	2·8	3·4
Resorcin...	,,	0·0	0·0	0·5	0·7	0·8	0·7
Ammonium benzoate...	,,	0·0	0·0	0·0	0·2	0·4	0·9
Ethyl iodide...	,,	39·5	41·5	42·7	43·4	44·0	45·2
Iodoform...	,,	0·0	0·0	0·0	0·0	0·1	0·1
[1] Carbolate of ammonia...	,,	0·9	1·2	1·9	2·4	3·1	3·9

[1] An aqueous solution was employed, containing equivalent proportions of phenol and ammonia. The figures represent proportions of PHENOL lost by the mixture during its evaporation at the several temperatures.

INDEX.

ACIDUM benzoicum, 195
— boracicum, 187
— carbolicum, 175
— gallicum, 173
— lacticum, 159
— salicylicum, 185
— tannicum, 171
— hydrocyanicum dilutum, 223
Aconitum napellus, 215
Actea racemosa, 222
Air, atmospheric, 107
— cold, 109
— warm, 111
— compressed, inhalation of, 112
— — expiration into, 114
— rarefied, inspiration of, 114
— — expiration into, 115
— compressed, inhalation of, in chamber, 117
— rarefied, inhalation of, in chamber, 128
— rarefied, expiration into, 130
— naturally rarefied, inhalation of, 131
— spray producer, 18, 51
— — — loss of medicaments with, 18
— — — improved, 52
Allen and Hanbury, 174
Aldehydum dilutum, 210
Alkaram, 230
Alumen, 166
Ammoniæ benzoas, 195
— liquor, 152
— carbonas, 152
Ammonii chloridum, 153
Amygdalæ oleum, 151
Amyl nitris, 210
Andral, 273
Aniseed, oil of, 199
Antimonium, 165
Aphonia, 265
Apthæ, 257
APPARATUS EMPLOYED IN INHALATION, 30
For Air and Gases, 30
cooled and warmed air, 31
Dr. Joscelyn Seaton's respirator, 31
Mr. Jeffrey's respirator, 31

APPARATUS EMPLOYED IN INHALATION—continued
For Compressed and Rarefied Air, 32
Waldenburg's, 33
Cube's double, 33
Tobold's, 33
Schnitzler's single, 34
— double, 36
Geigel and Mayer's, 35
Biedert's, 35
The Pneumatic Chamber, 35
Tabarié's chamber, 35
Lange's chamber, 36
Dr. G. von Liebig's chamber, 38
Simonoff's chamber, 38
Fontaine's chamber, 38
Limousin's oxygen receiver, 39
Burrough's chloride of ammonium inhaler, 40
Dr. Felton's ditto, 41
For the Inhalation of Medicated Vapours, 41
Oral and oro-nasal inhalers, 42
Dr. E. Blake's, 43
Dr. G. Hunter Mackenzie's, 43
Dr. Burney Yeo's, 44
Dr. W. Williams's, 46
Oral inhalers, 47
Nasal inhalers, 48
Dr. Feldbausch's, 48
Dr. Cousins's, 49
Dr. George Moore's, 49
Spray Producers or Nebulizers—
The air spray apparatus, 51
Improved air spray producer, 52
Dr. Sass's horizontal nebulizer, 53
Dr. Wright's atmonemeter, 53
Steam sprays, 54
Siegle's, 55
Dr. Cohen's modification, 55
Oertel's modification, 56
Mr. Benham's, 57
Dr. Lee's steam draft inhaler, 57
Action of spray producers, 58
The Sales Girons principle, 60

APP

APPARATUS EMPLOYED IN INHALA-
TION—continued
For the Inhalation of Warm Vapours, 60
Mackenzie's eclectic inhaler, 61
Dr. Spencer Thompson's inhaler, 61
For the Inhalation of Fumes, 61
For the Inhalation of Powders, 62
Rauchfuss's insufflator, 62
Mr. Bryant's insufflator, 62
Dr. Andrew Smith's insufflator, 62
Clinton Wagner's insufflator, 62
Rabiersky's insufflator, 62
New and Improved Forms of Apparatus, 63
Chamber inhaler and disinfector, No. 1, 63
Chamber inhaler and disinfector, No. 2, 66
The Globe oro-nasal inhaler 67
The water-bath for same, 70
The Globe oral inhaler, 72
Inhaler for concentrated vapours, 73
Apparatus for tar vapours, 74
Mr. L. Kay-Shuttleworth's inhaler, 75
Argenti nitras, 170
Arsenicum, 200
Assafœtida, 198
Asthma, 281
Atropa belladonna, 217

BALSAMUM Peruvianum, 196
— tolutanum, 197
Barthez, 338
Bataille's experiment, 8
Beddoes, 134, 140
Begbie, Dr., 289
Benham's modification of Siegle's steam spray producer, 57
Bennett, Dr. Hughes, 298
Benzinum, 182
Bergson's tubes, 52, 53
Bert, P., 120, 121, 129, 133
Berthold, 5, 187
Betulæ oleum pyroligneum, 184
Beverley, Dr. W. H., 285
Biedert's apparatus, 35
Blackley, 236
Blake's, Dr. E., inhaler, 43, 05
Bliss's cure, 219
Bosworth, 234
Braithwaite, Dr., 318
Brinton, 317
Bromide of ammonium, 224
— of potassium, 224

CLO

Bromum, 223
Bronchi, inflammation of, 268
— plastic inflammation of, 275
Bronchial tubes, dilatation of, 273
Bronchiectasis, 273
Bronchitis, 268
— acute, 268
— capillary, 268
— chronic, 269
Bronchorrhœa, 270
Browne's, H. Langley, apparatus, 88
Brown, Dr. J., 333
Brown-Séquard, 217
Brunton, Dr. Lauder, 211, 310
Bryant's insufflator, 62, 244
Buchholz, 189, 197
Bumstead, 205
Burrough's chloride of ammonium inhaler, 40

CADINUM oleum, 184
Calcis liquor, 156
— chloratæ liquor, 11
— saccharatus liquor, 156
— phosphas, 200
— hypophosphis, 199
Calx caustica, 156
— chloratæ, 148
Camphorum officinarum, 193
Cannabis indica, 220
Caraway, oil of, 199
Carbolic acid, experiments with, 7, 24, 27
Carbonic acid, 139
Carburetted hydrogen, 140
Caricæ papayæ succus, 158
Cassia, oil of, 199
Catarrhus senilis, 272
Cephaëlis ipecacuanha, 163
Chamber inhaler and disinfector, No. 1, 63
— No. 2, 66
Chandelon's process for carbolic acid, 7
Chinolinum, 188
Chlori vapor, 143
Chloral hydras, 209
Chlorine, 143
— experiments with, 145
Chloride of ammonium inhaler, 40
Chloroformum, 208
Churchill, Dr., 318
Cimifuga racemosa, 222
Cinnamon, oil of, 199
Clark, Sir Andrew, 297
Claviceps purpurea, 174
Clayton, F.C.S., Edwy G., 21, 219, 28, 357
Cloez, 192

INDEX. 361

CLO

Cloves, oil of, 199
Cocaine, 200
Cocculus indicus, 311
Codman and Shurtleff, 55
Cohen, Dr. Solis, 2, 31, 92, 138, 140, 157, 166, 218
— modification of Siegle's spray producer, 55
Collins, Mr. James, 96
Commerge, 141
Coniæ vapor, 12, 99
Conium maculatum, 214
Consumption, 293
— pulmonary, 295
— bronchial, 295
— laryngeal, 295
— pharyngeal, 295
— intestinal, 295
Corbyn and Stacey, 153, 177, 208
Corrigan, Dr., 273
— vaporizer, 88, 144
Cousins', Dr., nasal inhaler, 49
Creasoti vapor, 12, 99
Creasotum, 180
— experiments with, 8
Cresoline, 182
Croup, 267
— false, 264
— catarrhal, 264
— true, 344
— spasmodic, 345
Crum Brown, 214
Cube's double apparatus, 33
Cubeba officinalis, 193
Cupri sulphas, 171
Cuxon & Co., Messrs., 88
Cynanche trachealis, 344

Da Costa, 5
Datura stramonium, 218
— tatula, 218
Daly, Dr., 236, 241
Demarquay, 133, 134, 140
— experiments, 2
Deville, 197
Dewar, Dr., 134, 142, 143, 272, 290, 328
Digitalis purpurea, 225
Diphtheria, 337
— benignant, 337
— typical, 336
— inflammatory, 337
— malignant, 337
— chronic, 338
— micrococcus of, 335
Dodson, 140
Dogwood, Jamaica, 222
Dorema ammoniacum, 197
Drake, Dr. C., 111

GEI

Eclectic inhaler, Mackenzie's, 61
Emmerich, 278
Emphysema, 290
— lobular, 290
— lobar, 290
Entrance of medicaments into the organs of respiration, 1
— evidence of, furnished by sputa, 4
— — by smell, 5
— — by urine, 6
Experiments with oro-nasal inhalers, 7
— carbolic acid, 7
— creasote, 8
— thymol, 8
— iodine, 9, 12
— eucalyptol, 11
— oil of turpentine, 11
— special apparatus, 11
— conclusions from experiments, 19
Epistaxis, 244
Ergot, 317
Ergotin, Bonjean's, 175
Erythroxylon coca, 200
Espic, cigarettes d', 285
Ether sulphuricus, 206
Ethyl iodidum, 207
Eucalyptus globulus, 192
— rostrata, 173
Euchlorine, 145

Fagi oleum pyroligneum, 184
Faust and Homeyer, 192
Feldbausch's nasal inhaler, 48
Felton's, Dr., inhaler, 41
Ferula galbaniflua, 198
Ferrier's, Dr., snuff, 240
Ferri perchloridi, 167
— sulphas, 168
Fieber's, Dr. F., experiments, 3
Fischer, Dr., 103
Fontaine, Dr S. A., 38
Foster, Mr., 334
Fothergill, Dr., 215, 285, 309
Fournié's experiments, 2, 3
Foxglove, 225
Fränkel, 122
Fraser, 214
Frémy, 197
Fumes, constitution of, 6

Gairdner, Dr., 273
Galbanum, 198
Garrod, Dr., 192, 198, 201, 212
Gases, 107
Gaultheria procumbens, 186
Geigel and Mayer's apparatus, 35
Geiger, Dr., 157

GEL

Gelsemium, 308
Globe oro-nasal inhaler and bath, 67, 70
— oral inhaler, 72
Glycerinum, 151
Gottstein's tampons, 238
Granville's, Dr. J. Mortimer, powder, 240, 285
Graves, Dr., 141, 309
Green mountain cure, 219
Grindelia robusta, 219

Hack, 236
Hadra, 122
Hamamelis virginica, 174
Hannay, 239
Harley, Dr. J., 217
Hay fever, 235
Hazeline, 174
Heinze's statistics, 29^5
Helfft, 139, 140
Helmholz, 240
Hemlock, 214
Hemming, Messrs., 348
Henbane, 213
Hicks', Dr. Braxton, apparatus, 90
Himrod's cure, 219, 285
Hopkins and Williams, 158, 215
Hoppe-Seyler, 176
Howe, Dr. John E., 309, 310
Huggins, 286
Hunt, Dr., 187
Hutchinson, 33, 119
Hydrargyrum, 204
Hydrogen, 140
— peroxide, 135
Hydrobromic acid, 224
Hydroquinone, 180
Hyoscyamus niger, 213

Indian hemp, 220
— tobacco, 220
Influenza, 326
Inhalation chambers, 76
— different kinds of, 76
— principles of construction of, 76
— experiments relating to, 77
— description of Author's, 83
    Mr. Robson's eucalyptus machine, 87
    Mr. H. Langley Browne's apparatus, 86
    Dr. Corrigan's vaporizer, 88
    Messrs. Savory and Moore's vaporizer, 89
    Messrs. Williams and Co.'s 'Queen Mab,' 353
    Dr. Neale's chemical lung, 89

INH

Inhalation chamber, description of— continued
    Dr. Braxton Hicks' apparatus for air disinfection, 90
Inhalation, method of, 4, 97
— frequency of, 104
— duration of, 104
INHALATION TREATMENT OF DISEASES OF THE ORGANS OF RESPIRATION.
Nasal catarrh, 227
Dry nasal catarrh, 232
Hay fever, 235
Hypertrophy of the nasal mucous membrane, 248
Bleeding from the nose, 244
Catarrh of the nasal pharynx, 247
Dry catarrh of the naso-pharynx, 248
Catarrh of the pharynx, 249
Tonsillitis, 249
Follicular disease of the throat, 254
Apthæ, 257
Relaxed sore throat, 258
Rheumatic sore throat, 260
Gouty sore throat, 261
Ulcerated sore throat, 261
Inflammation of the larynx, 262
Catarrhal croup, 264
Laryngitis stridulosa, 264
Aphonia, 265
Catarrh of the trachea, 267
Tracheitis, 267
Inflammation of the bronchi, 268
Acute bronchitis, 268
Capillary bronchitis, 268
Chronic bronchitis, 269
Bronchorrhœa, 270
Catarrhus senilis, 272
Dilatation of the bronchial tubes, 273
Plastic inflammation of the bronchi, 275
Inflammation of the lungs, 275
Interstitial or fibroid pneumonia, 276
Lobar pneumonia, 277
Lobular pneumonia, 277
Pleuro-pneumonia, 277
Broncho-pneumonia, 277
Epidemic or endemic pneumonia, 277
Intermittent pneumonia, 277
Asthma, 281
Emphysema, 290
— lobular, 290
— lobar, 290
Consumption, 293
— pulmonary, 295
— bronchial, 295

INDEX.

INH

INHALATION TREATMENT OF DISEASES OF THE ORGANS OF RESPIRATION—*continued*
Consumption, laryngeal, 295
— pharyngeal, 295
— intestinal, 295
— acute tuberculosis, 296
— acute phthisis, 296
— acute tuberculo-pneumonic phthisis, 296
— catarrhal phthisis, 296
— fibroid phthisis, 297
— scrofulous phthisis, 298
— hæmorrhagic phthisis, 298
— chronic tubercular phthisis, 298
Inhalers, oro-nasal, 42
— oral, 47
— nasal, 48
— for concentrated vapours, 73
Iodine, experiments with, 9
Iodum, 202
Iodoformum, 207
Isambert, 338

JABORANDI, 310
Jacobson, 119
James, Mr. J. Brindley, 45, 99, 105
Jeffreys' respirator, 31
Jenner, Sir William, 258, 337
Jones, Dr. Macnaughten, 100, 153, 208
Jones, Dr. Talfourd, 284
Joy's cigarettes, 219
Juniper, oil of, 199
Juniperus oxycedrus, 184

KABIERSKY's powder-blower, 356
Klikowitsch, S., 138
Koch, Dr., 102
Kohlschütter, 137, 138
Kopp, E., 197
Kröhne and Sesemann, 54, 356
Küchenmeister, 156

LADENBURG, 217, 218
Laennec, 145, 204
Lange, 36, 119, 120, 121, 122
Langenbeck, Prof., 111
Laryngeal consumption, 295
— phthisis, 320
Laryngismus stridulus, 346
Laryngitis, 265
— stridulosa, 264, 346
Larynx, inflammation of, 262
Lee's, Dr., steam draft inhaler, 16, 57, 58
Lee, 205

MED

Lefferts, 241
Lemaire and Plügge, 176
Leptothrix buccalis, 335
Letzerich, 334, 335
Lewin, 4, 189, 256
Lewinstein, 122
Liebig, Dr. G. von, 38, 120, 121, 161
Limousin, apparatus of, 39, 733
Liquor carbonis detergens, 179, 184
Lobelia inflata, 220
Louis, 145
Lungs, inflammation of, 275

MACKENDRICK, 134
Mackenzie, Dr. G. Hunter, 43, 208, 319
Mackenzie, Dr. Morell, 61, 62, 225, 284, 286, 238, 240, 242, 248, 255, 256, 295, 322, 335, 338, 344
Maladies suitable for treatment by compressed air of pneumatic chamber, 124
Marcet, Dr., 324
Marjory, oil of, 199
Martindale, 211, 220, 223, 289
Martindale and Westcott, 180, 181, 188, 207
Matthieu, 3
Mauthner, J., 158
Maw, Son, and Thompson, 32, 63
Mayer, 13
Mayer and Meltzler, 7, 57, 88
Meadow sweet, 185
Medicaments, quantities of, 91, 100
— method of inhaling, 97
MEDICAMENTS EMPLOYED IN INHALATION, 107
Gases—
Atmospheric air, 107
Cold air, 109
Warm air, 111
Inhalation of compressed air, 112
Expiration into compressed air, 114
Inspiration of rarefied air, 114
Expiration into rarefied air, 115
Inhalation of compressed air in the pneumatic chamber, 117
Inhalation of rarefied air in the pneumatic chamber, 128
Expiration into rarefied air, 130
Inhalation of naturally rarefied air, 131
— oxygen, 132
— ozone, 134
— peroxide of hydrogen, 135
— nitrogen, 137
— nitrous oxide gas, 138

MED

MEDICAMENTS EMPLOYED IN INHALATION—continued
Gases—continued
  Inhalation of carbonic acid, 136
  — hydrogen, 140
  — carburetted hydrogen, 140
  — sulphuretted hydrogen, 141
  — sulphurous acid, 142
  — chlorine, 143
Diluents and Refrigerants—
  Cold water, 149
  Iced water, 149
Warming and Soothing remedies—
  Hot water, 149
  Solution of gum, 150
  — of vegetable extractive matters, 150
  Glycerinum, 151
  Oleum Olivæ, 151
  Oleum Amygdalæ, 151
  Verbascum thapsus, 152
Chemical Solvents—
  Liquor ammoniæ, ammoniæ carbonas, ammonii chloridum, soda caustica, sodii chloridum, sodæ carbonas, 154
  Liquor potassæ, potassæ carbonas, 155
  Liquor calcis, 156
  Tetramethylammonium hydroxide, 157
  Tetraethylammonium hydroxide, 157
  Neurin, 158
  Succus caricæ papayæ, 158
  Acidum lacticum, 159
Refrigerants—
  Potassæ chloras, 159
  Potassæ nitras, 160
Solvents, Alteratives, and Depurants
  Mineral waters, 161
  Sulphides, 162
Nauseants and Emetics—
  Cephaëlis ipecacuanha, 163
  Antimonium, 165
Mineral Astringents—
  Alumen, 166
  Liquor ferri perchloridi fortior, 167
  Ferri sulphas, 168
  Plumbi acetas, 168
  Zinci sulphas, 169
Caustic Astringents—
  Argenti nitras, 170
  Cupri sulphas, 171
Vegetable Astringents—
  Acidum tannicum, 171
  — gallicum, 173
  Eucalyptus rostrata, 173
  Hamamelis virginica, 174

MED

MEDICAMENTS EMPLOYED IN INHALATION—continued
Vegetable Astringents—continued
  Claviceps purpurea, 174
Non-oxidizing Antiseptics—
  Acidum carbolicum, 175
  Resorcin, 180
  Hydroquinone, 180
  Creasotum, 180
  Cresoline, 182
  Benzinum, 182
  Pix liquida, 183
  Salicinum, acidum salicylicum, sodæ salicylas, 185
  Acidum boracicum, 187
  Sodæ biboras, 187
  Chinolinum, 188
Oxidizing Antiseptics—
  Potassæ permanganas, 188
Stimulants, Balsamics, and Antiseptics—
  Thymus vulgaris, 189
  Oleum terebinthinæ, 190
  Terebena, 191
  Eucalyptus globulus, 192
  Cubebæ officinalis, 193
  Camphora officinarum, 193
  Acidum benzoicum, sodæ benzoas, ammoniæ benzoas, 195
Balsamic remedies—
  Balsamum Peruvianum, 196
  — tolutanum, 197
  Assafœtida, 198
  Galbanum, 198
  Olea essentialia, 198
Nervine stimulants—
  Calcis hypophosphis, sodæ hypophosphis, 199
  Calcis phosphas, 200
  Cocaine, 200
Alteratives—
  Arsenicum, 200
  Iodum, 202
  Hydrargyrum, 204
Anæsthetics—
  Æther sulphuricus, 206
  Ethyl iodidum, 207
  Iodoformum, 207
  Chloroformum, 208
  Chloral hydras, 209
  Aldehydum dilutum, 210
  Amyl nitris, 210
  Nitroglycerinum, 241
Anodynes and Narcotics—
  Papaver somniferum, 212
  Hyoscyamus niger, 213
  Conium maculatum, 214
  Aconitum napellus, 215
  Atropa belladonna, 217
  Datura stramonium, 218

INDEX.

MED

MEDICAMENTS EMPLOYED IN INHA-
LATION—continued
Anodynes and Narcotics—continued
  Datura tatula, 218
  Grindelia robusta, 219
  Cannabis indica, 220
  Lobelia inflata, 220
  Nicotiana tabacum, 222
  Cimifuga racemosa, 222
  Piscidia erythrina, 222
  Acidum hydrocyanicum dilutum, 223
  Bromum, 223
Vascular Sedative—
  Digitalis purpurea, 225
Powders—
  Various medicinal, 226
Meigs, Dr., 347
Meissner, 198
Merrill, Dr., 202
Micrococci, 278
Mitchell, Dr. Weir, 224
Moncorvo, M., 180
Monk's-hood, 215
Moore, Dr. George, 239, 241
— nose inhaler, 49
Muguet, 257
Mullein, the Great, 152
Murray, Dr. J. Carrick, 98, 105
Murrell, Dr., 164, 185, 224, 309
Myroxylon pereiræ, 196
— toluiferum, 197

NARTHEX assafœtida, 198
Nasal catarrh, 227
— dry, 232
Nasal inhalers, 48
Nasal mucous membrane, hypertrophy of, 243
Nasal pharynx, catarrh of, 247
— dry, 248
Neale's, Dr., chemical lung, 89
Neurin, 158
Nicotiana tabacum, 222
Nitrogen, 187
Nitroglycerinum, 211
Nitrous oxide gas, 138
Nose, bleeding from, 244
— tuberculosis of, 325

O'CONNELL, Dr. M. D., 239, 243
Oertel, 4, 31, 37, 113, 115, 116, 117, 124, 125, 129, 138, 140, 155, 158, 159, 161, 183, 195, 196, 256, 335, 348
Oertel's modification of Siegle's steam spray-producer, 5, 6
Oidium albicans, 257

PNE

Olea essentialia, 198
Oleum pini sylvestris, 191
Olivæ oleum, 151
Opium, 212
Oral and oro-nasal inhalers, defects of, 42
— loss of medicaments with, 94
Oral inhalers, 47
Oxygen, 132
— receiver, 39
Ozone, 134

PAGET, Dr., 237
Panum, 119, 120, 121
Papaver somniferum, 212
Papayotin, 158
Parkes, 149, 176
Partajas, Dr. J. G., 155
Peart, Dr., 351
Peri-pneumonia notha, 272
Pertussis, 329
Pfeiffer, 120
Pharynx, catarrh of, 249
— tuberculosis of, 325
Pharyngitis sicca, 254
Phenol, 7
Phillips, 241
Phthisis, acute, 296
— acute tubercular pneumonic, 296
— catarrhal, 296
— fibroid, 297
— scrofulous, 298
— hæmorrhagic, 298
— chronic tubercular, 198
— laryngeal, 320
Picrotoxine, 311
Pilocarpine, 310
Pinus palustris, 190
— pinaster, 190
— sylvestris, 190
Piorry, M., 204
Piscidia erythrina, 222
Pix liquida, 183
Plumbi acetas, 168
Pneumatic chamber, 35
— Tabarié's, 35
— Lange's, 36
— Dr. G. von Liebig's, 38
— Simonoff's, 38
— Fontaine's, 38
Pneumonia, 275
— acute lobar, 276
— croupous, 276
— interstitial or fibroid, 276
— lobar, 277
— lobular, 277
— pleuro-, 277
— broncho-, 277
— epidemic or endemic, 277

PNE

Pneumonia, intermittent, 277
Pneumonococci, 278
Pond's extract, 174
Poplar, 185
Potassæ liquor, potassæ carbonas,
— chloras, 159
— nitras, 160
— permanganas, 198
Potassium bromide, 224
Powders, 226
Preston, Dr., 246
PRINCIPLES CONCERNED IN VOLATILIZATION AND INHALATION, 21
  Experiments, table i., 22
  — table ii., 23
  Temperature, 23
  Humidity, 23
  Movement of air, 24
  Augmentation of surface, 24
  Extension of surface and movement of air, 27
  Augmentation of surface with increase of temperature, 28
  Effect of volatile solvents on rate of evaporation, 27
  Method of inhalation, 29
Ptychotis ajowan, 189

QUAIN, 300
Quincke, 120
Quinlan, Dr., 152
Quinsy, 250

RABIERSKY'S insufflator, 62
Ramadge, Dr., 111, 301
Raspail, M., 194
Rauchfuss's insufflator, 62
Resorcin, 180
Ringer, Dr., 5. 142, 164, 165, 185, 194, 201, 204, 205, 209, 210, 216, 218, 221, 222, 224, 241, 284, 286, 288, 289, 309, 351
Roberts, Dr. Lloyd, 159
Robinson Beverley, 248
Robson's eucalyptus machine, 87
Roe, Dr. John E., 236, 241
Rosemary, oil of, 199
Rossbach, 158
Rusci oleum pyrolignum, 184
Ruspini's styptic, 316

SAGE, oil of, 199
Sales Giron, 93
— principle, 60
Salex caprea, 185
Salicinum, 185
Salter, Dr. Hyde, 242

SUL

Sandahl, 120. 122
Sanderson, Dr. J. Burdon, 38, 119, 127, 161, 162
Sanson, Dr., 178
Sargent, 334
Sass's, Dr., horizontal nebulizer, 53
Savory and Moore's vaporizer, 89
Scarlatina, 349
— simplex, 349
— anginosa, 349
— maligna, 349
Scarlet fever, 348
Scharling, 197
Schnetzler, M., 104
Schnitzler and Störk's experiments, 3
— single apparatus, 34
— double apparatus, 36
Schüll, Dr., 103
Seaton's, Dr. Joscelyn, respirator, 31
Sedgwick, Dr. Leonard, 159, 231
Senator, 335
Shuttleworth's, L. K., inhaler, 75
Siegle's steam spray producer, 17, 55
— loss of medicaments with, 17, 95-97
Simonoff, 38, 122
Simpson, Dr., 187
— Sir James, 140
Snake root, black, 222
Smallpox, 350
Smith's, Dr. Andrew, insufflator, 62
Smith, Dr. R. Shingleton, 208, 319
— Dr. C. Solomon, 103
Soda caustica, sodæ carbonas, sodii chloridum, 154
Sodæ hypophosphis. 199
— chloratæ liquor, 143
— benzoas, 195
— biboras, 187
— salicylas, 185
Sormani, Dr. J., 103, 208, 319
Speck, 133
Spendler, 160
Sprays, action of. 58, 59
Spray producers, 51
Sprengler, 140
Squire, 195
Starkey, Dr., 5
Steam sprays, 54
Stembo, 121
Stokes, Dr., 145, 273
Stone, Prof., 145
Stuart, Dr. Grainger, 273
Strychnine, 310
Sulphides, 162
— ammonium, 162
— calcium, 162
— potassium, 162
— sodium, 162
Sulphurous acid, 142
Sulphuretted hydrogen, 141

INDEX.

Sweet flag, oil of, 199
Syphilis of organs of respiration, 353
— of nose, 354
— of pharynx, 354
— of larynx, 355
— of lungs, 355

TABARIÉ, 35
Tanner, Dr., 289
Tar, 183
— coal, 183
— wood, 183
— vapours, apparatus for, 74
Tavernier's experiment, 3
Terebena, 191
Terebinthinæ oleum, 190
Tetraethylammonium hydroxide, 157
Tetramethylammonium hydroxide, 157
Thompson's, Dr. Spencer, vapour inhaler, 61
Thorn apple, 218
Thorowgood, Dr., 287, 319
Throat, follicular disease of, 254
— relaxed, 258
— rheumatic, 260
— gouty, 261
— ulcerated, 261
Thrush, 257
Thymol, experiments with, 8
Thymolis vapor, 15
Thymus vulgaris, 189
Tobold's apparatus, 33
Tonsillitis, 249
Toogood, 61
Toulmouche, 145
Trachea, catarrh of, 267
Tracheitis, 267
Treutler, 137, 138
Trousseau, 205, 285
Tubercle bacillus, 102, 103, 294, 297
Tuberculosis, acute, 296
Tymowski, Dr. de, 319
Tyndall, Prof., 103
Typhoid fever, 350

VAPOR acidi carbolici, 14

Vapor conii, 13
— creasoti, 13
— thymolis, 15
Vapours, physical composition of, 6
Vapour of hot water, 15
Verbascum thapsus, 152
Vivenot, von, 118, 119, 120, 121, 122, 128, 129
Vogel, Prof., 161

WAGNER's, Clinton, insufflator, 62
Waldenburg, 115, 116
Waldenburg's apparatus, 33, 35
Walford, Dr. W. G., 144, 353
Walshe, Dr., 156
Warfwinge, Dr., 334
Water, 149
— cold, 149
— iced, 149
— hot, 149
— mineral, 161
Watts, 183
Whooping cough, 329
Willemin, 140
Williams, Dr. C. T., 296, 298
Williams, Dr. W., 289
— oro-nasal respirator, 46
Williams, Messrs., and Co., 353
Williamson, Dr., 351
Willigk's statistics, 295
Willow, 185
Willmott, 189
Winter green, 186
Witch hazel, 174
Witt, Mr. Charles, 351
Wright's, Dr., atmonemeter, 53
Wright and Co., Messrs., 51, 57
Würtz and Bæyer, 158

YEO's, Dr. Burney, respirator, 44, 105

ZDEKAUER, case of hæmoptysis, 4
Zinci sulphas, 169
Zygodesmus fuscus, 335

22

www.ingramcontent.com/pod-product-compliance
Lightning Source LLC
Chambersburg PA
CBHW031418230426
43668CB00007B/354